Trends and Applications in Advanced Polymeric Materials

Scrivener Publishing
100 Cummings Center, Suite 541J
Beverly, MA 01915-6106

Publishers at Scrivener
Martin Scrivener (martin@scrivenerpublishing.com)
Phillip Carmical (pcarmical@scrivenerpublishing.com)

Trends and Applications in Advanced Polymeric Materials

Edited by

Sanjay K. Nayak, Smita Mohanty and Lakshmi Unnikrishnan

Scrivener Publishing

WILEY

This edition first published 2017 by John Wiley & Sons, Inc., 111 River Street, Hoboken, NJ 07030, USA and Scrivener Publishing LLC, 100 Cummings Center, Suite 541J, Beverly, MA 01915, USA

For more information about Scrivener publications please visit www.scrivenerpublishing.com.

Wiley Global Headquarters
111 River Street, Hoboken, NJ 07030, USA

For details of our global editorial offices, customer services, and more information about Wiley products visit us at www.wiley.com.

Library of Congress Cataloging-in-Publication Data

Names: Nayak, Sanjay K., editor. | Mohanty, Smita, 1976- editor. |
Unnikrishnan, Lakshmi (Engineer), editor.
Title: Trends and applications in advanced polymeric materials / edited by
Sanjay K. Nayak, Smita Mohanty and Lakshmi Unnikrishnan.
Description: Hoboken, NJ : John Wiley & Sons, 2017. | Includes index. |
Identifiers: LCCN 2017036474 (print) | LCCN 2017041073 (ebook) | ISBN
9781119364788 (epub) | ISBN 9781119364160 (pdf) | ISBN 9781119363637
(cloth)
Subjects: LCSH: Polymers--Industrial applications. | Nanostructured materials.
Classification: LCC TA455.P58 (ebook) | LCC TA455.P58 T68 2017 (print) | DDC

620.1/92--dc23
LC record available at https://lccn.loc.gov/2017036474

Cover image: Pixbay.Com
Cover design by Russell Richardson

Set in size of 11pt and Minion Pro by Exeter Premedia Services Private Ltd., Chennai, India

10 9 8 7 6 5 4 3 2 1

Contents

Preface

Today, polymers rule the world with their diverse range of applications, fulfilling requirements ranging from the domestic to industrial and defense sectors. The changing world has seen a revolution created by polymers, the most versatile material ever discovered by human kind, owing to their lightweight, ease of processing and manufacturing, moldability into intricate shapes, and cost-effectiveness. They can easily be filled with a range of reinforcing agents like fibers, particulates, flakes and spheres in micro/nano sizes and compete with conventional materials in terms of performance, properties and durability. Polymers have become a multidisciplinary brand which is equally sought after by chemists, physicists, engineers, scientists, doctors and industrialists. More and more research is being carried out to further explore the material and modify it to meet the increasing demand.

"*The more important reason is that the research itself provides an important long-run perspective on the issues that we face on a day-to-day basis.*" These words of Ben Bernanke, the former U.S. Federal Reserve Chairman, have inspired us to discuss and share research on many advanced but widely used applications of polymers with techno-socio-economic importance.

Presented herein are the trends in polymer research and applications of many novel technologies. The chapters in this book feature major original works of young researchers along with a few reviews on innovative techniques and utilities. Primarily, use of polymers for advanced applications like coatings, solar energy harvesting, conducting polymers, sensors, electrolytes, drug delivery, hydrogels, batteries and high-barrier polymers has provided a framework in which to design the contents of this book. The chapters are well crafted with an abstract, introduction to the specific topic, theoretical and experimental techniques wherever necessary, a discussion of detailed results and a concise conclusion about the findings.

To start with, Chapters 1 and 2 on polymer nanocomposite coatings describe the potential of coatings beyond aesthetics. Specialized coatings are discussed wherein polymer matrices are reinforced with nanofillers for imparting specified properties like scratch/barrier resistance, electrical/thermal conductivity, as well as mechanical strength/durability. In Chapter 3, the book proceeds with a discussion of low smoke materials made for polymer nanocomposites from an industrial point of view. The authors describe the effect of various additives on the flame retardancy and smoke tolerance of the nanocomposites in detail. Polymer nanocomposites for solar energy harvesting, a topic of huge interest, is discussed in Chapters 4 and 5. While Chapter 4 deliberates the usefulness, challenges and current status of organic solar cells in the form of a review, Chapter 5 highlights the photocatalytic activities of highly efficient metal nanocrystals embedded in polymer matrix. The electrical sensitivity of various conducting polymer nanocomposite formulations, which are mainly used in sensors, is described in Chapters 6 through 8; and the luminescent characteristics are described in Chapter 9. A discussion on the gas transport phenomena in polymeric membranes filled with nanoparticles is incorporated as Chapter 10, which explains the underlying aspects and factors affecting the mobility of gas molecules through the membranes. Furthermore, the mechanism of ion transport across the membranes and the effect of nanomaterials in developing a continuous conducting path are elucidated in Chapter 11. The development of biocompatible hydrogels for drug delivery application is dealt with in Chapters 12 and 13. Factors like drug entrapment, release rate, compatibility with biological systems, etc., are clearly explained. Chapter 14 presents the advanced use of siloxane modified multilayered polyethylene for oxygen barrier packaging application. The improvement in oxygen diffusion for enhanced shelf life of packaged food items is discussed in detail. Finally, Chapter 15 highlights the advantages and opportunities in the field of rechargeable batteries, with specific emphasis on lithium-based batteries with polymer electrolyte membranes.

This book thus proposes to unveil the research prospects of polymers for indigenous development. The research discussed herein is resourceful in terms of its contents because the selected chapters cover the most recent developments and societal demands. We hope our venture will expressly encourage young people to streamline their research, industrialists to invest in advanced developments, and academicians and scientists to take their research further.

We are thankful to all the authors who have contributed and shared their expertise. Credit also goes to all members who have directly or indirectly helped us in the process of compiling this book. Special thanks go to Wiley-Scrivener for providing us with the opportunity to publish our book. Last, but not the least, we bow before the Almighty, without whose blessings we would not have been able to frame this book.

Sanjay K Nayak
Smita Mohanty
Lakshmi Unnikrishnan
Bhubaneswar, India, August 2017

1

Polymer Nanocomposites and Coatings: The Game Changers

Gaurav Verma

Dr. Shanti Swarup Bhatnagar University Institute of Chemical Engineering and Technology (formerly Department of Chemical Engineering & Technology), Panjab University, Chandigarh, India
Centre for Nanoscience and Nanotechnology (U.I.E.A.S.T), Panjab University, Chandigarh, India

Abstract
In recent years, polymer nanocomposites and coatings have caught the attention of the research world due to their versatile properties and widespread applications. The availability of new nanoscale fillers and additives provide polymer scientists with materials and a lot of options to modify the properties of the polymeric matrix and hence widen the ambit of their applications. Increasing usage of polymer nanocomposites and their exploration for more potential applications lead to game-changing solutions to many engineering and technological problems. The issue with nanoscale fillers is their stability and compatibility with polymeric matrix. A lot depends on the processing protocols used for fabricating the nanocomposites. This chapter provides an overview of the structure-property-processing relationship for polymer nanocomposites and coatings.

***Keywords*:** Polymer, nanoscale fillers, structure-property-processing, performance

1.1 Introduction

Polymer nanocomposites are composed of a polymeric matrix and a nanoscale filler which have at least one dimension in the nanometer range (usually 1–100 nm is accepted but nowadays many fillers up to a size of 500 nm have also been considered as nanofiller). The wide variation in

Corresponding author: gauravverma@pu.ac.in, gauravvermas@gmail.com

Sanjay K. Nayak *et al.* (eds.) Trends and Applications in Advanced Polymeric Materials, (1–22) 2018 © Scrivener Publishing LLC

shapes and sizes of these nanoscale fillers accounts for wide ranging structural and property modifications of the polymeric matrix, as shown in Figure 1.1. Although polymers themselves are versatile materials and can be easily tailored to imitate metals, natural materials and even biomaterials,

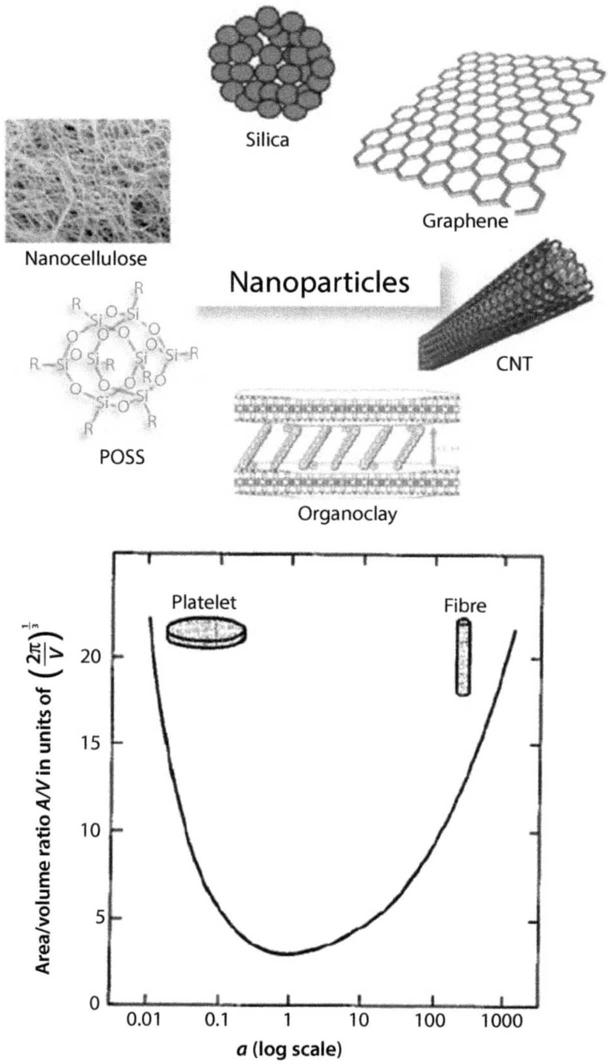

Figure 1.1 (Top) Different shapes and types of typical nanoparticles (nanoscale fillers) used to reinforce polymeric matrices. (Bottom) Particle shapes and surface area/volume (A/V) versus aspect ratio (a) variation for nanoscale structures which can reinforce polymer matrices [1].

the ever widening scope of new materials has yet to catch up with the latest technology, thus requiring constant research and updating.

Depending on how many dimensions of the particles are in nanometer range, nanoparticles are mainly categorized into three types. In the first type, called isodimensional nanoparticles, all dimensions are in the order of nanometers (0D) (e.g., spherical silica nanoparticles, some nanoclusters, etc.). In the second type, called nanotubes or nanowhiskers, two dimensions of the particles are in the nanometer range and the third one is larger, usually forming an elongated structure (1D) (e.g., carbon nanotubes, cellulose nanowhiskers). The third type of nanoparticles is characterized by only one dimension in the order of nano range (2D). In this case the particles are present in the form of sheets one to a few nanometers thick to hundreds to thousands of nanometers long (e.g., layered silicates [LS]).

The 21st century applications are far reaching, as scientists have started to explore more of outer space, planets and other celestial bodies. Aspirations on earth are also much more technologically advanced as compared to a decade ago. Better materials are needed to fabricate new types of automobiles like driverless cars, flying drones or supersonic aircrafts. There are two options for making materials to cater to these needs. One is to invent or discover totally novel materials with distinct properties, new structure and hence new properties. This approach is promising but may not be that helpful as there are shortcomings in a single material. Getting the best property and structural combinations is usually possible by modifying materials. So the second apprxoach is to use nanoscale fillers to tailor materials with better property combinations without compromising their inherent characteristics.

Using nanoscale fillers to modify conventional materials like polymers is an up-and-coming and very promising technique. By using only a very small quantity of nanoscale fillers like less than 10 weight percent, huge property benefits can be achieved. In comparison to conventional microscale filled composites, the reduction in weight percentage of the filler used is about 10 times or more, while the property improvements almost double. Sometimes the properties which couldn't be imbibed in conventional composites (Figure 1.2) can now be easily induced in nanocomposites. These advantages and other structural, morphological and physical improvements have led to increased interest in scientific and commercial communities. The only issue facing the use of nanocomposites is the ultimate control over nanofiller size, which still needs to be attained. Processing of nanocomposites, especially coatings, is also a far cry from realization; hence; the objective of this chapter is to instigate the research world into formalizing the processing protocols. By using examples of

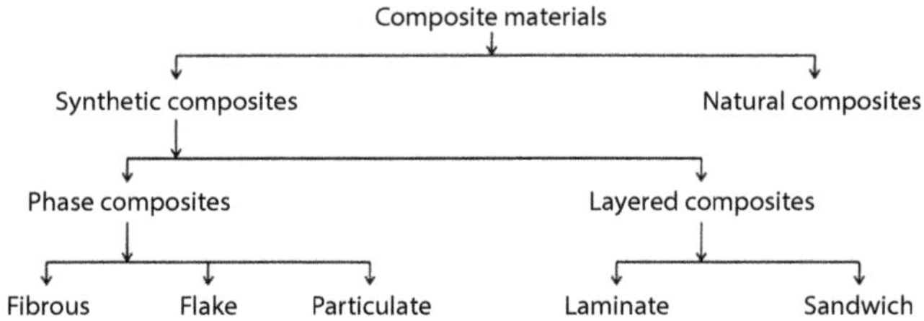

Figure 1.2 Types of composites.

polymer-based nanocomposites and coatings, this chapter presents useful information on structure-property-processing of nanocomposites. Some of their applications will also be briefly discussed.

1.2 Polymer Nanocomposites

1.2.1 Types of Polymer Nanocomposites: Processing

The possibility of combining nanoscale fillers with polymers is enormous, as polymers themselves are a huge class of materials with versatility in their chemistry and physical structure. The tunability of polymeric structures results in various forms like hard plastics, soft foams, coatings and even cellular structures and biomaterials. Inspired by many natural structures and composites like bone, teeth and nacre, many hierarchical structures of polymer nanocomposites have been built. Depending upon the use of either thermoplastic or thermosetting polymers, various processing techniques are adopted to fabricate these materials. For example, commercially developed large-scale thermoplastic composites of polymers can be processed using melt processing mechanical methods like extrusion and injection molding (Figure 1.3). Composite processing machines ranging from lab-scale customized extruders to pressurized plunger-type injection-molding machines are sometimes well suited for thermoplastic polymer and 1D nanofillers like nanotubes and nanofibers. Although these nanofibers/nanotubes significantly depend on their orientation in the shear developed during melt processing, many times a solution or solvent precasting with polymer is done for blending them with certain matrices. Shear is generated through twin-screw extruding for some matrices like polyether ether ketone (PEEK) in some cases to disperse nanofibers up to 1 weight percent (wt%) [2, 3]. The intrinsic viscosity of thermoplastic polymers contributes to the high shear which is generated inside the barrel

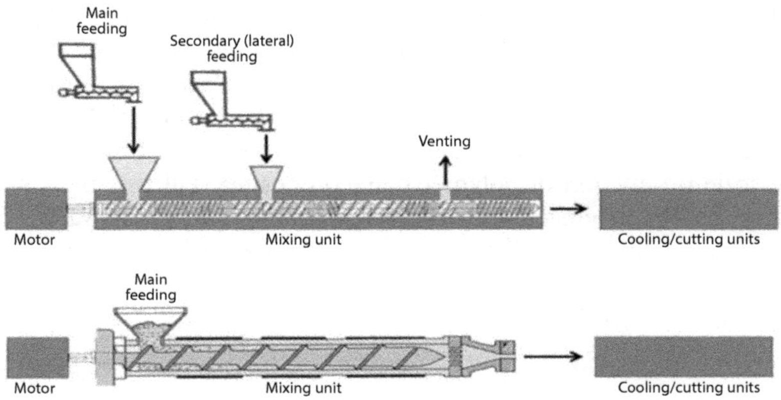

Figure 1.3 Processing of thermoplastic nanocomposites.

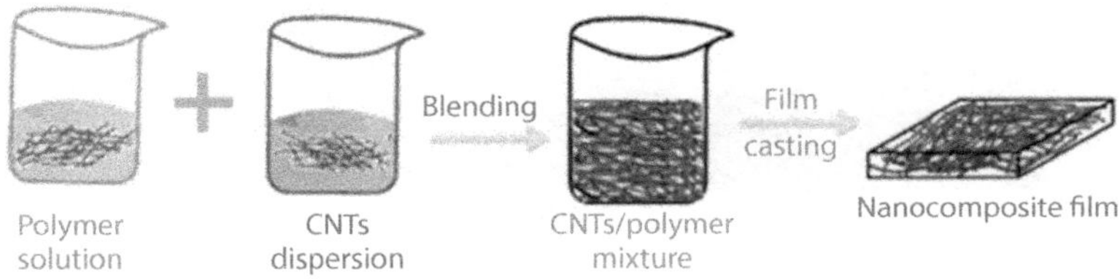

Figure 1.4 Processing of thermosetting nanocomposites.

during processing and helps in dispersing the carbon nanotubes/carbon nanofibers (CNTs/CNFs). The orientation of the fibers can be an issue when using such techniques. Certain composites may require directional properties, others might not. Depending upon the requirement of the product, the suitability of the technique may be decided. Some scientists have also used preprocessing or a precursor technique like ball milling or sonication prior to final or intermediate processing to achieve better aspect ratio of the nanofillers [4].

Thermosetting polymers like epoxy and polyurethane and their variants have to undego a solution/solvent-based processing protocol when reinforced with nanofillers like carbon nanotubes (CNTs) (Figure 1.4). Solvent-based processing requires separate dispersion or dissolution of CNTs directly in polymer mixture and then casting into a given mold, unlike thermoplastic processing techniques of extrusion, where *in-situ* mixing of nanofillers are carried out.

But the new nanoscale fillers demand greater technological advancements for dispersing and homogenizing them into polymer matrices. As needs have changed, an upgrade of conventional processing equipment used in the polymer and rubber industries is required. Parameters like stability and control over shape, particle size and surface area can only be attained if the equipment in use caters to their manipulation and alteration. New designs may soon be entering the industry for commercialization and large-scale production of polymer nanocomposites. Till that time, small batches may be produced with prototypes being used at laboratory and research scale. Shown in Figures 1.5–1.7 are three such essential types of equipment used by our laboratory to produce polyurethane-clay nanocomposite coatings [5–10].

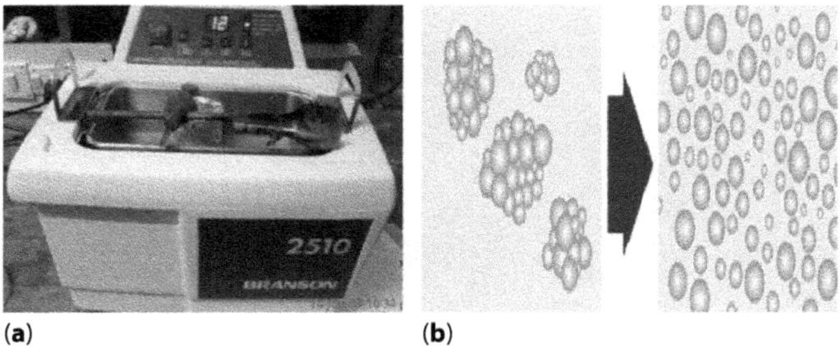

Figure 1.5 (a) View of ultrasonic bath used for sonication with time control. (b) Schematic representation of the deagglomeration of particles by sonication with the help of cavitation.

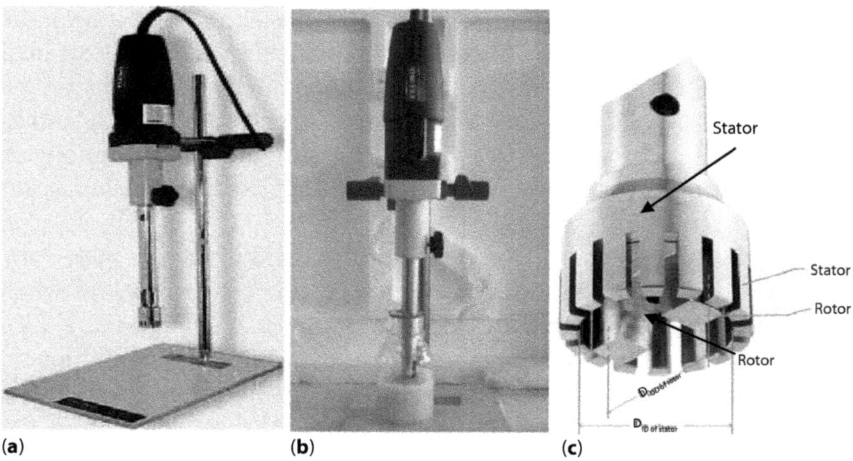

Figure 1.6 (a) Oblique and (b) front view of high shear homogenizer and (c) its stator-rotor.

Figure 1.7 (a,b) Stator-rotor system. (c,d) Action of high shear on sample and the vortex motion generated (white liquid streak marked with arrows) during sample processing.

Table 1.1 Equipments and its specifications.

Equipment	Function	Specifications
Ultrasonic bath	Breaks down agglomerates	Branson 3510; 240V; 40kHz
High shear homogenizer	Homogenizes, disperses	Fluka FA-25; 500W; 220V; 50Hz
Mechanical stirrer	Mixes and blends	Perfit; 220V, 250 rpm

1.2.1.1 Equipment and Processing

For preparation and processing of polymer-nanomaterial/nanofiller nanocomposites and coating formulations, a combination of three different types of equipment can be used. Table 1.1 is a concise list of the types of equipment and their functions.

The ultrasonic apparatus (Figure 1.5a) disperses and deagglomerates the nanomaterial/nanofiller using the principle of ultrasonic cavitation (Figure 1.5b). The sound waves that propagate in the liquid (organic solvents and binder) result in alternating high-pressure (compression) and low-pressure (rarefaction) cycles. This applies mechanical stress on the attracting electrostatic forces (e.g., van der Waals forces) between the individual particles.

Ultrasonic cavitation in liquids causes high speed liquid jets of up to a maximum of 1000 km/h. Such jets press liquid at high pressure between the particles and separate them from each other. Smaller particles are accelerated with the liquid jets and collide at high speeds, thus effectively milling the micron-size and sub-micron-size particles, as shown in Figure 1.5b.

The high shear homogenizer consists of a high speed rotor housed inside a fixed stator (Figure 1.6). The high speed rotation of the rotor generates strong centrifugal force by which materials are sucked into the narrow gap between the precision-engineered stator chamber and rotor from the top and bottom end of the agitator (Figure 1.7c,d). With the intense mechanical and hydraulic shear generated in the narrow gap between the stator and rotor, material is dispersed and emulsified in the first step (Figure 1.7a,b). With the function of high linear velocity (min = 15 m/s; max = 40 m/s) generated by the high speed rotating rotor, strong hydraulic shearing force (Equation 1.1) and other mechanical forces, the material is completely dispersed, emulsified and homogenized.

The final material is ejected from the slanted slot in the stator (Figure 1.7b). Material is continuously ejected at high speed from the slanted slots, while it faces internal material friction and wall resistance. Meanwhile, the flow direction of material is changed with the strong suction force of the high speed rotating rotor and a vortex motion is generated (Figure 1.7c,d). After several cycles, materials are thoroughly dispersed, emulsified and homogenized.

The shear stress exerted on the nanomaterials/nanofillers as per Cao *et al.* [11] can be calculated using Equation 1.1:

$$\tau = \mu\dot{\gamma} \tag{1.1}$$

where τ is the shear stress, $\dot{\gamma}$ is the shear rate, and μ is the solution viscosity. The shear rate of the homogenizer is calculated by Equation 1.2:

$$\dot{\gamma} = 2\pi(OD)R/(ID - OD)60 \tag{1.2}$$

where *OD* is the outside diameter of rotor, and *ID* is the internal diameter of the stator, as shown in Figure 1.8. *R* is the revolutions per minute of the stator-rotor system of the homogenizer.

The equipment discussed in the above section is effective for small-sized laboratory samples but may need a constant scale-up to cater to industrial needs.

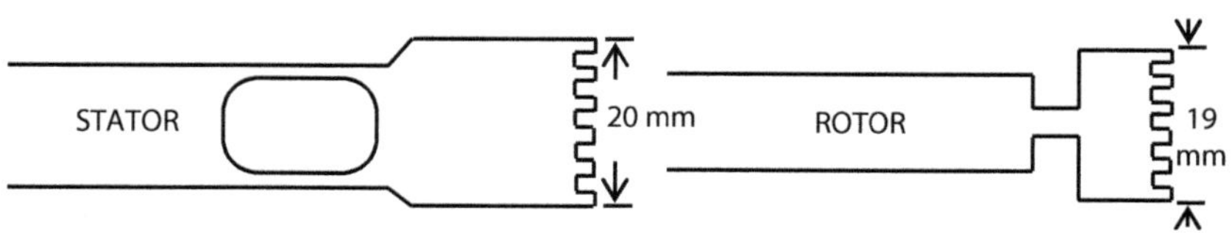

Figure 1.8 Schematic drawing of stator and rotor showing *ID* and *OD*. Dimensions shown in mm are not to scale.

1.2.2 Polymer Property Enhancements

A variety of polymer-nanofiller combinations, such as nylon, polypropylene, polyethylene, polybutene, polyurethanes, etc., have been fabricated and tested for property improvements. After the pioneering work done by the Toyota group [12], an enormous amount of work has been carried out on preparation of polymer-based nanocomposites. Now the focus is moving towards polymer nanocomposites, which are light, possess good mechanical properties, and are eco-friendly (i.e., biocompatible/degradable), for mainly two reasons: environmental concerns and the realization that our petroleum resources are finite. In this regard, several research groups have developed different nanocomposites based on biocompatible/degradable polymers which are obtained from fossil sources as well as renewable resources. Particularly, more attention has been paid to polymers from renewable resources because they allow adding value to agricultural products, which is economically important for many countries. So far, different biocompatible/degradable polymers have been used such as poly caprolactone (PCL), poly lactic acid (PLA), poly hydroxybutyrate (PHB), poly butylene succinate (PBS), natural rubber, starch, cellulose, etc. Several researchers have also reported nanocomposites based on blends of biocompatible/degradable polymers or obtained blending of these polymers with non-degradable polymers. Usuki *et al.* [12] from the Toyota research group were the first to prepare commercial polymer nanocomposites by solution polymerization of caprolactam in the clay galleries. Later on, this product was marketed by UBE Industries and Bayer. Currently, these nylon 6-based nanocomposites are used to make belts for Toyota car engines and also for the production of packaging film [12].

Table 1.2 briefly lists the polymers, nanofillers and their respective property improvements. Shown at a glance are some of the common polymer-based nanocomposites and coatings.

Table 1.2 Various polymers, nanofillers and expected property improvements.

Sl. no.	Polymer	Nanofiller	Property enhancement	Ref.
1	Polyurethane thermoset PU coatings	Clays	Scratch resistance, Barrier resistance	[5–10]
2	PU thermoplastic	Clay	Mechanical properties	[13–14]
3	PU	CNT	Tensile Strength, Electrical	[15]
4	TPU	MWCNT	Thermal, Electrical conductivity	[16]
5	PP	CNT	Tensile modulus	[17]

1.2.3 Polymer Nanocomposite Structure and Morphology

The long-chain molecular structure of polymers undergoes transformation depending on the kind of interface of polymeric chains with the included nanofiller. There are three main material constituents in any composite: the matrix, the reinforcement (fiber), and the so-called interfacial region. The interfacial region is responsible for communication between the matrix and filler and has conventionally ascribed properties different from the bulk matrix because of its proximity to the surface of the filler [18].

Figure 1.9 depicts the expected morphology of the polymer in the presence of a 2D nanofiller like clay or graphene. If the plate-like structures are not dispersed and delaminated at nanoscale then the resulting composite is a microcomposite. If the polymer chains happen to shear past the "d-spacing" of the 2D nanofiller an intercalated nanocomposite is formed. An exfoliated nanocomposite is formed when the polymer chains not only enter into the gallery spacing of the 2D nanostructure but also are able to separate/disperse and completely delaminate the clay/graphene into single platelets. A partially intercalated and exfoliated sample forms a mixed composite (Figure 1.9). This structure of polymer-nanofiller combination affects the final properties of the nanocomposite and is also responsible for its performance.

The surface and bulk morphology is governed by the nanofiller type, size and composition. For instance, the nanoclay imbibed into thermoset two-pack polyurethane significantly changes its hard-soft segmental

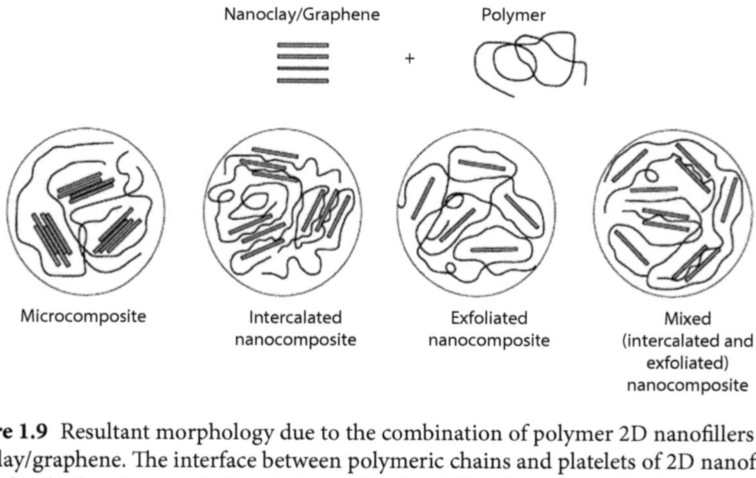

Figure 1.9 Resultant morphology due to the combination of polymer 2D nanofillers like clay/graphene. The interface between polymeric chains and platelets of 2D nanofiller controls whether the morphology is intercalated, exfoliated or a mix of both.

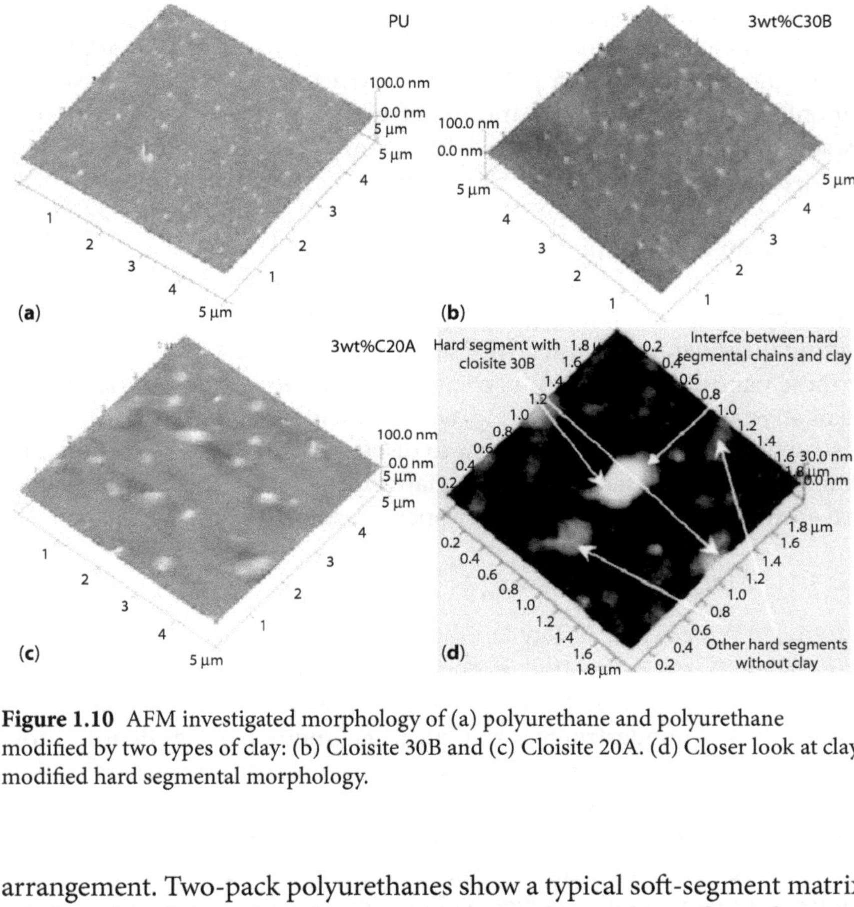

Figure 1.10 AFM investigated morphology of (a) polyurethane and polyurethane modified by two types of clay: (b) Cloisite 30B and (c) Cloisite 20A. (d) Closer look at clay modified hard segmental morphology.

arrangement. Two-pack polyurethanes show a typical soft-segment matrix reinforced by flakes of hard segments. The preferential interface of certain clays, like Cloisite 20A and Cloisite 30B, which have an OH group in their organic modifier with hard segments of polyurethanes, forms an orb-like or tubular surface morphology, as seen in Figure 1.10 [9]. This morphology is a result of the chemical and physical interactions [5]. These interactions are dependent on H-bonding and other primary bondings, and also the size of the nanoplatelets achieved in the formulation. The particles if fine or ultra-fine control the level of interactions and subsequently the morphology of the hard segments.

1.2.4 Characterization of Polymer Nanocomposites

Characterization and testing are vital tools to understand and apply new materials like polymer nanocomposites. It not only helps to apply materials for novel and established applications but also to further improve materials. For nanofilled composites it is essential to analyze and characterize the intricate changes in polymeric matrices. By estimating the size of nanofillers and nanomaterials more control of properties may be achieved. Establishing structure-property relationships in these composites lets engineers, scientists and those in the industry have supreme control over final properties and modulate the material characteristics according to the environment in which they are expected to perform. Moreover, service life predictions and other tests involving durability and specifically mechanical, thermal, optical, and chemical properties must be established before putting these nanocomposites into suitable applications. The performance of these nanocomposites in a particular application is judged based on all qualitative and quantitative testing, which can be either destructive or nondestructive in nature. It is pertinent to discuss some of these techniques of characterization. Based on what is analyzed the techniques can be classified as morphological, spectral or property testing.

1.2.4.1 Morphological Testing

1.2.4.1.1 Wide-Angle X-Ray Diffraction (XRD)

Wide-angle X-ray diffraction is one of the most dependable techniques to quantify dispersion and nanostructure in polymer-nanofiller nanocomposites. It is a nondestructive analytical technique for the identification and quantitative determination of long-range order in various crystalline compounds. X-rays are electromagnetic radiation generated when an electron beam accelerated through a high voltage field hits a metal, which acts as an anode. The wavelength (λ) of X-rays is characteristic of the target anode material used and is given by Equation 1.3:

$$\lambda = hc/E \qquad (1.3)$$

where h is the Planck's constant, c is the velocity of light and E is the energy of the photon. X-ray is used to interpret the morphologies of nanocomposites and to measure the d-spacing (Figure 1.11). X-rays are generated when high-energy electrons impinge on a metal target, in this case copper (Cu). With a Cu target, the characteristic spectrums, produced under the normal operating voltage of 30–50 kV between the cathode and anode, consist of K_α and K_β lines. In most polymer studies, only the K_α line, which is the stronger of the two characteristic lines, is used. The K_β lines are removed by filtering through monochromatization. The K_α line itself contains doublet lines, the $K_{\alpha1}$ (λ = 1.5405 Å) and $K_{\alpha2}$ (λ = 1.5443 Å), and is usually unresolved. It is therefore regarded as a single line of weighted average wavelength of 1.5417 Å. Crystal lattice can be considered as a family of planes extending through the atoms of the lattice and is defined by a triplet (hkl) of integral numbers that are called Miller indices. X-rays of wavelength λ are only diffracted from a system of parallel equidistant lattice planes of spacing d_{hkl} if Bragg's law (Figure 1.11) as given by Equation 1.4 is satisfied:

$$n\lambda = 2d_{hkl}\sin\theta \tag{1.4}$$

where n, a positive integer, denotes the order of reflection, λ is wavelength, θ is the angle between incident ray and reflecting plane, and d_{hkl} is the distance between adjacent planes in the crystals [19]. Of particular interest is the characterization of 2D nanofillers like graphene and clays (layered silicates), as the intergallery spacing or d-spacing are quite easily deciphered along with the Miller indices (hkl) or planar indices.

Although XRD offers a convenient and simple method to determine the average interlayer spacing of the silicate layers in the pristine layered silicates and in the intercalated nanocomposites, the above technique cannot determine the spatial distribution of the silicate layers as well as any structural non-homogeneity in the observed nanocomposites, namely the simultaneous occurrence of both intercalated and exfoliated structures. Thus, nanocomposite structures should be studied not solely by XRD patterns, but also by using transmission electron microscopy (TEM). The results from both can be well corroborated for reaching a good estimation of dispersion and its extent in the nanocomposites.

1.2.4.1.2 Transmission Electron Microscopy (TEM)

Transmission electron microscopy allows a qualitative evaluation of the internal structure and spatial distribution of the various phases through

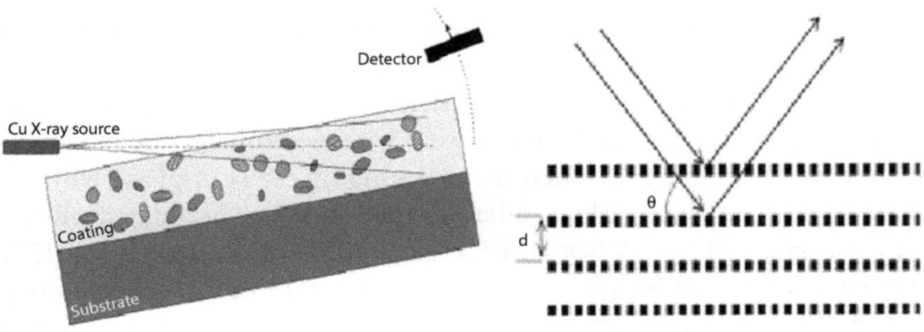

Figure 1.11 X-ray diffraction beam projected on (a) nanocomposite coating sample and (b) a schematic for Bragg's law.

direct visualization. In the case of 2D nanofillers like clay, besides the occurrence of intercalated and exfoliated nanocomposites, the above technique can also prove the formation of flocculated structures. In typical TEM images clays are seen to be randomly dispersed, with some intercalation of the platelets in the PU matrix [14, 20, 21]. Verma *et al.* [8, 9] observed delaminated, exfoliated and homogeneously dispersed platelets in some regions of cross-linked polyester clay nanocomposites, while in some regions there were just agglomerated structures. These representative regions may be indicative of the "little intercalation" and "large intercalation" corresponding to the primary and secondary characteristic XRD peaks of PU-clay nanocomposites [22].

In fact, images from TEM help to know the affect of polymerization on the extent of exfoliation. In a specific investigation, Solarski *et al.* [23] found that during the chain growth of PU some cross-linking reactions occurred and this reaction prevented the exfoliation of the layers. The clarity of the pictures facilitates a better understanding of nanostructures. Large domain sizes (100 to 500 nm) for the intercalated silicates can be easily analyzed [24]. A rather intricate study by Wang *et al.* [25] demonstrated that the intercalated silicate domains had a collection of 5–10 nearly parallel layered silicates (shown by dark lines) with d-spacings of about 5–8 nm, 6–12nm and 15–25 nm. Hence, TEM is a complimentary technique to XRD for characterizing dispersion of organoclay in PU nanocomposites.

1.2.4.2 Spectral Testing

1.2.4.2.1 Fourier Transform Infrared Spectroscopy-Attenuated Total Reflectance (FTIR-ATR)

Chemical structures of the organoclays, components of polyurethane and their nanocomposite coatings are examined by FTIR-ATR [26]. The chemical bonds and their assignments and different molecular structures can be identified by analyzing IR spectra. The molecular vibrations result in a change in the bond dipole moment, as a consequence of change in the electron distribution in the bonds. It is possible to stimulate transitions between energy levels by interaction with electromagnetic radiations of appropriate frequency. When the vibrating dipole is in phase with the electric vector of the incident radiation the vibrations are enhanced and there is transfer of energy from the incident radiation to the molecule [27]. The detection of this energy absorption constitutes infrared spectroscopy. Figure 1.12 shows how the spectral transitions are detected by scanning through the frequency whilst continuously monitoring the transmitted light intensity [28].

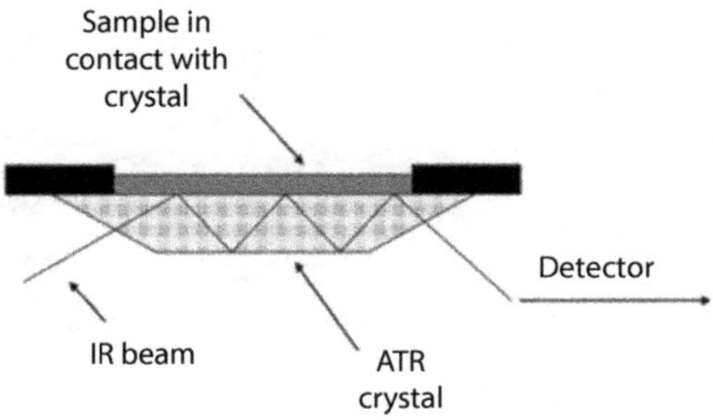

Figure 1.12 Schematic for detection of energy absorptions in FTIR-ATR.

The simplified relationship between the intensities and the dipole moment is given by Equation 1.5 [28]:

$$I \sim \left(\partial \mu / \partial l\right)^2 \tag{1.5}$$

where I is the intensity of the peak, μ is the weak dipole moment, and l is the distance between the atoms in the bond under assessment [27].

1.2.4.2.2 Color-Eye Spectrophotometer

The Color-Eye spectrophotometer has programmable adjustments; a motorized control automatically adjusts the UV content found in natural D65 daylight. A motorized specular insert measures color shifts as a result of gloss or texture in coatings. It gives precise readings on diffuse or direct transmission samples, including thin films, by using the removable sample holder called a cuvette. The machine measures the percentage transmittance of polymer suspensions in the wavelength region of 350–750 nm. These are particularly useful for polymeric coatings.

1.2.4.3 Testing

1.2.4.3.1 Rheological and Flow Characterization

Flow characteristics are of principal importance to polymers and their coatings. From storage to application of wet formulation till the formation of final dry film, rheology is a vital parameter. Each step involves varying shear rate ranges, and the polymer viscosities are specific to that range.

1.2.4.3.2 Mechanical Testing

Mechanical tests like the ones performed on a universal testing machine measuring tensile strength and compressive strength are of prime importance apart from impact testers and flexural or bend testers. These provide a fair idea of the mechanical strength of polymeric nanocomposites and

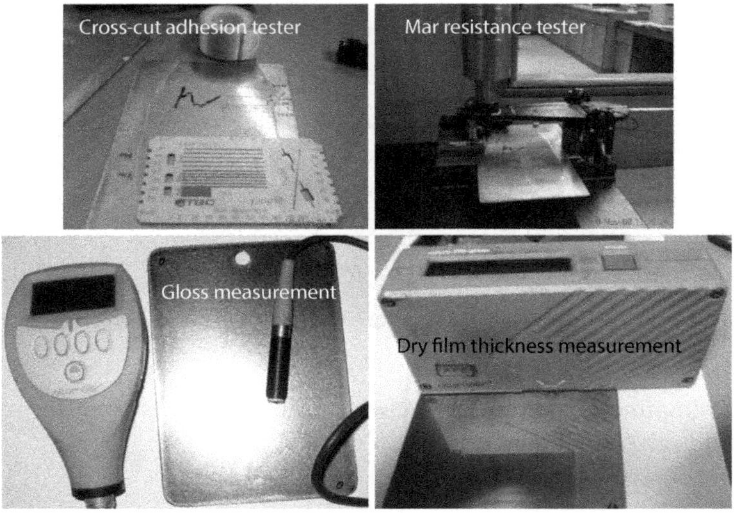

Figure 1.13 Various surface and mechanical testing for coatings on coated mild steel substrates.

their coatings. Other primary tests are Hardness testing (Shore A & D) and wear or tribological tests like pin on disk or Taber abrader. These help to estimate the dynamic wear or abrasion resistance properties. Most of these tests are destructive in nature. For coatings, mar resistance, scratch resistance and adhesion tests are necessary for knowing its quality (Figure 1.13).

1.2.5 Applications

Nanocomposite science and technology is a corollary of the field of engineering plastics, and has potentially entered into the areas of coatings/thin films, structural materials, medical/biomedical product packaging, and electronic and photonic devices. These materials with versatile properties have also attracted the attention of researchers in the field of dielectrics and electrical insulation. Since new properties are brought about from the interactions of nanofillers with polymer matrices, mesoscopic properties are expected as an outcome, which could be of interest to both scientists and engineers [29].

The possibility of fabricating electronic devices is immense by use of techniques like soft lithography, lamination, spin casting and solution casting. Polymers are intrinsically insulating except for a few of them like polyaniline, etc., which could be conducting; by use of certain semiconducting nanofillers like CdS, CdSe, CuS, ZnS, and Fe_3O_4 there is a total turnaround of conducting properties. So these combinations create greater opportunities for creating new and peculiar devices with versatile functions suitable for novel and new-age applications.

Of all the polymer nanocomposites, polyurethane-clay has evinced a lot of interest because of its wide chemistry and compatibility with wide applications in areas like coatings.

1.2.5.1 *Nanocomposite Coatings: Focus PU-Clay Coatings*

A wide variety of fillers, whiskers and fibers, as well as clay and wollastonites, are being used in PU formulations to reinforce the PU matrix for improved hardness, strength and stiffness [30]. In general, the properties of PU composites depend on filler shape, average diameter and interfacial coupling. To meet modern technological demands in the coatings arena, the size of the materials should be reduced to the nanometer scale. The high specific surface area of nanomaterials, which results in high reactivity and a strong tendency towards agglomeration and rapid grain growth, also needs to be controlled as per the application [31–33].

Nanomaterials can be classified into nanocrystalline materials and nanoparticles. The former are polycrystalline bulk materials with grain sizes in the nanometer range (less than 100 nm), while the latter refers to ultrafine dispersive particles with diameters below 100 nm. Nanoparticles are generally considered as the building blocks of bulk nanocrystalline materials. Nanoparticles synthesized from several routes may have different internal structures that would affect the properties of materials consolidated from them. As unique properties of nanocrystalline materials derived from their fine grain size, it is of crucial importance to retain the microstructure at a nanometer scale during consolidation to form the bulk materials.

The search for nanostructured coatings with unique physicochemical properties is driven by the improvement in coating technologies and the availability of various kinds of synthesized nanopowders, which results in the incorporation of an interface with high cohesive strength to the PU matrix; good mechanical, thermal, barrier and flame-retardant properties; higher specific heat and electrical resistivity; and improved hardness and fracture toughness [34–36]. Since the past decade, favorable applications have been found for hard and wear-resistant ceramic coatings in industrial sectors [37]. However, polymer-clay nanocomposite coatings in which clay platelets (e.g., MMT and its organically modified forms or even micas and their variants) at the nanoscale are dispersed in a polymeric matrix possess several more advantages such as low clay content (between 1 and

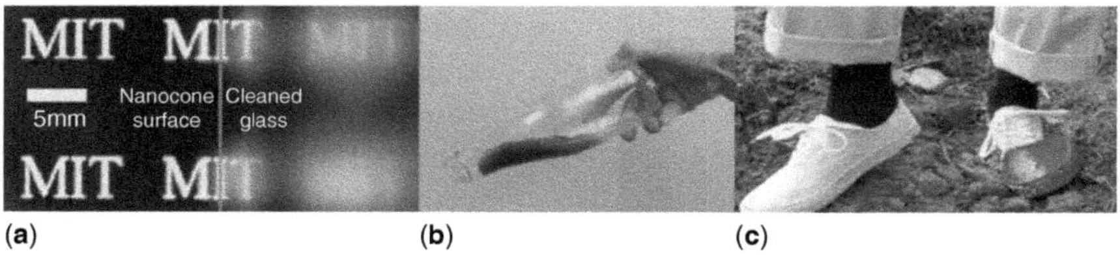

Figure 1.14 Applications of nanocomposite coatings (a) Anti-reflective, anti-fogging surface, (b) superhydrophobic surface inside a sauce bottle, and (c) left shoe with self-cleaning coating [42].

5 wt%), lighten the weight and are economical; and also possess enhanced mechanical behavior (i.e., higher stiffness and strength) [38, 39]; show better permeation-barrier properties due to increase in tortuous movement of water, oxygen or any other penetrating molecule [14, 36, 40, 41]; and improve thermal properties such as fire retardancy and heat distortion temperature (HDT) [35, 38].

Recently, researchers at MIT, Cambridge, Massachusetts, have controlled nanotextured surfaces of coatings to develop anti-reflective with super-transmittive properties [42] (Figure 1.14). These nanotextured surfaces may be tuned into various useful applications like super hydrophobic bottles and self-cleaning dirt-resistant shoes (Figure 1.14b,c). Also, PU-clay nanocomposite coatings have found application as superhydrophobic coatings with high adhesion strength and temperature and humidity resistance [43, 44]. Superhydrophobic coatings can potentially be extended into anti-fogging, self-cleaning and glare-free coatings for applications in solar panels, automotive windshields, house and building windows, smart phones and television screens. Some of these coatings have organically modified variants of montmorillonite (MMT) clays. Montmorillonite actually belongs to the general class of materials known as layered silicates, which incidentally have been an integral part of paint formulations [30].

1.3 Conclusions

Over the span of the last decade or more, polymer nanocomposites have come up with promising applications which are slowly replacing conventional materials. Due to such a wide spectrum of applications to which these materials are able to cater, it is pertinent that we call polymer nanocomposites and coatings as game changers for the new age. The scientific society has been eager to take up new challenges in developing novel materials using this combination. Soon large-scale commercialization by scaling up laboratory-based technologies and methods will take place with the entry of these products into consumer markets. The scope is huge and the potential is infinite once we understand how to specifically control particle or nanofiller size and thus tame the mysterious interfaces formed between polymeric chain molecules and nanofiller surfaces. Once the interfacial mystery is solved, technologists will have total control over polymeric nanocomposites and the properties will be tailored as per their suitability for an application.

Acknowledgments

The author acknowledges the projects under TEQIP-II, SAP and PURSE grants for supporting most of the work undertaken in polymer-clay nanocomposites. This chapter is an extension of the actual work.

References

1. Fischer, H., Polymer nanocomposites from fundamental research to specific applications. *Mater. Sci. Eng. C.* 23, 763–72, 2003.
2. Patel, P., Stec, A.A., Hull, T.R., Naffakh, M., Diez-Pascual, A.M., Ellis, G., Safronava, N., and Lyon, R.E., Flammability properties of PEEK and carbon nanotube composites. *Polym. Degrad. Stab.* 97, 2492–2502, 2012.
3. Díez-Pascual, A.M., Naffakh, M., Gómez, M.A., Marco, C., Ellis, G., Martínez, M.T., Ansón, A., González-Domínguez, J.M., Martínez-Rubi, Y., and Simard, B., Development and characterization of PEEK/carbon nanotube composites. *Carbon* 47, 3079–3090, 2009.
4. Ma, P.C., Siddiqui, N.A., Marom, G., and Kim, J.K., Dispersion and functionalization of carbon nanotubes for polymer-based nanocomposites: A review. *Compos. Part A* 41, 1345–1367, 2010.
5. Verma, G., Kaushik, A., and Ghosh, A.K., Interaction of nano-sized materials with polymer chains in polymer nanocomposite thin films: An AFM perspective. *AIP Conf. Proc.* 1393, 377–378, 2011.
6. Verma, G., Kaushik, A., and Ghosh, A.K., Preparation and characterization PU-clay nanocomposites, in: *Emerging Paradigms in Nanotechnology*, pp. 57–69, Pearson: New Delhi, 2012.
7. Verma, G., Kaushik, A., and Ghosh, A.K., Developing and characterizing polyurethane-nanoclay coatings for better scratch and mar resistance. *Adv. Mater. Res.* 585, 473–477, 2012.
8. Verma, G., Kaushik, A., and Ghosh, A.K., Preparation, characterization and properties of organoclay reinforced polyurethane nanocomposite coatings. *J. Plast. Film Sheeting* 29, 56–77, 2012.

9. Verma, G., Kaushik, A., and Ghosh, A.K., Comparative assessment of nanomorphology and properties of spray coated clear polyurethane coatings reinforced with different organoclays. *Prog. Org. Coat.* 76, 1046–1056, 2013.
10. Verma, G., Kaushik, A., and Ghosh, A.K., Nano-interfaces between clay platelets and polyurethane hard segments in spray coated automotive nanocomposites. *Prog. Org. Coat.* 99, 282–294, 2016.
11. Cao, X., Lee, L.J., Widya, T., and Macosko, C., Polyurethane/clay nanocomposites foams: Processing, structure and properties. *Polymer* 46, 775–783, 2005.
12. Usuki, A., Kojima, Y., Kawasumi, M., Okada, A., Fukushima, Y., Kurauchi, T., and Kamigaito, O., Synthesis of nylon 6-clay hybrid. *J. Mater. Res.* 8, 1179–1184, 1993.
13. Maji, P.K., Guchhait, P.K., and Bhowmick, A.K., Effect of the microstructure of a hyperbranched polymer and nanoclay loading on the morphology and properties of novel polyurethane nanocomposites. *ACS Appl. Mater. Interfaces* 1, 289–300, 2008.
14. Maji, P.K., Das, N.K., and Bhowmick, A.K., Preparation and properties of polyurethane nanocomposites of novel architecture as advanced barrier materials. *Polymer* 51, 1100–1110, 2010.
15. Shamsi, R., Mahyari, M., and Koosha, M., Synthesis of CNT-polyurethane nanocomposites using ester-based polyols with different molecular structure: Mechanical, thermal, and electrical properties. *J. Appl. Polym. Sci.* 134, 44529, 2017.
16. Wang, W., Jiang, F., Jiang, Y., Lu, Y., and Zhang, L., Preparation and properties of polyurethane/multiwalled carbon nanotube nanocomposites by a spray drying process. *J. Appl. Polym. Sci.* 126, 789–795, 2012.
17. Jose, M.V., Dean, D., Tyner, J., Price, G., and Nyairo, E., Polypropylene/carbon nanotube nanocomposite fibers: Process–morphology–property relationships. *J. Appl. Polym. Sci.* 103, 3844–3850, 2007.
18. Wagner, H.D., and Vaia, R.A., Nanocomposites: Issues at the interface. *Mater. Today* 7, 38–42, 2004.
19. Cullity, B.D., and Stock S.R., *Elements of X-ray Diffraction*, 3rd ed., Prentice-Hall, 2001.
20. Maiti, M., and Bhowmick, A.K., New insights into rubber–clay nanocomposites by AFM imaging. *Polymer* 47, 6156–6166, 2006.
21. Lee, H.T., and Lin, L.H., Waterborne polyurethane/clay nanocomposites: Novel effects of the clay and its interlayer ions on the morphology and physical and electrical properties. *Macromolecules* 39, 6133–6141, 2006.
22. Kim, D.S., Kim, J.T., and Woo, W.B., Reaction kinetics and characteristics of polyurethane/clay nanocomposites. *J. Appl. Polym. Sci.* 96, 1641–1647, 2005.
23. Solarski, S., Benali, S., Rochery, M., Devaux, E., Alexandre, M., Monteverde, F., and Dubois, P., Synthesis of a polyurethane/clay nanocomposite used as coating: Interactions between the counterions of clay and the isocyanate and

incidence on the nanocomposite structure. *J. Appl. Polym. Sci.* 95, 238–244, 2005.
24. Tien, Y., and Wei, K.H., The effect of nano-sized silicate layers from montmorillonite on glass transition, dynamic mechanical, and thermal degradation properties of segmented polyurethane. *J. Appl. Polym. Sci.* 86, 1741–1748, 2002.
25. Wang, W.J., Chin, W.K., and Wang, W.J., Synthesis and structural characterizations of [chromophore]+-saponite/polyurethane nanocomposites. *J. Polym. Sci., Part B: Polym. Phys.* 40, 1690–1703, 2002.
26. Almeida, E., Balmayore, M., and Santos, T., Some relevant aspects of the use of FTIR associated techniques in the study of surfaces and coatings. *Prog. Org. Coat.* 44, 233–242, 2002.
27. Stuart, B., *Infrared Spectroscopy: Fundamentals and Applications*, John Wiley & Sons, Inc., 2005.
28. Campbell, D., and White, J.R., *Polymer Characterization: Physical Techniques*, Chapman and Hall: London, 1989.
29. Tanaka, T., Montanari, G.C., and Mulhaupt, R., Polymer nanocomposites as dielectrics and electrical insulation-perspectives for processing technologies, material characterization and future applications. *IEEE Trans. Dielectr. Electr. Insul.* 11, 763–784, 2004.
30. Paul, S. (Ed.), *Surface Coatings. Science and Technology*, John Wiley: New York, 1985.
31. Baer, D.R., Burrows, P.E., and El-Azab, A.A., Enhancing coating functionality using nanoscience and nanotechnology. *Prog. Org. Coat.* 47, 342–356, 2003.
32. Perera, D.Y., Effect of pigmentation on organic coating characteristics. *Prog. Org. Coat.* 50, 247–262, 2004.
33. Fernando, R.H., and Sung, L.-P. (Eds.), *Nanotechnology Applications in Coatings*, vol. 1008, American Chemical Society, 2009.
34. Advani, S.G., *Processing and Properties of Nanocomposites*, World Scientific Pub., 2007.
35. Ke, Y.C., and Stroeve, P., *Polymer-Layered Silicate and Silica Nanocomposites*, Elsevier, 2005.
36. Alexandre, M., and Dubois, P., Polymer-layered silicate nanocomposites: Preparation, properties and uses of a new class of materials. *Mater. Sci. Eng. R-Rep.* 28, 1–63, 2000.
37. Tjong, S.C., and Chen, H., Nanocrystalline materials and coatings. *Mater. Sci. Eng. R-Rep.* 45, 1–88, 2004.
38. Chattopadhyay, D.K., and Raju, K.V., Structural engineering of polyurethane coatings for high performance applications. *Prog. Polym. Sci.* 32, 352–418, 2007.
39. Tjong, S.C., Structural and mechanical properties of polymer nanocomposites. *Mater. Sci. Eng. R-Rep.* 53, 73–197, 2006.
40. Chou, C.S., LaFleur, E.E., Lorah, D.P., Slone, R.V., and Neglia, K.D., Aqueous nanocomposite dispersions: Processes, compositions, and uses thereof, US Patent 6838507, assigned to Rohm and Haas Company, 2005.

41. LeBaron, P.C., Wang, Z., and Pinnavaia, T.J., Polymer-layered silicate nanocomposites: An overview. *Appl. Clay Sci.* 15, 11–29, 1999.
42. Park, K.C., Choi, H.J., Chang, C.H., Cohen, R.E., McKinley, G.H., and Barbastathis, G., Nanotextured silica surfaces with robust superhydrophobicity and omnidirectional broadband supertransmissivity. *ACS Nano* 6, 3789–3799, 2012.
43. Steele, A., Bayer, I., and Loth, E., Adhesion strength and superhydrophobicity of polyurethane/organoclay nanocomposite coatings. *J. Appl. Polym. Sci.* 125, E445–E452, 2012.
44. Yeong, Y.H., Steele, A., Loth, E., Bayer, I., De Combarieu, G., and Lakeman, C., Temperature and humidity effects on superhydrophobicity of nanocomposite coatings. *Appl. Phys. Lett.* 100, 053112, 2012.

2

DGEBA Epoxy/$CaCO_3$ Nanocomposites for Improved Chemical Resistance and Mechanical Properties for Coating Applications

Manoj Kumar Shukla[1]*, Archana Mishra[2], Kavita Srivastava[3], A K Rathore[4], and Deepak Srivastava[5]*

[1]*Faculty of Engg. & Tech., Mewar University, Chittorgarh, Rajasthan, India.*
[2]*Department of Plastic Tech., Govt. Polytechnic College, Kota, Rajasthan, India.*
[3]*Department of Chemistry, V. S. S. D. College, Kanpur, U.P., India.*
[4]*School of Chem. Tech., Department of Chemical Engineering, H.B.T.U., Kanpur, U.P., India*
[5]*School of Chem. Tech., Department of Plastic Technology, H.B. T. U. , Kanpur, U.P., India*

Abstract

Epoxy/$CaCO_3$ nanocomposites were prepared by mechanical mixing using two different grades of diglycidyl ether of bisphenol A (DGEBA) epoxy resins (i.e., EEW 180 and EEW 220) and nano $CaCO_3$ (1–10 wt%) followed by sonication for 30 min. Furthermore, these mixtures were used for casting films of 100 μm and subjected to curing at 110 °C for 1 h and post curing at 80 °C for 5 h. The prepared nanocomposite films were exposed to acids, alkalies, artificial seawater and solvents for six months. It was observed that the films of epoxy resins, EEW 180 and EEW 220, containing 5 and 1 wt% of nano-$CaCO_3$ respectively, showed resistance to these chemicals for maximum period of exposure. It was also observed that 5 wt% and 1 wt% loaded nano-$CaCO_3$ composite films respectively, showed maximum resistance to impact, scratch hardness, adhesion and flexibility as compared to other compositions of nanocomposites and neat resin matrices.

Keywords: Nanocomposites, resin matrix, DGEBA epoxy resin, coating, nano-$CaCO_3$

**Corresponding authors*: mkshukla.10@gmail.com; dsri92@gmail.com

Sanjay K. Nayak *et al.* (eds.) Trends and Applications in Advanced Polymeric Materials, (23–44) 2018 © Scrivener Publishing LLC

2.1 Introduction

Epoxy resins are the most commonly used thermosetting resins. The commercial epoxy resins contain aliphatic, cycloaliphatic or aromatic backbones. The most widely used ones are epichlorohydrin and bisphenol-A derived resins. The outstanding performance characteristics of these resins are conveyed by bisphenol-A, which provide toughness, rigidity and elevated temperature performance. The ether linkages and hydroxyl groups provide chemical resistance and adhesive properties respectively [1]. The resins have a variety of applications, such as coatings, adhesives, floorings, in aerospace, electrical and electronics, due to their excellent properties, viz., chemical resistance, outstanding adhesive properties, good mechanical properties and thermal stability [2–3]. The applications of surface coating with epoxy matrix are not only limited to the metal substrate but also extended to wood, furniture, paper and plastic coatings [4]. The presence of hydroxyl groups in the resin offers a variety of curing agents for curing reaction ranging from room temperature to elevated temperature as per the process requirement. The curing reaction can result in a three-dimensional bonding network in the resins, whose properties are dependent on the extent of cross-linking. However, because of its highly cross-linked structure, the resin showed brittleness and crack initiation, followed by propagations and lack of toughness after being cured [5–6]. These shortfalls restrict the resins for wider applications by showing poor resistance to chemicals and coating properties. These limitations can be bridged by the incorporation of various nanofillers.

The incorporation of inorganic nanofillers, such as nano $CaCO_3$, TiO_2, SiO_2 and clay (particle size $\leq$ 100 nm), in resin matrix is attractive for improving the mechanical and chemical properties of resins [7]. The particle size of nanofillers has a major impact on the improvement of properties of nanocomposites. The smaller size nanoparticles penetrate further into the amorphous region or free space available between the chains of macromolecules than the larger ones and provide rather increased toughness and rigidity to nanocomposites [8]. But a major drawback related to these nanofillers is the difficulty of uniform dispersion within the matrix because they have a tendency to agglomerate. To avoid the agglomeration of nanoparticles and enhance the interface adhesion between the polymeric chains and nanoparticulates, surface treatment with dispersants or coupling agents is required [9–12]. The preparation methods also have great effect on the properties of nanocomposites [13]. Basically, three commonly used methods are available to prepare nanocomposites, they are: solution casting, melt blending, and *in-situ* polymerization. In solution casting, polymer, solvent and nanofiller are thoroughly mixed by ultrasonication. In melt mixing, extruder or internal mixtures are used to mix the resins and the nanofillers. The *in-situ* polymerization method includes the blending of monomer and the reinforcement. The monomer is allowed to intercalate between the layers of nano-reinforcement and get polymerized [14]. Kumar *et al.* [15] prepared the epoxy/Al_2O_3 nanocomposites by the solution process and observed that the filling of nanoparticles changed the structure of nanocomposites. Nanocomposite was prepared by Mishra *et al.* [16] by mechanical mixing of synthesized (by *in-situ* deposition technique) nano-$CaCO_3$ and epoxy resin and they found better dispersion in the case of nanoparticulates than commercial ones. To achieve moderate exfoliation of nanoparticles in polymer matrix, common dispersion methods, such as ultrasonication, planetary centrifugal mixing and magnetic dispersion, are used [17].

The prepared epoxy/nano-$CaCO_3$ mixture was used to cast the films on the different panels for characterization of various mechanical properties such as impact strength, toughness, adhesion, flexibility, scratch hardness, chemical resistance and cupping properties. Many researchers have proved that the incorporation of nanoparticles could improve not only the toughness but also the mechanical strength and chemical resistivity of nanocomposite coatings. This might be due to phase interaction between the nanoparticles and the matrix by hydrogen bonding and van der Waals forces as well as the strong covalent or ionic covalent bond developed between organic and inorganic phases [18].

The epoxy resin/nano-$CaCO_3$ composites obtained through *in-situ* and inclusion polymerization showed well-dispersed nanofiller and their tensile property, cupping property and neutral salt spray corrosion resistance were distinctly improved [19]. The impact strength of the resultant nanocomposites was improved due to the nanoparticles, which act as stress concentrators in the nanocomposites to initiate massive crazes and are capable of increased load sharing [20]. This toughening mechanism is responsible for improvement in the impact strength. The impact strength and tensile strength of modified nanoclay reinforced epoxy resin nanocomposites increased by 87.7 and 20.9 percent, respectively, when the nanofiller loading was 3 wt% [21]. This could possibly be due to the deviation and branching of the crack front, forcing a higher energy absorption in the composite [22]. The energy can also be consumed at the particle/matrix interface under strong bonding conditions and therefore the addition of nanofiller can simultaneously improve the stiffness and impact energy. The toughness of nanocomposite was also found to be better than conventional material filled epoxy composites [23].

The epoxy nanocomposites also exhibit improved chemical resistance as compared to conventional composites due to the presence of nanoparticles between the polymer chains and the compact three-dimensional cross-linked structures of the nanocomposites and longer diffusion path of the systems, which restrict the penetration of the chemical reagents, viz., alkalies, acids and solvents. The chemical resistance behavior of epoxy/MMT (with varying amounts; i.e., 0, 1, 3, 5 wt%) nanocomposite samples against different acids, alkalies and solvents was studied by Pashaei *et al.*, [24] who found excellent resistance to chemicals by all the samples against all the corrosive chemical reagents. The performance of epoxy nanocomposite under a corrosive environment was also studied by Abacha *et al.* [25], who evaluated the water diffusion and sulfuric acid corrosion. The diffusion of the environmental solution was measured by noting the increase in weight of samples as a function of immersion time at 80 °C. They found that with the addition of organoclay there was better flexural strength retention under immersion in sulfuric acid.

In the present study, the nanocomposite films were prepared by mechanical mixing and followed by sonication of two series of epoxy resin matrices and nano-$CaCO_3$, subsequently casting the films on different panels. The effect of nanosize $CaCO_3$ on the mechanical properties, viz., impact hardness, scratch resistance, adhesion and flexibility of films were studied. The chemical resistance of nanocomposite films against various acids, alkalies, artificial seawater and solvents was also studied during the investigation.

2.2 Experimental

2.2.1 Preparation of Epoxy/$CaCO_3$ Nanocomposites

The DGEBA epoxy resins having epoxy equivalent weight (EEW) 180 and 220 were mixed separately with varying amounts (i.e., 0, 1, 2, 3, 5 and 10 wt%) of nano-$CaCO_3$. The mechanical mixing was done at high rpm (4000 rpm) for 30 min using an electrical stirrer followed by sonication using a probe-type sonicator (model no. 750F, PCI Analytics Pvt. Ltd., Mumbai) for a further 30 min. The stoichiometric amount of hardener polyamide was mixed thoroughly into the resin mixtures. The mixtures were applied on glass and mild steel panels of size 150 × 50 × 1.25 mm and 150 × 100 × 1.25 mm through a film applicator to form a 100 micron film for each composition. These nanocomposite film coated panels were cured in an air oven for 1 h at 110 °C and post cured at 80 °C for 5 h. The samples prepared by mixing 0, 1, 2, 3, 5 and 10 wt% of nano-$CaCO_3$ into epoxy resins (EEW 180 and EEW 220) were designated as EP^1C-0, EP^1C-1, EP^1C-2, EP^1C-3, EP^1C-5, EP^1C-10, EP^2C-0, EP^2C-1, EP^2C-2, EP^2C-3, EP^2C-5 and EP^2C-10.

2.2.2 Preparation of Panels

Different panels of mild steel of 150 × 100 × 1.25 mm size and glass of 150 × 50 × 1.25 mm size were used, as per IS-101:1986, to evaluate the mechanical and chemical properties of epoxy/nano-$CaCO_3$ composite.

2.2.3 Preparation of Reagents for Chemical Resistance

All of the acids and alkalies taken for chemical resistance testing were prepared by diluting them with distilled water to make a 10% dilution solution, as per IS-101 standard.

2.2.3.1 Artificial Seawater (ASW)

The artificial seawater was prepared in lab; the salts and weight taken to prepare the ASW are given in Table 2.1.

Table 2.1 Artificial seawater prepared as per IS 1404:1970 standard.

Salt	Weight (g)
Sodium chloride	23.476
Magnesium chloride	4.981
Sodium sulphate	3.917
Calcium chloride	1.102
Potassium chloride	0.664
Sodium bicarbonate	0.192
Potassium bromide	0.096
Boric acid	0.026
Strontium chloride	0.024
Sodium fluoride	0.003

2.2.4 Preparation of Films

The films of epoxy/$CaCO_3$ nanocomposite systems were cast on the mild steel panels for evaluation of the mechanical properties and on the glass panels for the evaluation of the chemical resistance to acids, alkalies, artificial seawater and solvents using a Bird film applicator (Sheen Instruments Ltd, U.K.). A dry film thickness of about 100 microns was maintained on all the panels; the films were then cured as per curing schedule, i.e., at 110 °C for 1 h and post curing at 80 °C for 5 h. After the post curing, the film coated mild steel and glass panels were used for the characterization of mechanical and chemical properties respectively.

2.3 Characterization of Epoxy/$CaCO_3$ Nanocomposite

2.3.1 Fourier Transform Infrared (FTIR) Spectra

To confirm the extent of curing reaction and the functional groups present in the epoxy resin matrix, we generally used Fourier transform infrared spectroscopy (FTIR). As we added the curing agent into the resin matrix, the curing reaction started via consumption of hydroxyl groups and the oxirane groups present in the polymeric chain. The disappearance of absorbance band of these two groups in the FTIR spectrum revealed an idea of the extent of curing reaction in the polymeric matrix. The FTIR test of cured and uncured epoxy resin matrix was performed on a VERTEX series FTIR spectrometer (Bruker Corporation, U.K.) at room temperature. The FTIR spectra were recorded in the range of 400–4000 cm^{-1} at a resolution of 2 cm^{-1} with 10 scans. The background spectra were taken in the empty chamber before measurements to eliminate the influence of moisture and carbon dioxide in air.

2.3.2 Mechanical Properties

The mechanical properties of the cured nanocomposite cast films were characterized for the impact resistance, scratch hardness, flexibility and adhesion test or mandrel bend test.

2.3.2.1 Impact Resistance

The impact hardness of the cured films was evaluated as per IS 101 standard using a tubular impact hardness tester (Khushboo Scientific, Mumbai, India). The apparatus consists of a steel block of 500 gm weight which slides vertically between two guides. The block was allowed to fall under gravity onto the die block with a hole at the center. The test panel between the die blocks was impacted by the steel block or indenter. The depth of the indentations was varied by inserting mild steel washers of known thickness between the indenter and tool block. The failure of coating films was shown by cracking and the loss of adhesion at the deformed portions of the test panels.

2.3.2.2 *Scratch Hardness*

The scratch hardness of the cured films of epoxy/$CaCO_3$ nanocomposite was evaluated as per IS 101 standard by using an automatic scratch hardness tester (Sheen Instruments Ltd, U.K.). In this test a hard hemisphere needle of the diameter of 1 mm was allowed to run on the mild steel test panels at a rate of 30–40 mm/sec. A specific load was placed on the top of the needle and the panels were examined for a sign of bare metal.

2.3.2.3 *Adhesion and Flexibility Test*

The adhesion and flexibility of cured films were evaluate as per IS 101 standard. The test was performed on a cylindrical mandel (model 809; Sheen instruments Ltd., U.K.). The mandrel of diameters 1/12 to 1.3 of an inch were taken. Cylindrical bars of different diameters were passed through this test panel. At specified cylindrical bar, the panel was examined in signs of cracking, flapping and detachment from the substrate, after the sample had been bent through 180°.

2.3.2.4 *Chemical Resistance Test*

The chemical resistance of nanocomposite cured films was tested as per IS 101 standard by immersing the glass panels coated with different compositions of nano-$CaCO_3$ in the epoxy matrix into the alkalies, acids, artificial seawater and solvents for six months. The monitoring was carried out for each month. The coated glass panels were half-dipped into the chemical reagents and the remaining portion was left unexposed to compare the film for corrosive properties.

2.3.2.5 *Morphological Properties*

Scanning electron microscopic studies were conducted to investigate the surface morphology of the fractured samples. They give an idea of the dispersion of nanofillers in the resin matrix, that is, whether methods used to prepare the nanocomposites were value added or not. The heterogeneous dispersion led to deterioration of the properties of nanocomposites rather than improving them as expected. The scanning electron microscope (SEM) images of the fractured surface of nanocomposite samples were obtained by using high resolution and low vacuum SEM equipment (model no. Quanta 200 FEG; FEI Company, U.S.A.) to investigate the microstructure and fractured surfaces of nanocomposites. The samples were mounted on aluminium stubs using carbon tape. The samples were coated with a thin layer of gold to prevent charging before observation by SEM.

2.4 Results and Discussion

2.4.1 FTIR Spectroscopic Analysis

The FTIR spectra of uncured neat epoxy matrix, 1 wt% $CaCO_3$ loaded EEW 180 and EEW 220 resins are given in Figures 2.1, 2.2 and 2.3 respectively. The absorption bands at around 3500 cm^{-1} and 913 cm^{-1} were due to the –OH and oxirane group stretching of neat epoxy matrix, respectively. Similar absorption bands also appeared in the case of 1 wt% $CaCO_3$ loaded epoxy matrices. It revealed that nano-$CaCO_3$ remained unreacted in both resin matrices. Hence, it can be concluded that the nanoparticles were physically blended in the matrix systems.

The IR spectra of neat, 1 wt% $CaCO_3$ loaded cured EEW 180 (EP^1C-1) and EEW 220 (EP^2C-1) resin matrices are given in Figures 2.4, 2.5 and 2.6 respectively. In all cured matrices the hydroxyl group formed due to the reaction of oxirane group of epoxy resin and primary amine, which was identified from the appearance of bands near 3355 cm^{-1} and 3500 cm^{-1}. During the curing reaction of epoxy resins with polyamide curing agent,

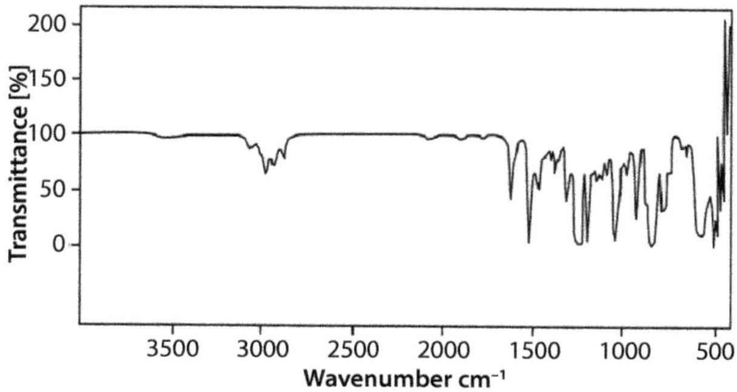

Figure 2.1 FTIR spectrum of uncured neat epoxy matrix.

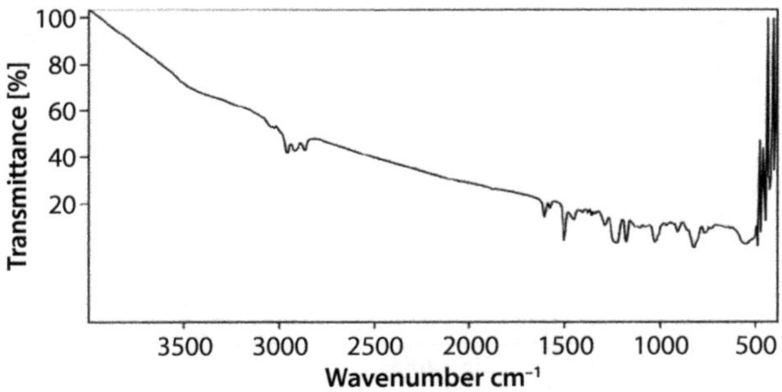

Figure 2.2 FTIR spectrum of uncured EP[1]C-1 sample.

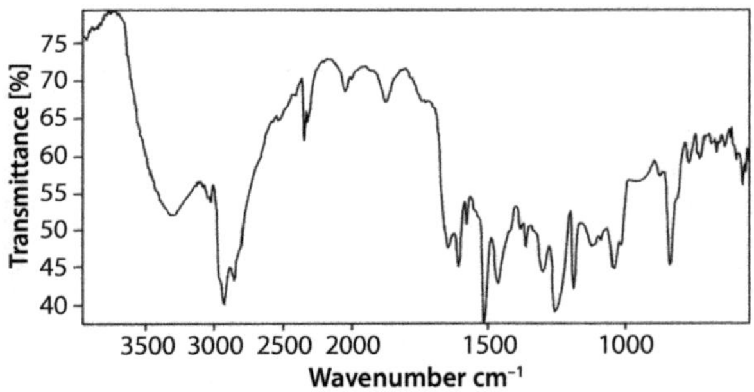

Figure 2.3 FTIR spectrum of cured EP[1]C-0 sample.

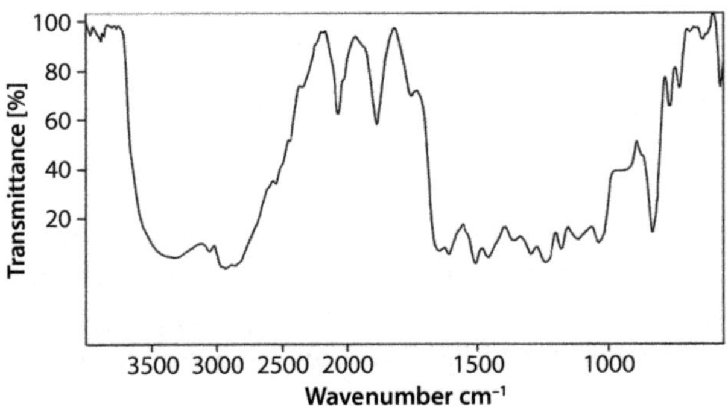

Figure 2.4 FTIR spectrum of cured EP[1]C-1 sample.

primary amine converts into secondary amine and further to tertiary amine. This conversion led to the evolution of –OH groups. A broad absorption band near 3300 cm^{-1} might indicate an increase in the number of –OH groups due to the reaction between oxirane groups and the

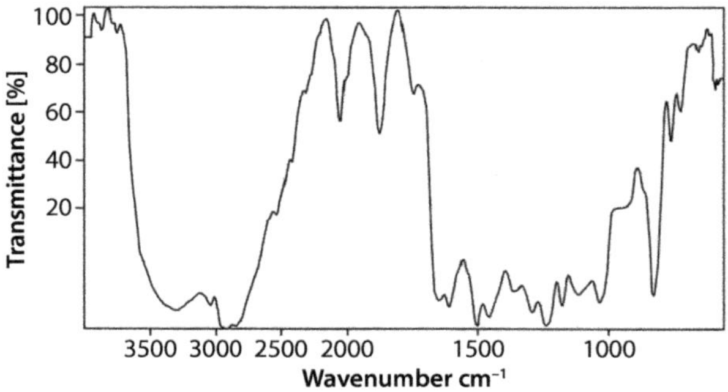

Figure 2.5 FTIR spectrum of EP^2C-1 cured sample.

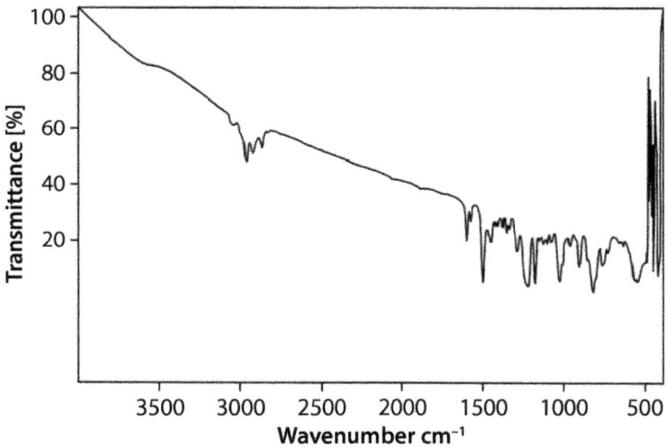

Figure 2.6 FTIR spectrum of uncured EP^2C-1 sample.

amines. Similar bands also appeared for 1 wt% $CaCO_3$ loaded cured epoxy matrices. This showed that the nano-$CaCO_3$ in the epoxy matrix remained unreacted. The oxirane group cleavage during the curing reaction was also confirmed by the absence of bands around 913 cm^{-1} in both of the cured matrices.

2.4.2 Studies on Mechenical Properties

2.4.2.1 Impact Resistance

The experimental results of the impact resistance of nanocomposite films with various concentrations of $CaCO_3$ are shown in Table 2.2. It is clear from

Table 2.2 The experimental data sheet for mechanical properties of coating films.

Sl. no.	EEW	Sample name	Nano $CaCO_3$ (%)	Impact height (cm)	Scratch hardness (kg)	Adhesion & flexibility (Bend test)
1	180	EP^1C-0	0	6	4.2	Fail
2	180	EP^1C-1	1	7	4.3	Fail
3	180	EP^1C-2	2	10	4.5	Fail
4	180	EP^1C-3	3	20	4.8	Fail
5	180	EP^1C-5	5	30	5.2	Pass
6	180	EP^1C-6	10	25	4.8	Fail
7	220	EP^2C-0	0	15	4.0	Fail
8	220	EP^2C-1	1	20	4.5	Pass
9	220	EP^2C-2	2	15	4.3	Fail
10	220	EP^2C-3	3	12	4.2	Fail
11	220	EP^2C-5	5	11	4.1	Fail
12	220	EP^2C-10	10	10	4.0	Fail

the table that as the concentration of nanoparticles increased the impact strength also increased and was found maximum for samples EP^1C-5 and EP^2C-1 respectively. Thereafter, there was a decrease in impact strength upon further loading of nanoparticles. This might be due to agglomeration of nanoparticles in resin matrices, which causes the deterioration in toughness, while the nanoparticles exhibit reinforcing capacities when added in small amounts [26]. The improvement in the impact resistance might also be contributed by load sharing via nanoparticles and the force of attraction developed between macromolecules and nanoparticles due to the hydrogen bond, which results in the enhancement of toughness in matrices at fully cured stage [27–28]. These results are supported by Jain *et al.* [29].

The improvement in the impact strength could also be explained on the basis of energy conservation law, i.e., the applied sock should transfer the energy to the sample upon impact and convert the kinetic energy into heat or deformation. Since impact is applied onto the sample for a very short duration, the energy transferred is measured with respect to time. The response of sample to this effect is to behave like it is brittle to occurrence of fracture at the impact site as per the theory of time-temperature superposition [30–32]. Figures 2.7 and 2.8 showed that as the filler concentration increased in both the epoxy matrices, the impact height increased up to nearly 5 wt% and 1 wt% loading of nanoparticulates respectively. It is obvious that the absorbed energy was also shared by the nanoparticles and led to the prevention of fracture or crack initiation during the impact [33].

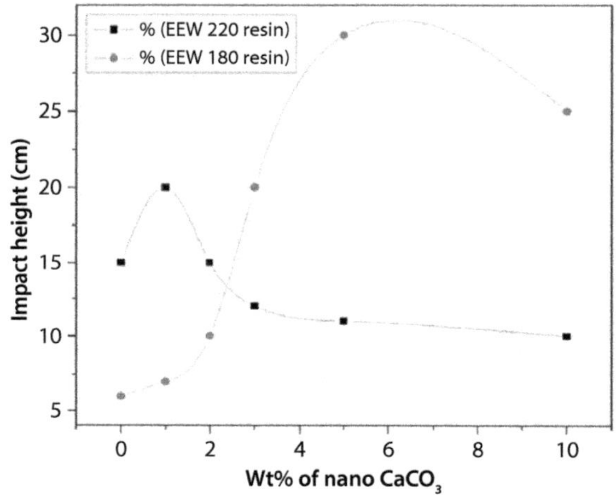

Figure 2.7 Impact resistance of nanocomposite films.

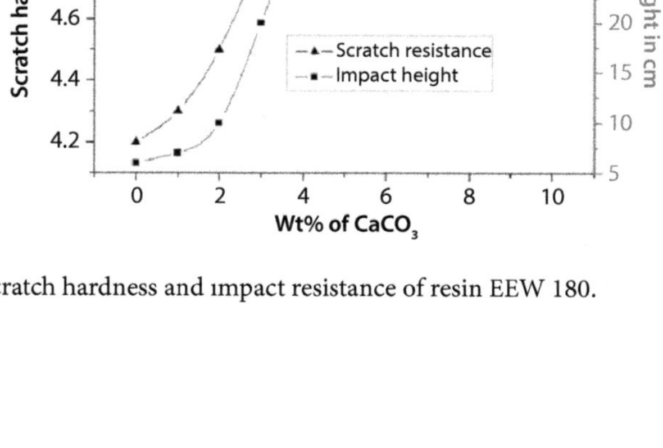

Figure 2.8 Scratch hardness and impact resistance of resin EEW 180.

It was also observed that the loading of nano-$CaCO_3$ and maximum impact height value were found higher in the case of EEW 180 compared to EEW 220 matrix. The higher impact height might be due to the fact that more nanoparticles were present in the amorphous region, which shared more load in the case of the EEW 180 than EEW 220 resin.

2.4.2.2 Studies of Scratch Hardness

The scratch hardness of nanocomposite films is given in Table 2.2, which shows that the scratch resistance of nanocomposite films increased as nano-$CaCO_3$ concentration increased in the matrices and was found maximum for EP^1C-5 and EP^2C-1 sample, respectively; and decreased thereafter. This might be due to aggregation of nanoparticles, which caused a reduction in toughness and bonding between matrix and nanoparticles in both matrices. It can also be observed from Figures 2.9 and 2.10 that more nanoparticle loading is required for maximum scratch hardness in the case of EEW 180 than EEW 220 resin because the first resin system has more free space available than the second one. Since the hardness value is directly related to the proportion of cross-linking between epoxide groups and curing agents, the higher the cross-linking the higher the hardness [34].

The degree of cross-linking also contributes to the surface hardness of neat matrix systems. As soon as we incorporated the nanoparticles into the matrices, the nanocomposite films showed improved surface

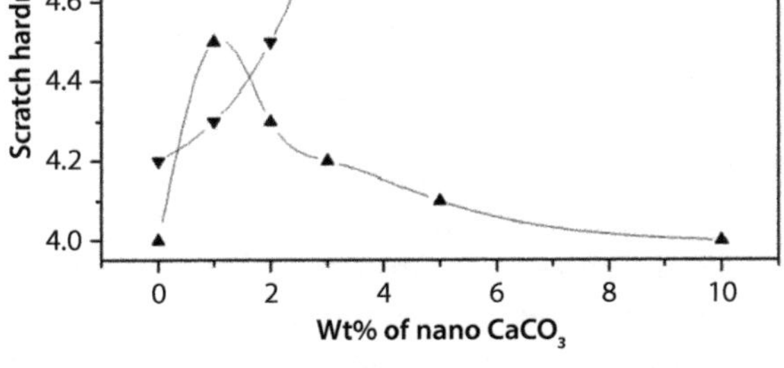

Figure 2.9 Scratch hardness of nanocomposite films vs. content of nano-$CaCO_3$.

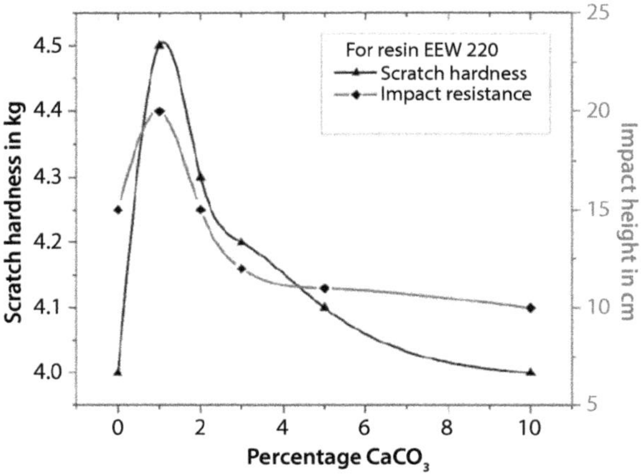

Figure 2.10 Scratch hardness and impact resistance of EEW 220 resin.

hardness and higher scratch resistance as compared to the neat epoxy matrix. The scratch hardness is also affected by the polarity of hydroxyl and oxirane groups present in the backbone of polymeric chains of epoxy matrix [35]. Since nano-$CaCO_3$ particles are coated with stearic acid which also has carboxylic polar groups, the polar groups of matrices and the nanoparticles formed a hydrogen bond, which led to enhanced surface hardness of nanocomposite coating films [36]. Hence, it can be concluded that a drastic improvement in scratch resistance was found, up to a certain extent, after the incorporation of nanofiller into the matrices.

2.4.2.3 Adhesion and Flexibility Test (Mandrel Bend Test)

The data related to mandrel bend test nanocomposite films are given in Table 2.2, which indicate that all film samples showed poor flexibility except 5 wt% and 1 wt% of nano-$CaCO_3$ loaded in EEW 180 and EEW 220 epoxy matrix respectively, and no visible cracking was found. The failure samples may be attributed to higher cross-link density and heat curing at elevated temperature, which led to lower toughness. Whereas passed sample showed higher toughness due to uniform distribution of nano-$CaCO_3$ in the resin matrices. The SEM images, as discussed later in this paper, clearly showed a uniform dispersion of nanoparticles which led to more crazes being generated in the nanocomposites than in the neat matrix, which enhanced the toughness in the nanocomposite samples.

2.4.3 Studies on Chemical Resistance

The chemical resistance of nanocomposite samples is shown in Figures 2.11 and 2.12. It was observed that the sample EP^1C-5 remained unaffected for the maximum periods of 5, 3, 4, 5, 4, 5, 4, 5, 5, 5 and 5 months and sample EP^2C-1 remained unaffected for the maximum periods of 5, 3, 4, 5, 4, 5, 4, 5 and 5 months against artificial seawater, HCl, H_2SO_4, CH_3COOH, NaOH, NH_4OH, xylene, CH_3OH and toluene respectively as compared to neat epoxy cured films and other compositions of nanocomposite films.

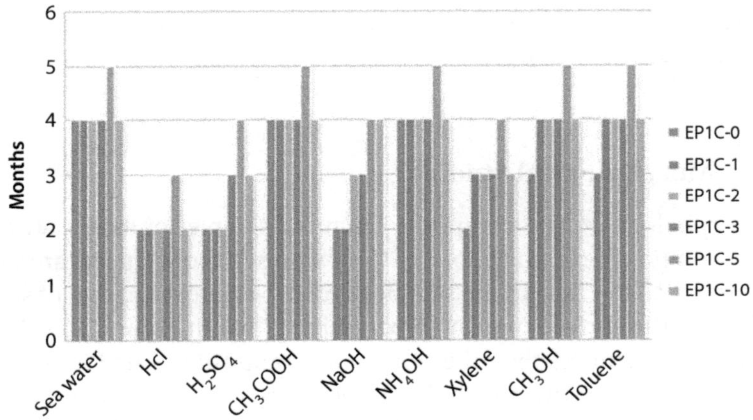

Figure 2.11 Chemical resistance of nanocomposite films of EEW 180 resin.

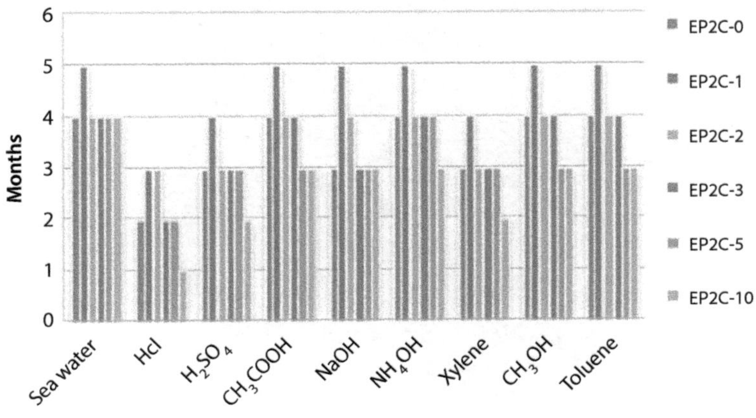

Figure 2.12 Chemical resistance of nanocomposite films of EEW 220 resin.

The improvement in chemical resistance might be due to the blocked free space available between the chains of matrices by increasing nanoparticles up to 5 wt% and 1 wt% for EEW 180 and EEW 220 matrices respectively; hence, the penetration of these chemicals remained restricted. The compact three-dimensional cross-linked structures of the epoxy nanocomposites and longer diffusion path of the systems also resulted in improvement in the chemical resistivity [37].

We also found that the unaffected duration in the case of HCl was less than other chemicals in both samples EP^{1}C-5 and EP^{2}C-1. It showed that the hydroxyl groups present in resin matrix are more reactive towards HCl than other chemicals. After comparing both matrices, it was found that because of the higher molecular weight of EEW 220 matrix, the maximum unaffected period was attained at only 1 wt% loading of nano-$CaCO_3$.

2.4.4 Morphological Studies

The morphology of nanocomposite fractured samples of neat matrix and various composition with nano-$CaCO_3$ of both series of matrices are shown in Figures 2.13 to 2.19. All the samples showed different degrees of surface roughness. The surface of neat epoxy resin shown in Figure 2.13 showed less surface roughness than the other nanocomposites. The lower surface roughness indicates the possibility of more crack initiation and propagation. However, it is obvious that as the loading of nano-$CaCO_3$ increased in

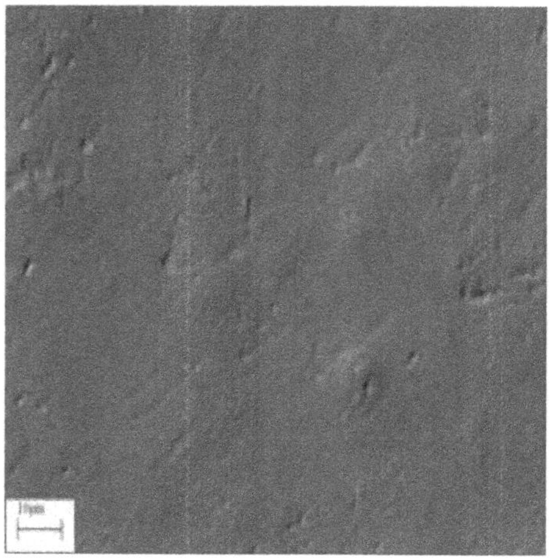

Figure 2.13 SEM image of EP^{1}C-0 sample.

Figure 2.14 SEM image of EP[1]C-1 sample.

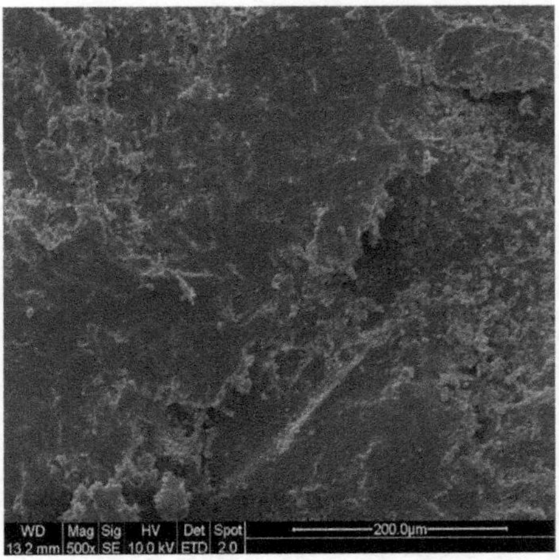

Figure 2.15 SEM image of EP[1]C-3 sample.

the epoxy matrices the roughness of the fractured surface also increased. The increase in surface roughness increased the surface area to deflect the crack initiation and propagation in the samples, which also increased the absorbed energy of fracture by the crack length during the deformation. The higher energy absorption also enhanced the fracture toughness, which

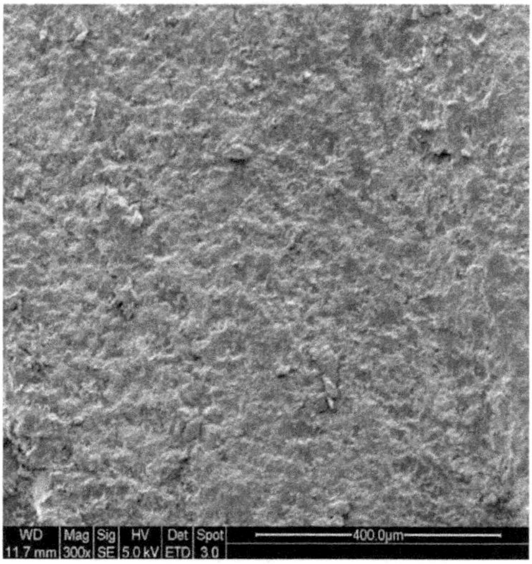

Figure 2.16 SEM image of EP^{1}C-5 sample.

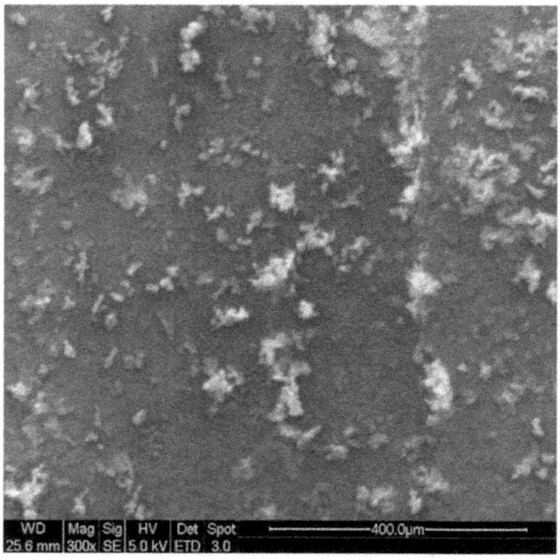

Figure 2.17 SEM image of EP^{1}C-10 sample.

is clear from Figure 2.16. Hence, it is concluded that samples EP^{1}C-5 and EP^{2}C-1 have the uniform exfoliation of nanofiller and showed maximum surface roughness as compared to other nanocomposite samples and neat epoxy resin matrices. This investigation has been corroborated by Alamri and Low [38].

Figure 2.18 SEM image of EP^2C-0 sample.

Figure 2.19 SEM image of EP^2C-1 sample.

2.5 Conclusion

Varying amounts of nano calcium carbonate were mixed in the epoxy resin matrices using polyamide as a curing agent to optimize the formulation of nanocomposite systems for different properties of coatings. It has been found that the uniform dispersion of nanoparticles in the resultant nanocomposite films greatly influenced the properties of coatings. The different mechanical properties, viz., scratch resistance, impact resistance and mandrel bending, were described. While the chemical resistance of nanocomposite films were characterized after immersion into various acids viz., HCl, H_2SO_4, CH_3COOH; alkalies viz., NaOH, NH_4OH; solvents viz., xylene, methyl alcohol, toluene; and artificial seawater, for up to six months duration. The scratch resistance and impact resistance test results reported that as the nanofiller loading increased in the formulation, the resistance to scratch and impact also increased accordingly, and was found maximum for samples EP^1C-5 and EP^2C-1. The mandrel bend test failed for all samples except EP^1C-5 and EP^2C-1 respectively. The chemical resistivity of nanocomposite films was also found maximum for the same samples (5 months duration) towards all chemical reagents as compared with neat matrices and other formulations of nanocomposite films. As far as the loading of nanofillers in matrices is concerned, resin EEW 180 had more loading capacity than resin EEW 220, i.e., the former resin having more amorphous region than the latter one.

References

1. Lee, H., and Neville, K., *Handbook of Epoxy Resins*, McGraw Hill: New York, 1967.
2. Weatherhead, R.G., *FRP Technology: Fibre Reinforced Resin Systems*, Applied Science Publisher Ltd: London, 1980.
3. Misev, T.A., and Linde, R.V.D., Powder coating technology: New technology at the turn of century. *Prog. Org. Coat.* 34, 160, 1998.
4. Richler, K., and Steiger, R., Thermal Stability of wood-wood and wood-FRP bonding with polyurethane and epoxy. *Adv. Eng. Mater.* 7, 419–422, 2005.
5. Brydson, J.A., *Plastics Materials*, D. Van Nostrand Company, 1966.
6. Goodman, S.H. (Ed.), *Handbook of Thermoset Plastics*, pp. 133–182 Noyes Publications, 1986.
7. Shi, Q., Wang, L., Yu, H., Jiang, S., Zhao, Z., and Dong, X., A novel epoxy/$CaCO_3$ nanocomposite and its mechanism of toughness improvement. *Macromol. Mater. Eng.* 291, 53–58, 2006.
8. Fu, S.-Y., Feng, X.-Q., Lauke, B., and Mai, Y.-W., Effects of particle size, particle/matrix interface adhesion and particle loading on mechanical properties of particulate-polymer composite. *Compos. Part B* 39, 933–961, 2008.
9. Zebarjad, S.M., Sajjadi, S.A., Tahani, M., and Lazzeri, A., A study on thermal behaviour of HDPE/$CaCO_3$ nanocomposites. *J. Achiev. Mater. Manuf. Eng.* 17, 173–176, 2006.
10. Spanoudakis, J., and Young, R.J., Crack propagation of a glass particle filled epoxy resin: Effect of paticle-volume fraction and size. *J. Mater. Sci.* 19, 473–86, 1984.

11. Moloney, A.C., Kausch, H.H., Kaisher, T., and Beer, H.R., A Review: Parameters determining the strength and toughness of particulate filled epoxide resin. *J. Mater. Sci.* 22, 93–381, 1987.
12. Yong, V., and Hahn, H.T., Processing and properties of SiC/vinyl ester nanocomposites. *Nanotechnology* 15, 1338, 2004.
13. Yu, H., Wang, L., Shi, Q., Jiang, S., and Jiang, G., Preparation of epoxy resin/ $CaCO_3$ nanocomposites and performance of resultant powder coatings. *J. Appl. Polym. Sci.* 101, 2656–2660, 2006.
14. Anandhan, S., and Bandyopadhyay, S., Polymer nanocomposites: From synthesis to applications, in: *Nanocomposites and Polymers with Analytical Methods*, Cuppoletti, J. (Ed.), pp. 3–28, InTech, 2011.
15. Sudheer Kumar, N., Venkateshwar Rao, K., Chakraverti, G., and Shilpa Chakra, C.H., Tribological properties of epoxy/Al_2O_3 nanocomposites. *IJETT* 4, 2095–2098, 2013.
16. Mishra, S., Sonawane, S., and Chitodkar, V., Comparative study on improvement in mechanical and flame retarding properties of epoxy-$CaCO_3$ nano and commercial composites. *Polym. Plast. Technol. Eng.* 44, 463–473, 2005.
17. Agubra, V.A., Owuor, P.S., and Hosur, M.V., Influence of nanoclay dispersion methods on the mechanical behavior of e-glass/epoxy nanocomposites. *Nanomaterials* 3, 550–563, 2013.
18. Hussain, F., Hojjati, M., Okamoto, M., and Gorga, R.E., Review article: Polymer-matrix nanocomposites, processing, manufacturing, and application: An overview. *J. Compos. Mater.* 40, 1511–1575, 2006
19. Yu, H.J., Wang, L., Shi, Q., Jiang, G.H., Zhao, Z.R., and Dong, X.C., Study on nano-$CaCO_3$ modified epoxy powder coatings. *Prog. Org. Coat.* 55, 296–300, 2006.
20. Weon, J.-I., Gam, K.-T., Boo, W.-J., Sue H.-J., and Chan, C.-M., Impact toughening mechanism of $CaCO_3$ reinforced polypropylene nanocomposite. *J. Appl. Polym. Sci.* 99, 3070–3076, 2006.
21. Zhang, K., Wang, L., Wang, F., Wang, G.,and Li, Z., Preparation and characterization of modified-clay-reinforced and toughened epoxy-resin nanocomposites. *J. Appl. Polym. Sci.* 91, 2649–2652, 2004.
22. Wetzel, B., Haupert, F., and Zhang, M.Q., Epoxy nanocomposites with high mechanical and tribological performance. *Compos. Sci. Technol.* 63, 2055, 2003.
23. Ng, C.B., Schadler, L.S., and Siegel, R.W., Synthesis and mechanical properties of $Ti0_2$-epoxy nanocomposites. *Nanostruct. Mater.* 12, 507–510, 1999.
24. Pashaei, S., Siddaramaiah, and Syed, A.A., Investigation on thermal, mechanical and morphological behaviours of organo nanoclay incorporated epoxy nanocomposites. *ARPN J. Eng. Appl. Sci.* 5, 76–86, 2010.
25. Abacha, N., Kubouchi, M., Tsuda, K., and Sakai, T., Performance of epoxy-nanocomposite under corrosive environment. *Express Polym. Lett.* 1, 364–369, 2007.
26. Mir, M.A.N., Khanmohammadi, M.R., Garmarudi, A.B., and Haghshenas, M., Nanosilica reinforced epoxy floor coating composites: Preparation and thermophysical characterization. *Curr. Chem. Lett.* 1, 13–20, 2011.

27. Kumar, M.A., Reddy, C.R., Reddy, G.M., Kumar, S.A., and Rafee, D., Epoxy/WGFR/macro/nanocomposites on chemical resistance. *Int. J. Nanomater. Biostruct.* 1, 25–27, 2011.
28. Zhang, W., Dehghani-Sanij, A.A., and Blackburn, R.S., IR study on hydrogen bonding in epoxy resin-silica nanocomposites. *Prog. Nat. Sci.* 18, 801–805, 2008.
29. Jain, R., Narula, A.K., and Choudhary, V., Studies on epoxy/$CaCO_3$ nanocomposites. *J. Appl. Polym. Sci.* 114, 2161–2168, 2009.
30. Jin, F.-L., and Park, S.-J., Thermal stability of trifunctional epoxy resin modified with nanosized calcium carbonate. *Bull. Korean Chem. Soc.* 30, 334–338, 2009.
31. Hiemenz, P.C., and Lodge, T.P., *Polymer Chemistry*, 2nd ed., Taylor and Francis: Florida, 2007.
32. Li, R., Time-temperature superposition method for glass transition temperature of plastic materials. *Mater. Sci. Eng.* A278, 36–45, 2000.
33. van Gurp, M., and Palmen, J., Time-temperature superposition for polymeric blends. *Rheol. Bull.* 67, 5–8, 1998.
34. Bhattacharya, A., Rawlins, J.W., and Ray, P. (Eds.), *Polymer Grafting and Cross-Linking*, John Wiley and Sons: New Jersey, 2009.
35. Khan, R., Samad, U.A., Alam, M.A., Boumaza, M., and Saeed, M.A.-Z., Effect of ZnO nano powder on mechanical properties of epoxy/polyaminoamide adduct coatings. *IJACST* 2, 25–28, 2013.
36. Al-Zubaidi, A.B., Al-Tabakh, A.A., Al-Qassey, H.A., and Al-Kaseey, R.N., Mechanical thermal wear characteristics of polymer composite material reinforced with calcium carbonate powder, presented at: 4th International Scientific Conference on Nanotechnology, Advanced Materials and Their Applications, ICNAMA, 3–4 Nov, 2013.
37. Kahraman, R., Effect of aluminium filler content on moisture diffusion in to epoxy adhesives in distilled water. *J. Appl. Polym. Sci.* 98, 1165, 2005.
38. Alamri, H., and Low, I.M., The effect of water absorption on mechanical properties of nano-filler reinforced epoxy nanocomposites. *J. Mater. Design* 42, 214–222, 2012.

3

An Industrial Approach to FRLS (Fire Retardant Low Smoke) Compliance in Epoxy Resin-Based Polymeric Products

Hari R[1]* and Sukumar Roy[2]*

[1]*Insulation and Chemical Sciences Laboratory, Bharat Heavy Electricals Limited, Corporate Research and Development, Hyderabad, India*
[2]*Ceramic Technological Institute, Bharat Heavy Electrical Limited, Corporate Research & Development, Prof. CNR Rao Circle, Bangalore, India*

Abstract
An industrial approach to modifying conventional epoxy resin-hardener polymeric system by incorporating nanostructured alumina additive material along with various other additives (e.g., calcium carbonate, aluminum phosphate, zinc borate, etc.) in optimized proportions has been developed. The thus-modified epoxy resin hardener material, which was fabricated in the form of an industrial product, confirms the latest compliance norms of FRLS (fire retardant low smoke) properties under the classifications of UL 94 and IEC 60076-11 respectively. As per the process, the nanostructured alumina material with hollow morphology (in-house synthesized through a patented process) along with other additive materials (e.g., calcium carbonate, aluminum phosphate, zinc borate, etc.) with defined weight ratio in the formulation was successfully incorporated in epoxy resin-hardener polymeric system. Furthermore, various laboratory specimens were fabricated first for bench-scale testing and then in the form of a high voltage insulation prototype for product-level testing thereof. All the additive-modified epoxy resin-hardener materials were cured by maintaining similar conditions (i.e., in the temperature range of 120–140 °C), which resulted in cured epoxy resin polymer composites without any deterioration of their associated mechanical, thermal and chemical properties as compared to virgin epoxy resin-hardener material. Moreover, the desired FRLS properties were achieved in the derived composites. The FRLS testing results of the additive-modified epoxy resin/hardener polymer

**Corresponding authors*: harir@bhel.in; sroy@bhel.in

Sanjay K. Nayak *et al.* (eds.) Trends and Applications in Advanced Polymeric Materials, (45–58) 2018 © Scrivener Publishing LLC

composites confirmed the desired FRLS properties in the new composite material as per the latest UL 94 classifications and IEC 60076-11 respectively.

Keywords: FRLS compliance, epoxy resin-hardener system, high voltage insulation products

3.1 Introduction

Epoxy resin is a polymer material that has been widely used for several industrial and household applications over the last 60 years due to its excellent characteristics such as toughness, chemical, mechanical and electrical resistance, low shrinkage upon cure and high adhesion to many substrates. Epoxy resin-hardener systems are also widely used for numerous surface coatings, castings, laminates, adhesives, composites, potting and painting materials, especially in application areas that comprise electronic, electrical, transportation industries, etc. A variety of epoxy resin-hardener systems are available commercially. From the perspective of application and new environmental compliances, one major disadvantage of epoxy resins, like other organic polymers, is their flammable property. Over the years, different strategies have been adopted in order to address this issue, such as, usage of inherently flame retardant polymers, modification of the polymer backbone or incorporation of flame retardants into polymers, etc.

While inherently flame retardant polymers can lead to higher manufacturing costs, the modification of the base epoxy polymeric system by incorporating suitable additive material system could balance out the enhanced cost. It is quite evident that modification of the polymer backbone by inclusion of P, Si, B or N elements often provide good fire resistant properties. The incorporation of such flame retardant in polymers is a simple and cost-effective way to achieve flame retardant polymers and is widely used. Very different mechanisms are involved in the reaction to fire of polymers containing flame retardant particles, and their effects are often mixed. Physical actions are encountered when an insulating protective layer is formed on top of the polymer during combustion (intumescent systems), when the degradation of the additive is an endothermic reaction ("cooling" effect; e.g., with metallic hydroxides) or when inert gases are released by the additive upon degradation (e.g., calcium carbonate), etc. All of these protective actions are associated with the so-called barrier effect. Furthermore, chemical effects are also involved, for example, when the flame retardant or its degradation products disturb the radical mechanism occurring in the flame that leads to its extinction (halogen/phosphorus-based flame retardants).

Although there are certain flame retardant epoxy products available commercially, these are reported to primarily contain high concentrations of halogen compounds [1], which are often combined with antimony trioxide as a synergist. The problem with these compounds is that, though they have fire retardant properties, they have other associated anti-environmental properties that can affect human health. For example, antimony-trioxide is listed as a carcinogenic chemical, the use of which should be avoided due to safety concerns. Also, upon thermal decomposition, halogen compounds (bromine based) split into bromine radicals and hydrogen bromide, which are highly corrosive and toxic in the presence of air or oxygen [2]. This reaction poses a great risk to human health in the long run. Besides the above, disposing of bromine is a challenging task with regard to environmental pollution.

Therefore, the effort to enhance the retardant reaction to fire of epoxy resin-hardener system has been further developed and innovative solutions were looked for. The focus of this chapter is on recent developments in this field, i.e., bringing FRLS compliance in epoxy resin-hardener material, particularly the usage of epoxy resin-hardener material system as high voltage electrical insulation products in electrical power and allied industries. This chapter is further dedicated to the growing use of nanoparticles with defined morphology and agglomerate structure as potential additive material for enhancing the retardant reaction of thermally-cured epoxy polymeric products to fire.

3.1.1 Incorporation of Additives

Epoxy-based resin-hardener system is widely used as an insulation material in many electrical apparatus because of its excellent electrical and mechanical properties. It is usually mixed with appropriate fillers to enhance the electrical and mechanical properties. With the development of nanotechnology, polymer nanocomposites have drawn much attention because they are expected to have improved properties over pure polymers or conventional polymer composites by the addition of a few parts of nanofillers.

Not all polymers with combinations of inorganic additives will form nanocomposites. The compatibility and interfacial properties between polymer matrix and inorganic additives significantly influence the essential characteristics of materials. Generally, inorganic additives have poor compatibility with the polymer matrix, except for water-soluble polymers. Therefore, inorganic additives must be organically modified using organic surfactants to improve compatibility. The organic surfactants in the inorganic additives play the important role of lowering the surface energy of the inorganic host, improving the wetting characteristics and miscibility with the polymer matrix.

The route for incorporating inorganic additives into fire retardant polymers, in particular to epoxies, has always been favored. It is generally a cheap and easy way of achieving sufficient levels of flame retardancy. However, the traditional additive flame-retarding solutions tend to have limits. As discussed, the use of halogen-containing flame retardants is being questioned; besides which, the possibility of bioaccumulation and hence the long-term effects on workers handling such substances should also be of concern. Metallic hydroxides, such as aluminum or magnesium hydroxide, are efficient at very high loadings (50 wt% or more), but their usage resulted in degraded mechanical properties besides difficulties in processing. Therefore, the incorporation of nanoparticles in epoxy resin hardener system has attracted much interest in the last ten years, in which the level of incorporation of additives is pretty low because of the enhanced surface area-weight ratio of the nanoparticles. Indeed, apart from providing fire retardant properties to the polymer(s), a lower level of incorporation of nanoparticles does not alter or deteriorate the other counter properties (i.e., mechanical, thermal, electrical properties, etc.), but rather often enhances some of these properties as well.

There is one technical approach to introduce FRLS properties into commercial epoxy resin-hardener polymeric system. The primary technical aspects of bringing FRLS properties into epoxy-based resin-hardener polymeric system are the following:

i. The curing process for the conventional epoxy resin-hardener system is chemically very sensitive. The fabricated product does not cure properly in the presence of any foreign substances in the formulation of the reacting components, thereby resulting in inferior products at the end with deteriorated mechanical, thermal and electrical properties. So, if the additive route is to be followed, the primary criteria is to ensure a complete curing process in the presence of the additives in a manner that obtains a defect-free product with desired properties at the end of the process.
ii. Other associated technical challenges in an industrial scenario would be: a) not altering either the processing conditions or processing equipment drastically, and b) not changing the counter raw materials in the epoxy resin-hardener system, such as accelerators, flexibilzers, etc., all of which are commercially available in several sources at fairly low cost or standardized cost.

Hence, the above technical factors can be translated into the following ground rules: i) not to replace epoxy resin-hardener system with other polymer(s) which is/are already FRLS compliant and ii) to introduce additive material(s) into the formulation in a manner which does not affect the curing process, hence with no deterioration of other counter mechanical, thermal and electrical properties of the fabricated product at the end, and furthermore iii) not to change the processing conditions along with counter additive materials, like accelerators, flexibilzers, etc., and preferably the processing equipment as well, in order to make the product economical.

3.2 Experimental

Conventional epoxy resin (biphenyl-A) hardener (acid anhydride)-based polymeric system along with specific counter materials, like accelerators, flexibilzers, etc., suitable for the epoxy resin system, which are available commercially with more than one source were used in this study.

A special type of nanostructured alumina material was synthesized by utilizing an in-house developed and patented process, i.e., so-called LPG (liquid petroleum gas) fired spray pyrolysis system, and was used as an additive material in the FRLS formulation. The additive nanostructured alumina material is composed with hollow spherical morphology with 70–100 nanometer-sized primary particles in micron-sized agglomerates of the particles. The material is further associated with very low tap density in the range of 0.02–0.03 g/cc (very high volume-to-weight ratio) that also has an amorphous structure in the X-ray powder diffraction pattern. This specific material was one important component in the additive formulation in the conventional epoxy resin-hardener system that was primarily responsible for bringing the FRLS properties in the product. Importantly, the material also did not interfere in the curing process of epoxy resin-hardener system because of the chemical inertness of the particles with reference to all the components present in the epoxy resin-hardener polymeric system.

Besides the said additive material [3], other additive materials (e.g., calcium carbonate, aluminum phosphate and zinc borate, etc.) in optimized proportions were also used in the formulation in a manner that did not alter the curing schedule and conditions; and counter-reacting additive materials like flexibilizers, hardener, etc., along with the machineries, etc., in the conventional epoxy resin-hardener polymeric system also made the conventional epoxy resin-hardener system [4] FRLS compliant in the end.

Specimen samples of thus-derived additive-modified epoxy resin hardener composite materials were fabricated with dimensions of 200 × 10 × 3 mm in a lab model epoxy casting machine using a vacuum level of about 3 mbar. The samples were cured at a temperature in the range of

120–140 °C for a period in the range of 6–8 h. The flame retardant properties of these cured samples were evaluated by various specified testing.

Limiting oxygen index (LOI) measures the minimum concentration of oxygen in a flowing mixture of oxygen and nitrogen that will just support flaming combustion. The LOI (ASTM D2863) provides information about the flammability of the samples (Figure 3.1). LOI values remain low for samples with virgin epoxy polymer (blank epoxy polymer without any additive) as compared to the additive-modified epoxy composites and the differences between these two samples is significant (Table 3.1). The flame spread was evaluated by the horizontal burner test, in which the time required for burning of a fixed distance of 150 mm of the sample is recorded (Table 3.1). Therefore, the longer the time, the better is the FRLS behavior of the sample. As can be seen, the time required for burning the sample is drastically increased by the incorporation of the identified additives in the epoxy polymer matrix along with other ingredients. Furthermore, contrary to the unfilled (virgin) epoxy body, the additive-modified [5] epoxy composites do not drip.

Most of the thermoplastic materials have the tendency to burn, in which the burning process could be associated with melting and decomposition depending on the intensity of the heat. The resultant burnt polymer drips, thereby causing additional problems [6]. Therefore, the UL 94 standard captures this aspect of the polymers via dripping criteria. The vertical burning test for classifying materials as UL 94 was carried out according to the "Standard for Flammability Tests of Plastic Materials in Devices and Appliances" issued by the Underwriters Laboratories Inc. In order to achieve a UL 94 VTM-0 rating, there cannot be dripping particles or debris that causes cotton positioned 300 millimeters below the test subject to be ignited. For testing, the compositions as given in Table 3.1 are taken and molded into 125 × 13 × 3 mm (L × W × T) sample and tested as per UL 94

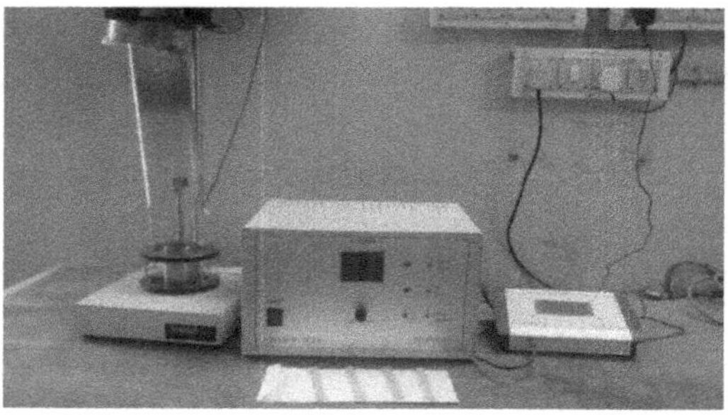

Figure 3.1 Limiting oxygen index (LOI) test apparatus and representative samples.

Table 3.1 Flammability characteristics of nanostructured alumina and other additive modified in epoxy resin-hardener system vs. unfilled (virgin) epoxy resin-hardener body.

Samples	Additive incorporated					LOI (%)	Horizontal burner test; time for burning 150 mm in horizontal position(s)
	Nano-structured alumina (by Vol%)	Calcium carbonate (by Wt%)	Aluminum phosphate (by Wt%)	Zinc borate (by Wt%)	500 mesh quartz (by Wt%)		
Unfilled (Virgin) epoxy	0	0	0	0	0	18.5	260
Set 1	1–2	0.5–1.0	0.5–1.0	1.0–1.5	0	19.6	350
Set 2	2–3	0.5–1.0	0.5–1.0	2.0–2.5	0	19.8	380
Set 3	3–5	0.5–1.0	0.5–1.0	3.0–4.0	0	20.6	395
Set 4	1–2	0.5–1.0	0.5–1.0	1.0–1.5	1.0–2.0	23.2	420
Set 5	2–3	0.5–1.0	0.5–1.0	2.0–2.5	2.0–3.0	24.5	426
Set 6	3–5	0.5–1.0	0.5–1.0	3.0–4.0	3.0–5.0	25.6	435

and all the test samples qualified according to UL 94 VTM-0 protocol, as given in Table 3.2 of material classification.

For samples to obtain a VTM-0 rating, not only the flaming and glowing time has to be short, but also the flame should not drip and ignite the cotton.

After the FRLS properties were achieved in the "specimen level" (laboratory samples for purpose of optimization and bench-scale testing), a prototype for high voltage insulation product called "dry-type transformer" (DTT) was further fabricated by following the specified formulation optimized in the specimen level. FRLS testing of the fabricated prototype was conducted as per the standard procedure and FRLS compliance is also confirmed at the prototype level, in which all the test results have similar data points of all FRLS testing parameters to that of specimen level testing data.

Besides FRLS testing as per the classified protocol, other relevant testing was also conducted as per desired norms in each case, i.e., electrical testing (breakdown voltage, resistivity, dielectric permittivity, tan delta, etc.), mechanical testing (tensile, flexural, elastic moduli), and thermal testing (glass transition, thermal expansion, shrinkage profiles, etc.) respectively. The additive formulation(s) in the epoxy resin-hardener system was/were adjusted so that in each case there was no deterioration of any of those counter properties in the newly derived additive-modified epoxy composite body [7, 8].

Table 3.2 Material classifications as per UL 94 protocol.

Criteria/conditions	VTM-0	VTM-1	VTM-2
After flame time for each individual specimen t1 or t2	≤ 10s	≤ 30s	≤ 30s
Total after flame time for the set (t1 plus t2 for the 5 specimens)	≤ 50s	≤ 250s	≤ 250s
After flame plus afterglow time for each individual specimen after the second flame application (t2 + t3)	≤ 30s	≤ 60s	≤ 60s
After flame or afterglow of any specimen up to the holding clamp	No	No	No
Cotton indicator ignited by flaming particles or drips	No	No	Yes

3.3 Characterization, Results and Discussion

As previously mentioned, the bench-scale testing (specimen level testing) of the optimized formulation of additive-modified epoxy resin-hardener polymeric system is used to fabricate an industrial product prototype called dry-type transformer (DTT) to prove that the modified epoxy polymeric resin system is still FRLS compliant at the actual prototype level. It is to be mentioned here that such insulation products are often vulnerable to fire hazards because of unfavorable or limited atmospheric conditions, etc. while in service.

The prototype is further fire tested in a specially fabricated environmental chamber as per international standard IEC 60076-11, as shown in Figure 3.2a–c. All the specified parameters, including airflow (inlet and outlet), air temperature (inlet and outlet), cast coil temperature of HV and LV, magnetic core, and smoke density at the outlet, are monitored and recorded using appropriate testing instruments for the purpose. At the end of the fire test, the residual mass of the coils is measured in order to know the balance weight after burning for a period of one hour in order to determine the efficacy of FRLS properties in the derived additive-modified epoxy composite prototype over the conventional unfilled (virgin) epoxy resin-hardener polymeric body.

A complete transformer coil (HV+LV+Core), as shown in Figure 3.3, was placed in the specially designed environmental chamber as per IEC 60076-11, shown in Figure 3.4. The test was started when the alcohol (isopropanol) in the tank with initial level of 40 mm was ignited, simultaneously switching on the 24 kW radiant panel. Time duration for the testing was fixed at 60 minutes, which is in compliance with the standard. As the radiant panel and alcohol burning started simultaneously, the start time was recorded as zero minute, alcohol combustion was found to be finished at the end of the 18th minute and after the 40th minute, the radiant panel was switched off, as per the standard. During the 60 minute testing period, all the test parameters are recorded at 2 minute intervals, i.e., air inlet flow rate and temperature, outlet air flow rate and temperature, temperatures of LV, HV and core and smoke density respectively. In order to evaluate the role of additive formulations in the FRLS behavior of epoxy composite body, testing has been carried out both on additive-modified epoxy resin-hardener material coil as well as virgin epoxy resin-hardener coil under identical conditions, as shown in the Figure 3.4.

Figure 3.3 shows the actual profile of the additive-modified epoxy resin composite material fabricated product prototype, dry type transformer (DTT), which has FRLS properties (left), and another one without any FRLS properties (virgin epoxy resin-hardener body).

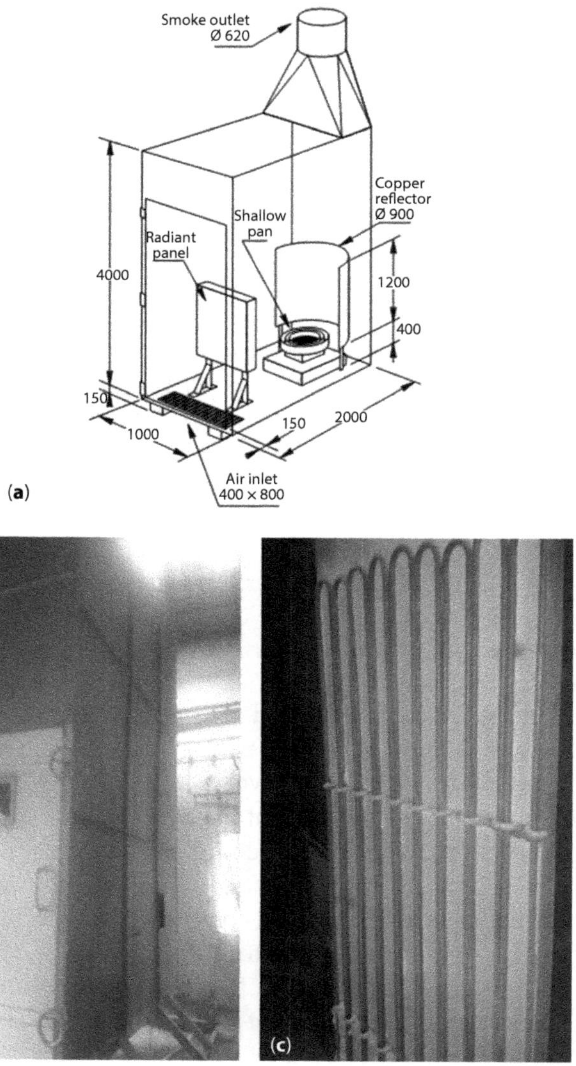

Figure 3.2 (a) Schematic drawing of a specially fabricated fire behavior environmental chamber with radiant panel as per IEC 60076-11. (b,c) Physical appearance of specially fabricated fire behavior environmental chamber with radiant panel as per IEC 60076-11.

Figure 3.4 shows the burning behavior of the FRLS-compliant additive-modified epoxy resin-hardener product prototype (product image on the left) and virgin epoxy product without any FRLS properties (image on the right) during testing as per IEC 60076-11 standard that show suppressed

Figure 3.3 Actual product prototypes of dry-type transformers (DTTs): i) FRLS compliant (left) additive-modified epoxy resin-hardener polymer composite as an electrical insulation material and ii) non-FRLS (right) virgin epoxy resin-hardener polymer material without any additive(s), before testing as per IEC 60076-11 standard.

Figure 3.4 Photos depicting the burning characteristics of the product prototype of a DTT having epoxy resin-hardener polymer composite material (image on the left, FRLS compliant) as an electrical insulation material during testing as per IEC 60076-11 standard and virgin epoxy resin-hardener non-FRLS (image on the right).

burning of the material with FRLS-compliant, as compared to non-FRLS virgin epoxy body.

The photos in Figure 3.5 depict post-burning of FRLS-compliant epoxy resin-hardener product versus non-FRLS epoxy product respectively, showing that FRLS-complaint material is not charred because of suppressed burning characteristics, thereby avoiding propagation of burning and hence fire, and also release of smoke, etc., in the process of burning.

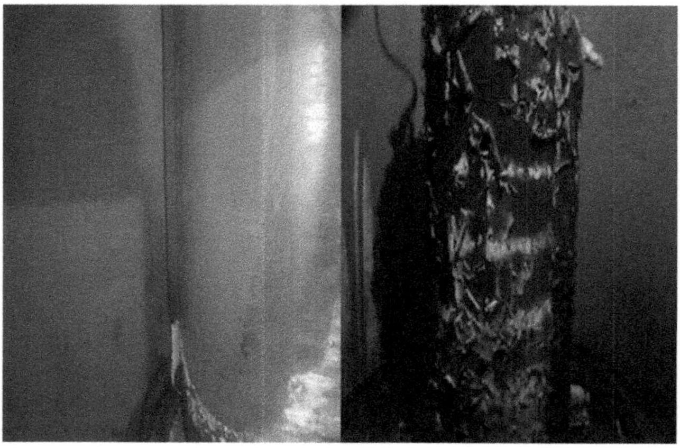

Figure 3.5 Photos of the product prototype of DTT having epoxy resin-hardener polymer as an electrical insulation material after testing per IEC 60076-11 standard with i) FRLS-compliant additive-modified epoxy resin-hardener composite prototype (image on the left) and ii) non-FRLS virgin epoxy resin-hardener material prototype (image on the right).

Table 3.3 FRLS compliant DTT coils (additive-modified epoxy body).

		Mass (Kg)		
Sl. no.	**Item description**	*Before test*	*After test*	**Mass loss (%)**
1	High Voltage Winding	206.52	199.54	3.07
2	Low Voltage Winding	119.5	116.46	
3	Total Assembly (HV, LV)	326.02	316	

Table 3.4 Non-FRLS DTT Coils (virgin epoxy body).

		Mass (Kg)		
Sl. no.	**Item description**	*Before test*	*After test*	**Mass loss (%)**
1	High Voltage Winding	36.68	28.76	17.00
2	Low Voltage Winding	16.62	15.48	
3	Total Assembly (HV, LV)	53.3	44.24	

After completion of the test, both (FRLS and Non-FRLS) prototype coils are measured for mass loss, as given in Table 3.3 and Table 3.4 respectively.

The scientific principle involved in optimization of additive formulation for achieving FRLS properties in epoxy resin-hardener polymeric system is to minimize the combustion (both rate and mass) of the composite material in the presence of air or oxygen, thereby creating a chemical environment which retards combustion and further prevents the propagation of combustion reaction.

The choice of additive material(s) and its combination is very important to achieve FRLS properties in epoxy resin-hardener polymers. One additive material, i.e., nanostructured alumina, which is chemically inert with respect to all the reacting components in the epoxy resin-hardener polymeric system, has hollow spherical morphology in which primary particles in the range of 80–100 nm comprise several micron-sized agglomerates in the particles. Such particles have significant porosity and hence a very high volume-to-weight ratio in the material. It is believed that the porosity in such nanostructured particles is associated with the counter surface profile in which chemistry, surface area-to-weight ratio and amorphous nature of the additive particles play a very crucial role in retarding the combustion process, thereby achieving FRLS properties in a combustible polymer, like epoxy resin-hardener system.

The selection of other additive materials (e.g., calcium carbonate, aluminum phosphate, zinc borate, etc.) along with their incorporation ratio in the formulation also influences the achievement of FRLS behavior that finally complies with the latest norms of UL 94 classifications.

Another important aspect that has been highlighted in the preceding section as well that the known FRLS-compliant epoxy resin-hardener polymeric system contains objectionable chemicals like halogen compounds or antimony compounds, or in some cases both, in contrast to the present research wherein no such compounds are used in the formulation for obtaining FRLS properties. Viewed this way, the new additive-modified epoxy resin-hardener composite material system brings new functionality of FRLS properties to conventional commercial grades of epoxy resin-hardener polymeric material system, in which any geometry, shape, or dimension for various products in the fields of electrical insulation and or allied areas could be fabricated by using the optimized formulation of the additives and the developed process.

3.4 Conclusion

This chapter has attempted to establish the FRLS compliance of epoxy resin-hardener polymer-based insulation products with fire safety and environmental hazard norms, which has a bearing on both the societal and industrial fronts. The present research work could also be viewed as how to introduce a new functionality (i.e., desired level) of fire retardant low smoke (FRLS) properties in a conventional commercial grade of epoxy resin-hardener-based polymeric material system by judiciously selecting and optimizing certain additive materials in which one nanostructured material (i.e., nanostructured alumina) with hollow porous morphology plays a significant role. The work also highlighted that by deriving the specific formulation of additive materials, the additive-modified epoxy resin-hardener polymeric system could bring the desired FRLS properties to the new composite material, and, more importantly, without deteriorating its associated mechanical, thermal and electrical insulation properties respectively.

Acknowledgments

The authors thank BHEL Management for their permission to publish this work. The authors also gratefully acknowledge the technical support and help from their colleague, Mr. G. Muralidharan.

References

1. Duquesne, S., Michel, L.B., Bourbigot, S., Delobel, R., Vezin, H., Camino, G., Berend, E., Lindsay, C., and Roels, T., Expandable graphite: A fire retardant additive for polyurethane coating. *Fire Mater.* 27, 103, 2007.
2. Sen, A.K., Mukheriece, B., Bhattacharya, A.S., Sanghi, L.K., De, P.P., and Bhowmick, K., Preparation and characterization of low-halogen and non-halogen fire-resistant low-smoke (FRLS) cable sheathing compound from blends of functionalized polyolefin and PVC. *J. Appl. Polym. Sci.* 43, 1673, 1991.
3. Liu, J., Gao, Y., Wang, F., and Wu, M., Preparation and characteristics of non-flammable polyimide materials. *J. Appl. Polym. Sci.* 75, 384, 2000.
4. Lee, H., and Neville, K., *Handbook of Epoxy Resins*, pp. 4–17, McGraw-Hill: New York, 1967.
5. Manias, E., Polizos, G., Nakajima, H., and Heidecker, M.J., Fundamentals of polymer nanocomposite technology, in: *Flame Retardant Polymer Nanocomposites*, Morgan, A.B., Wilkie, C.A. (Eds.), John Wiley and Sons: NY, USA, 2007.
6. Chen, D.Q., Wang, Y.Z., Hu, X.P., Wang, D.Y., Qu, M.H., and Yang, B., Flame-retardant and anti-dripping effects of a novel char-forming flame retardant for the treatment of poly(ethylene terephthalate) fabrics. *Polym. Degrad. Stab.* 88, 349–356, 2005.
7. Indian Patent Application number, 1103/Kol/2015, "A process of fabricating additive-modified epoxy resin cast body with targeted level of flame retardant and low smoke properties for high voltage insulation application" by Hari R., K.C. Muduli, KSR Prasad and S Roy. October 20, 2015
8. Indian Patent Application Number, 947/Kol/2011, "A nano-additive modified epoxy resin composition for high voltage insulation applications" by G. Muralidharan, S. Roy and R. Agarwal. July 15, 2011

4

Polymer-Based Organic Solar Cell: An Overview

Neha Patni*, Pranjal Sharma, Mythilypriya Suresh, Birendrakumar Tiwari and Shibu G. Pillai

Department of Chemical Engineering, Institute of Technology, Nirma University, Gujarat, India

Abstract

Solar energy is a highly abundant renewable source and in this age of depleting fossil fuels, its harnessing is paramount, which has led mankind to discover organic solar cells. Discovered and researched in the 21st century, this field offers valuable hope for resolving future energy requirements. The emergence of polymer solar cells and their working methodology and construction are included in this chapter. Developments pertaining to polymer solar cells and the conundrums related to them are discussed. Included is the life-cycle analysis technique used to indicate the environmental impact of organic solar cells. Synthesis techniques to prepare polymers for the cells are explained along with the subtle differences between them. Reasons for the deformation and degradation of the performance index of polymer solar cells are reported along with degradation mechanisms via physical and chemical means. The relevance of conjugated polymers over polymer solar cells and the effect of side chains on improving solar cell conductive performance are also presented in brief. Also given is a brief introduction to dyes and their utility in the increment of conduction performance. Natural dyes are briefly discussed and frequently used dyes are also indicated. Polyaniline (PANI) as polymer and the possibility of using and blending it with dye-sensitized solar cell (DSSC) are discussed.

**Corresponding author*: neha.patni@nirmauni.ac.in

Sanjay K. Nayak *et al.* (eds.) Trends and Applications in Advanced Polymeric Materials, (59–82) 2018 © Scrivener Publishing LLC

Keywords: Dye-sensitized solar cells, organic solar cells, polymer solar cells, dyes, polyaniline, natural dyes, ITO, TiO_2

4.1 Introduction

Today, the problem facing the world is not inadequate energy but instead the method of harnessing the abundant energy spread across the globe. Currently, the human race is highly dependent on energy generated from fossil fuels, which are now being depleted at an exponential rate due to the booming population. To meet rising energy demands we now need to devise other methods to harness various types of energy available on earth; from this notion comes the idea of utilizing renewable resources which promise a better and sustainable future. For years human beings have been in a dilemma about how to cater to their energy needs, until recently, when in the 19th century the strengths of renewable energy sources like wind, solar, geothermal, tidal, etc., were explored. Lately, humans have considered the surplus amount of solar energy acquired from the sun. The earth receives about 120,000 terrawatts in the form of solar energy, which is ten-fold of what we require to survive, but the conundrum with this source is, "How shall we efficiently harness this abundance of energy without any environmental degradation?".

This is where solar cells come into the picture. They were developed on the photovoltaic theory which grew up sometime in the mid-19th century and then the 20th century. Suddenly there was a surge of solar cells in the market as they were used in space crafts and space shuttles for energy production in space. Since then there have been huge developments in the field of silicon-based solar cells and by the 21st century the efficiency of these cells has increased up to 46% [1], though there were certain quandaries associated with these cells.

Later, solar cells were classified into three categories according to the usage of silicon in their structure: first generation, second generation and third generation solar cells. First generation solar cells utilize silicon wafers and provide an efficiency of about 15–20%; the benefits with this type are high stability and high performance, but their drawbacks are high energy demand during production and their rigidity [2]. Second generation solar cells utilize amorphous silicon and provide an efficiency of about 10–15%. The benefits of this type are that they are light and flexible, but this type also uses vacuum process and high temperature production procedures and scarce resources, which make this type costly [2].

Third generation solar cells use organic materials and are so-called organic solar cells. Polymer solar cells (PSCs) are a subcategory of organic solar cells. Currently, development is going on in this generation and performance efficiency of about 10% has been indicated, but the problems associated with this type are poor life cycle, instability and poor performance efficiency. This class offers a simple and highly inexpensive means to convert solar energy into usable energy. But, it lags behind first and second generation solar cells when considering the above-stated drawbacks. If a method to enhance stability and improve efficiency is devised in the near future we may have a new and highly sustainable source to generate energy. Until then, many advancements are required before their successful commercialization in the global markets.

Polymer solar cells are built from thin films of organic semiconductors including polymers, such as polyphenylene vinylene; small-molecule compounds like copper phthalocyanine; and carbon fullerenes and fullerene derivatives such as [6,6]-phenyl-C61-butyric acid methyl ester (PCBM). Recently, there have been extensive developments in dye-based solar cells. During the last two decades, conducting polymers (polyaniline [PANI], polyhexylthiophene, poly(4-ethylenedioxythiophene) [PEDOT] and polypyrrole [PPy]) have garnered considerable attention for the fabrication and development of solar cell applications. As dye-sensitized solar cells have become more efficient, more research is being done to increase the area of the cells [3–7].

4.2 Polymer Solar Cells: An Insight

Polymer solar cells, unlike any other class of solar cells, utilize photoelectric effect for generation of electricity, but instead of using silicon or any other semiconductor, they use conjugated polymers for electron transfer. The fundamentals of semiconductors are found to be the same for the functioning of conjugated polymers as they are for silicon or germanium.

In the year 2000, Alan J. Heeger, Alan MacDiarmid and Hideki Shirakawa discovered conjugated polymer solar cells [4] and were awarded the Nobel Prize in chemistry for that year. It was only then that the conducting behavior of polymers was observed by doping them with iodine [2, 8]. One of the greatest advantages of these solar cells is that they can be easily made using an organic solution. By coating this solution or printing it on a substrate we can produce solar cells in no time. And this fact has urged developers to research this field. Lately, a demonstration of about 10% performance efficiency with a shelf life of about several years has been indicated, which is one of the stark achievements for this class. Another benefit is that these solar cells utilize a small area as compared to first and second generation solar cells [2].

4.2.1 Why Polymer Solar Cells are Preferable

There is a general prospect that comes to mind for all those who want to contribute to the field of polymer solar cells, which is that they are quite lucrative. They have the following potential advantages:

1. Flexibility;
2. Processability;
3. Low material cost;
4. Reduction in usage of scare resources.

The drawbacks which are to be tinkered with in order to obtain a fully commercialized product are stability, efficiency and shelf life [2].

4.3 Layer Stack Constructıon of Polymer Solar Cells

Making a polymer solar cell is rather subtle. First the polymer is dissolved in the solvent and then, via certain printing or coating methodology, layers are coated over the substrate. The layer materials utilized for constructing polymer solar cells are:

- *Active layer:* This layer is divided into two parts: donor phase and acceptor phase. The donor phase is the one which absorbs the light and the acceptor phase is the one which extracts the electron from the exciting-bound electron hole, which results in an electron being transferred into the acceptor phase and also a hole being transferred into the donor phase. For the process to be efficient the transport of the charge carriers must be efficient towards the electrodes and the dissociation of electron-hole pair should be efficient.
- *Substrate:* There are two groups of substrate: glass and plastic groups. Transparent substrates are used for high efficiency. In general, a substrate should be able to hold the layer stack structure efficiently and should allow enough of an amount of sunlight onto active layer.

- *Electrodes:* The selection of electrodes having suitable energy level and electrodes providing enough transparency for the sunlight to pass through to penetrate are preferred. Indium tin oxide (ITO) is a great electrode since it has high optical transmission and low resistance. Another material which is gaining attention is poly(3,4-ethylenedioxythiophene) polystyrene sulfonate (PEDOT:PSS); when doped, this material becomes more flexible, tolerant and conductive.
- *Transport layers (ETL & HTL):* Transport layers are those which have the capability to primarily transfer either charged particles or photons.

These are inexpensive compared to silicon, as they require no expensive manufacturing steps. For instance, titanium dioxide (TiO_2) is already widely used as a paint base [9–11]. In the table in Section 4.9, various dyes are described that can be used in the cells.

4.4 Simple Working of a Polymer Solar Cell

Like all solar cells, polymer solar cells convert light into electricity by converting a flux of photon incident on the cell into a flux of charged particles. This conversion is acquired by using various types of materials which are utilized in the building of a polymer solar cell, where their optical and electrical characteristics are considered in production of the cell [2].

When a incident photon on a polymeric semiconductor exceeds the band gap binding energy, then this photon makes the electron jump to a higher unoccupied level and hence creates an electron-hole pair [4]. This electron-hole pair is subsequently separated over a built-in gradient in the electrochemical potential of the polymer solar cell. In the end, the electron and hole reach the opposite electrodes and are made to recombine after acquiring work from them for completing the circuit [2].

4.5 Life-Cycle Analysis (LCA)

Life-cycle analysis is a method of analyzing any product by considering various aspects and potential effects throughout the life cycle of the product. It can be used as a tool for recognizing any scope of improvement in a product, thereby reducing its impact on human life, the environment and resource depletion. The LCA for plastic photovoltaics can be understood as follows:

- *Goal and scope definition:* This step questions the actual need for product (objective), and the scope/boundaries in which this product can be utilized for human and environmental welfare. It questions the very need for PSC and to what extent it could be used.
- *Inventory analysis:* This step is a collection of data concerning the pollutants in the usage and production of the product (here, plastic photovoltaics). In this step one also collects all information regarding the material required for production.
- *Impact assessment:* In this step the impacts of the product on its surroundings and human lives are considered. The emissions concerned with this product and how they can be reduced, etc., are considered in this step.
- *Interpretation:* Finally, an output is presented in the fourth step termed "interpretation," which encapsulates all the processed information within a few statements of the steps prior to the fourth step.

So, LCA steps are followed and a well-designed inventory related to polymer solar cells is generated. And by following the impacts, inventory and goal analysis, proper interpretation regarding the materials to be used for the development of polymer solar cells are chosen/selected.

4.6 Current Condition of Polymer Solar Cells

Till now, we have only discussed polymer solar cells in basic terms; now the current situation regarding PSCs will be covered. Polymer solar cells have been investigated for about one and a half decades now and certain improvements have surged, which we cannot just disregard. With the great potential offered by this field, there are many problems which still remain unaddressed and we need to find foolproof answers [12].

Recently, the surge in growth of PSCs has made them a strong future contender for silicon-based solar cells. Due to several factors listed in the above sections, PSCs promise a much brighter future for mankind. About 8% improvement in the efficiency of the cells has been offered within the last decade and now PSCs are considered able to contend with polycrystalline silicon-based solar cells [12].

The questions which still remain unaddressed are the development of proper solvent, proper active layer and electrodes which do not use indium, which is a scarce resource [12].

The solvents around which our PSCs have grown are derivatives of chlorobenzene and xylene. The point on which we should shed some light is the fact that any change in solvent highly affects the performance of the PSC and using more benign solutions will require us to change the morphology of the entire material-solvent at nanolevel. Some highly efficient polymer solar cells are affected by water and oxygen and hence they are kept in a preserved environment in order to function, which is a severe drawback [12].

Currently, aluminum is used as a replacement for indium, but it is used in a vacuum deposition method and hence increases the production cost many folds. Processes to print aluminum in the form of ink over the substrate are under development and the efficiency of PSCs is observed under this new treatment [12].

Polymer solar cells are struggling with the same scientific drawbacks which silicon-based solar cells once suffered from; and there is a high possibility that, like silicon solar cells, polymer organic cells will also work their way through all these quandaries and reach the apex, thereby brightening all mankind.

4.7 Materials Used for Developing PSC

In this section we intend to deliver information about the basic materials which are used for processing a polymer solar cell. This section is further segregated into topics regarding the synthesis of the material to be used, conjugated polymer, side-chain influence in the polymers, and finally methods to process and purify polymers for their usage in polymer solar cells.

4.7.1 Synthesis of Polymer Materials

Polymerization is the basic process over which the entire synthesis depends. As is known, polymerization is a reaction, otherwise called a mechanism, which binds monomers into long chains which are called long-chained polymers. Basically, it is a process wherein the monomers are appended over and over to form a long-chain compound.

The synthesis of polymer material basically involves three methods which are of great importance in the reviewed subject. These methods are: Stille cross-coupling, Suzuki cross-coupling and direct acylation polymerization.

4.7.1.1 *Stille Cross-Coupling*

This is a coupling reaction wherein generation of new carbon-carbon bonds takes place, which is carried out by transmetalation of an organotin compound with a halogenide as a coupling partner. Here palladium is used as a catalyst in a suitable solvent. This coupling reaction takes place under mild conditions and no protective groups are required as the reaction is tolerant to mostly all functional groups [13]. This reaction procedure appears to be attractive as it is particularly stable and is not affected by moisture in the air [14]. Its only drawback is that the toxicity level of this reaction makes it unsuitable, as previously discussed in the section on Life-Cycle Analysis.

Illustration of Stille process [13–14]:

$$n(R)_3\,Sn{-}R^1{-}Sn(R)_3 + nX{-}R^2{-}R \rightarrow -[-R^1{-}R^2-]_n- \tag{4.1}$$

where R = alkyl; R^1 and R^2 = aryl alkenyl; X = Cl, Br, I, OTf; In presence of Pd(O)-Catalytic; Ligand.

4.7.1.2 *Suzuki Cross-Coupling*

This is another coupling reaction wherein new carbon-carbon bonds are generated. In this reaction an organoboron functionalized compound is partnered with an organic halide to produce the desired polymerization. Here again, palladium is used as a catalyst with suitable solvent. This reaction is also tolerant of mostly all functional groups and can be processed under mild conditions. The organoboron compounds used in this synthesis are less toxic as compared to organotin compounds and therefore suitable for industrial-scale use. One difference between the Stille and Suzuki process is the requirement of additional base in the Suzuki process as compared to that of Stille [9, 10]. This additional requirement is deemed to accelerate the transmetalation step (where ligands are transferred from one metallic compound to another) [15, 16].

Illustration of Suzuki process [15–16]:

$$n(R)_3\,B{-}R^1{-}B(R)_3 + nX{-}R^2{-}X \rightarrow -[-R^1{-}R^2-]_n- \tag{4.2}$$

where R = alkyl, OH, O-alkyl; R^1 and R^2 = aryl alkenyl; X = Cl, Br, I, OTf; In the presence of Pd(O)-Catalytic; Base; Ligand.

4.7.1.3 *Direct Arylation Polymerization*

This method is rather attractive compared to the other two methods listed above which are based on transition-metal-catalyzed cross-coupling polymerization techniques. This method does not involve a metalation step and thus eliminates the need to purify the monomers which are metaled, which is a more challenging problem. Another problem is elimination of toxic substances formed in other polymerization techniques; it has been indicated that this technique is free from such problems. Direct arylation polymerization is a technique which produces conjugated polymers with desired properties and structure. This reaction acquires precise control of structures by altering the reaction parameters. Currently this technique is used for small molecules and methods are being devised for its utilization for large molecules, as it provides a future alternative to the Suzuki and Stille techniques [17].

Illustration of Direct Arylation Polymerization process [17]:

$$nH{-}R^1{-}H + nBr{-}R^2{-}Br \rightarrow -[-R^1{-}R^2-]_n- \quad (4.3)$$

where R^1 and R^2 = aryl; In the presence of Pd(O)-Catalytic; Acid/Base.

4.7.1.4 Polymerization Rates

All the reactions mentioned above are examples of step reactions prepared from two monomers having di-functional units. Monomers initially react vigorously in a step reaction polymerization until a large amount of monomers are consumed and the polymerization rate tends to decrease. The reaction carries on until the concentration of functional groups is reduced to such an extent that the reaction production is insignificant or it carries on until the polymerized structure becomes immobile due to its increasing weight. Stoichiometry affects the step reaction, hence if long-chain/heavy polymers are to be formed, highly purified monomers need to be used [18].

4.7.2 Conjugated Polymers

On the basis of broad classification, almost everything in this universe can be classified as conductors, insulators, and existing somewhere in between these are semiconductors. The ability of a polymer to exhibit semiconductor properties is decided by its energy band gap (E_g). The energy band gap for a polymer is the energy difference between the highest occupied molecular orbital (HOMO) and lowest unoccupied molecular orbital (LUMO); which are analogous to the conduction band and valance band of a general silicon- or germanium-based semiconductor [19].

Now when a photon with light energy equivalent to or more than E_g is illuminated over a polymer, then excitation will occur and the electron will be ejected if the energy is higher than E_g, since in this fashion it will be able to release excess energy in the form of kinetic energy. For example, consider the polymer polyacetylene; now when a photon falls on the surface of this polymer, the pi electrons are transferred from HUMO to LUMO; and this is where the conjugation of polymers comes into the picture. The electron can be transferred from a pi to pi* state, i.e., from low energy state to excited state. Now if we use a conjugated polymer this delocalization of double bond varies, as the charge densities vary over the length of the polymer chain [20,21].

This variation of charge density is called the Peierls effect. This effect is responsible for the higher semiconductor resemblance of the polymer to the semiconductors [20, 22].

In such a case, the p_z orbitals are located in a perpendicular manner to the polymer chain length, which allows easy delocalization of pi electrons. This delocalization helps in the conduction mechanism of the conjugated polymer, hence conjugated polymer aids the conduction with alternating double bonds, which normal single-chain polymer fails to do [22].

4.7.3 Side-Chain Influence in Polymers

Organic solar cells are highly attractive due to their ability to be processed from just a coating of solvent over the layers. Now, this helps in the fast delivery of cheap, flexible substrates that are processed using available technologies. Now, to improve the solubility of the conjugated polymer solution in the applied solvent, conjugated polymers need to incorporate side chains which help in the stability and solubility. Some of the tested and appreciable side chains are alkanes, alkoxies, ketones and esters [23].

The side chains appreciably affect the final performance of the product as they influence the bulk heterojunctions of the active layer. It has been noted that changing the size, position and nature of the side chain affects the performance appreciably [24–28]. Research now states that, as a general rule, the side chains should be placed at such position where the steric hindrance effect is minimal [29,30]. In this manner the conjugated polymer is turned into a more planar form rather than a twisted form; the backbone tends to be more in a single plane, which convincingly influences the performance level [25, 31].

Another relation is that the number of carbons in a side chain should not be less than the number of carbons in the main chain of the conjugated polymer. This helps in the solubility of the polymer in the common organic solvent.

4.7.4 Purification

The monomers utilized in the polymerization reactions need to be highly pure and for that numerous techniques are taken into consideration. If the monomer is in a solid state, then filtration or recrystallization can be used; and if the monomer is a liquid, flash chromatography, distillation or dry column vacuum chromatography may be used.

4.8 Degradation and Stability of a PSC

In current usage, the performance of polymer solar cells degrades over time when operated. This degradation is due to both internal and external instabilities and poor coherence. However, external problems are mechanical (e.g., degradation due to interconnections, yellowing of encapsulation material, etc.) and can be remedied with simpler changes. Moving towards internal problems, which are of high scientific importance, many researchers are working towards remedies for these issues; for example, the stability of materials used in fabricating the PSC and also the level of acceptance among the layers and their affinity for each other, which tend to degrade PSCs.

4.8.1 Physical Degradation

Physical degradation is attributed to two main issues, namely:

- Morphological stability;
- Flexibility and delamination.

4.8.1.1 Morphological Stability

It was a breakthrough when bulk heterojunction layers were introduced in the science of solar cells. In contrast to the bilayer interface of acceptor and donor layers, in bulk layer heterojunction the acceptor and donor layers interweave, forming bi-contiguous and interpenetrating networks. This facilitates in the transfer of excited electrons; as the lifetime and diffusion of excitons is small, this increment in interfacial area improves the performance of the polymer solar cells exceptionally [32–34].

However, there happen to be factors which push this layer into hyperactivity, hence leading to it degrading/consuming itself. The major cause of this is that the morphology of the bulk layer is not in thermal equilibrium and when it is subjected to elevated temperatures it falls into imbalance [34].

Other factors deemed responsible for instability are donor-acceptor blend ratio, region regularity, molecular weight and processing solvents. Optimal stabilization of the bulk heterojunction morphology is required to attain proper functioning of the PSC [35]. Approaches have been made to immobilize the evolution of the bulk heterojunction layer due to elevated temperatures. Polymer side chains help to stabilize this agenda.

4.8.1.2 Flexibility and Delamination

The flexibility of polymer solar cell is one of its strong points and is counted as one of its most important traits. For a long time now, ITO has been used as a transparent conductor but has been limited by its flexibility, as it is a brittle inorganic material. ITO film tends to break/crack upon flexing. This break upon flexing increases the overall resistance of the ITO film, thus affecting conductivity. As the sheet resistance increases it does not allow the transfer of excited electrons into the external circuit, hence the performance of the polymer solar cell will degrade.

Another cause for degradation is delamination, which is caused by poor adhesion at one or more interfaces; when this happens the transfer of electrons will be hampered and thus the performance will degrade.

4.8.2 Chemical Degradation

Chemical degradation is attributed to these two factors:

- Polymer instability
- Photochemical degradation

4.8.2.1 Polymer Instability

Among the several reasons for the degradation of polymer solar cells, polymer instability is one of the prime causes. When moisture from atmosphere and oxygen diffuse within the polymer, several oxidation or degradation reactions take place, as shown in Figure 4.1.

Generally these reactions are positively affected by the sunlight to which the PSC is exposed; hence the oxidative degradation increases (degradation due to photooxidation). This degradation in its present state is not eliminated completely but can be reduced significantly by following some general rules. The very first rule is to never use exocyclic double bonds in the backbone of the polymer, as it aids or reacts actively towards photooxidation and hence "improves" the rate of degradation, which is undesirable.

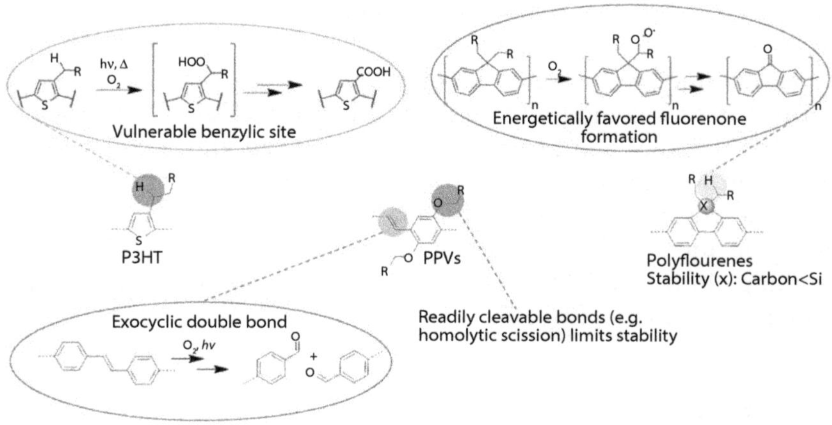

Figure 4.1 Figure illustrating the reason for instability in a polymer [2].

Secondly, the side-chain elements should be noted before being appended onto the backbone of the polymer. Though side chains are necessary for the improvement of the conductive performance of the PSC, they are not desirable from a degradation and stability point of view. Thus, side chains should be selected on such a basis, so that both the polymer stability is maintained and yet the conductive performance is superior. Also, a point to be noted here is that alpha carbons should be protected as they are highly vulnerable to photochemical reactions and thus degradation.

4.8.2.2 Photochemical Degradation

Large efforts are put into recognizing the degradation of polymer solar cells. Therefore, when such efforts are put forth one should never forget the very basic root of the structure, which in this case is the polymer [36, 37].

Polymers, as we know, are capable of degrading via atmospheric air and moisture. Though currently no essentially valid mechanism is available for a perfect understanding of this, to some extent it is known that if a polymer is exposed to sunlight then molecular oxygen tends to destroy the pi conjugation of the polymer and renders the absorbance slowly to nullity. In liquid state it is believed that singlet oxygen is the cause of degradation, whereas for solid-state polymers it is suggested that first allylic hydrogen is affected and then the side chains [38].

The above-stated process is responsible for degradation in solid state, as via this method the polymeric backbone is affected and the conjugation is lumped into oblivion, leading to bleaching. Here it is important to note that for solid-state polymers, the proposed mechanisms state that singlet oxygen intermediates are not the only cause of degradation but several intermediates form and lead to distortion of the backbone. This degradation is both photo- and thermal-oxidation directed, thus reinforcing the above sentence regarding oxygen. It has been observed that the mechanism proceeds in the following fashion: first, free radicals are formed via photogeneration, and then free radicals are formed via photolysis upon absorbing ultraviolet light; hence fissuring occurs in the conjugation and degradation continues [37, 39].

4.9 Dyes

Dyes are colored substances which have an affinity for the substrate over which they are applied. This basic definition defines any colored substance as dye. There are several types of dyes utilized for efficient working of polymer solar cells [40].

A dye should meet the following requirements for coating a polymer solar cell substrate:

- It should absorb all light below a wavelength of 920 nm;
- The affinity for the substrate should be very high;
- The excitation energy of the electron excited from the dye should have enough energy to reach the conduction band of the electrode;
- It should have a stable life;
- Regeneration may be employed using redox electrolyte potential [40].

Dyes can be classified into two categories, namely, synthetic and natural dyes. Synthetic dye is prepared by man-made means and may adopt some duplication techniques so as to enhance itself from its natural dye-based counterpart. And natural dye is generally obtained from the peels of fruits, and extracted from the leaves, barks and seeds of plants and trees. Generally, natural dyes can be obtained from almost all naturally occurring substances.

Currently dyes are used as photosensitizers in polymer solar cells, and synthetic dye-based cells have attained a performance efficiency of about 11% [41]. Natural dye-based solar cells lag in efficiency but are better substituents to synthetic dyes, as they are easily processable and yet easily

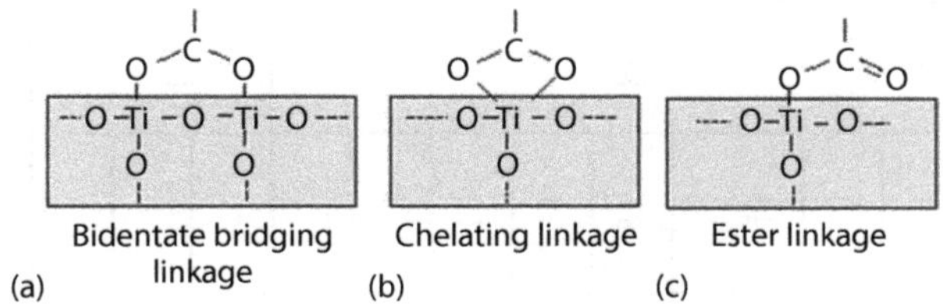

Figure 4.2 Illustration of a natural dye bonding with a TiO_2 layer: (a) bidentate bridging, (b) chelating and (c) ester linkages [41].

available around us in numerous forms, which makes them a very attractive research source.

Dyes have certain groups which help them to attach themselves to certain metal layers and these groups help in the proper transfer of electrons. Generally, dyes are used with TiO_2 nanoparticles as they are efficient at aiding dyes for transfer of electrons. Now for the dyes to be associated with the TiO_2, some groups need to enhance their associativity. These groups are generally carboxylic groups and their derivatives, i.e., acid halides, esters, etc. Hydroxyl groups also show the tendency of binding onto TiO_2 [41–43]. Figure 4.2 clearly shows how a dye can bond with a TiO_2 substrate where carboxylic acid is concerned.

4.9.1 Natural Dyes Used for Polymer Solar Cells

Natural dyes, as described in the above section, are one of the highest binding resources for PSC production as compared to synthetic dyes. Though synthetic dyes provide better efficiency, they are harmful to the environment. It should be taken into consideration that synthetic dyes may not be biodegradable, plus the metals used to make them are scarce, ruthenium being one of those elements. Therefore, in Table 4.1 some research results are shown which state the best possible efficiency for a particular dye and the characteristics of the cell when the dye was used.

Considering the information given in Table 4.1 and seeking its observations, it can be noted that spinach and beetroot dyes offer better efficiency. The reason for the use of spinach as dye gaining high efficiency may be due to the high chlorophyll concentration in the spinach leaves. It should also be noted that a significant amount of research has been put into improving the efficiencies of red beetroot, betalains and red turnip. Anthocyanins are preferred over other plant pigments because they offer a broader absorptive spectrum, running the gamut from red to purple. Secondly, chlorophyll

Table 4.1 Various dyes used in dye-sensitized solar cells.

Sl. no.	Dye	Band gap	Absorption range (nm)	Open circuit voltages V_{OC}	Short circuit photocurrent densities J (mA/cm^2)	IPCE (%)	Ref.
1	Spinach	3.14	300–700			15	[3]
	Beetroot	2.48				30	
	Spinach + Beetroot	2.45				45	
2	Teak	2.22	400–700	0.430–0.610	0.11–0.29	37	[4]
	Tamarind	3.07				33	
	Eucalyptus	2.47				12	
3	Olive		412–664	0.54–.058	0.35–0.85		[5]
	Lycium shawii		432–666	0.44–0.62	0.15–1.20		
4	Betalains	1.1	111–245				[6]
5	Carotenoids	1.42					
6	Chlorophyll	2.9					

is taken into consideration because it presents two absorptive peaks and thus enhances the energy absorption. Several other pigments from betalains offer good absorption. The following aspects are of great importance where polymer solar cells are concerned: Affinity of the dye for the TiO_2, degradation of the dye, absorptive pigment concentration, stabilization of the dye, and effect of polymer electrolyte or any other electrolyte over the efficiency of the PSC.

4.10 Performed Experiments

Chemicals required: Analytical grade potassium iodide, methylene chloride, acetone, acetic acid, nanocrystal line TiO_2 and ZnO powder were purchased from Piyush Chemicals, Ahmedabad. Materials like multimeter, clear plastic tapes, binder clips, mortar and pestle, and hot plate were also used. The ITO coated glass plates with dimensions 25 mm × 25 mm × 1.1 mm having resistivity < 10 were purchased from Techintrso, Nagpur.

Preparation of photoelectrode (TiO_2/dye): Acetic acid was slowly added in 1 ml increments to the TiO_2 and was ground well each time for about 30 min, followed by addition of 1 to 2 drops of detergent to act as surfactant. Preparation of PANI: Equimolar mixture of 2.6 ml aniline and 4 ml HCl were mixed to prepare aniline hydrochloride. To this mixture ammonium peroxydisulphate was added in the stoichiometric ratio of 1.25. The PANI thus prepared was kept in storage for two days after which the observations were obtained.

4.10.1 Experimental Setup 1

Coating the slides with TiO_2 solution: ITO sheets were cleaned using acetone and a multimeter was used to check which side was conductive, as shown in Table 4.2. Transparent tapes were used to tape the slides onto the table, leaving roughly 4 mm on each side. A drop of TiO_2 was put on one of the glass slabs and was spread evenly using a rod.

Annealing the Slides:

- First observation:
  The counter electrode was annealed for the first time on a burner and stored along with the TiO_2 coated electrode in vinegar solution but the coating dissolved as soon as it was immersed in vinegar solution.

Table 4.2 Materials and their resistances.

Material	Resistance
PANI	12.2 ohm
ITO Counter	250 ohm
ITO Electrode	170 ohm

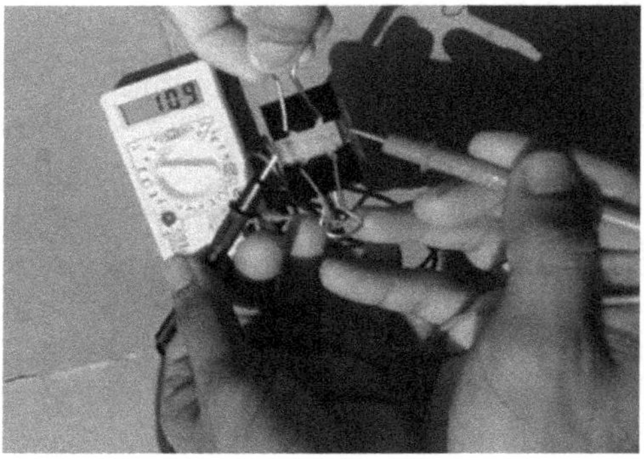

Figure 4.3 Experiment setup I.

- Second observation:
  The second time it was kept in a muffle furnace for about 2 h at 500 °C. To make the counter electrode, the second slide was held by metal tongs and was passed through the flame to get it carbon coated. This time the cells retained their coating. The counter electrode was coated with a drop of PANI solution and left out to dry. To complete the circuit the electrode and the counter electrode were placed on top of one another, as shown in Figure 4.3, and 1 and 2 drops of electrolyte solution were added in between the slides; this was held together using clips.
  Observation: 109 mV.
- Third observation:
  The TiO_2 coated sheets were left out to dry while the carbon coated slides were prepared using ITO sheets and then using glass slabs. The carbon coating was done by rubbing pencil on the slab. This was coated with prepared PANI solution. The TiO_2 coated slabs were then immersed in EBT dye for about 30 min to make a dye-sensitized cell. The electrode and the counter electrode were

then joined together to form a solar cell and a drop of electrolyte solution was put between them.
Observation: 320 mV

The PANI used was old and the graphene layer came off while applying PANI on the counter electrode.

4.10.2 Experimental Setup 2

ZnO coated slides were prepared by the same procedure as TiO_2 coated slides.

Coating the slides with ZnO solution: ITO sheets were cleaned using acetone and a multimeter was used to confirm the conductive side of the sheet. Transparent tape was used to tape the slides onto the table, leaving roughly 4 mm on each side. A drop of ZnO was put on one of the glass slabs and was spread evenly using a rod.

Annealing the Slides:

This time the ITO sheets were coated with ZnO and dried. The glass slab was coated with carbon by annealing on the burner and the extra coating was removed by using a tissue. It was then further coated by a few drops of PANI and kept in the muffle furnace for 30 min at 400 **°C.** The two cells were then combined with a drop of electrolyte between them and bound together using binder clips. The cell was then tested under sunlight at 5 pm (Figure 4.4).

Observation: 568 mV.

Figure 4.4 Experiment setup 2.

4.11 Summary

Organic cells are lightweight and flexible, meaning that they will be relatively easy to install. There are two main areas where organic cells need to be improved before they can be commercially viable. Their efficiency needs to continue to increase, but also the lifetime of these cells needs to be extended. As it stands, these cells degrade rather quickly; current commercial devices only last for a couple of years. Once the efficiency is competitive and these cells can last for a reasonable lifespan, then they will be a cost-effective, renewable power source. Herein we have experimented using titanium dioxide and zinc oxide applied on ITO coated glass slabs using polyaniline. The voltage readings of ITO coated glass slab consisting of a graphite layer on the plane glass slab were less as compared to that with carbon coated plane glass slab (counter electrode). The efficiency of polymer/organic solar cell is greatly affected by the type of metal oxide used and thickness of the oxide layer. Also, the selection of polyaniline is novel and the readings show the possibility of its exploration.

Today the world is in dire need of acquiring a renewable source of energy production as all the fossil fuels are being depleted at an exponential rate. In this chapter an attempt has been made to explain how we can search for renewable resources existing around us. The topic, polymer-based organic solar cells, suggests that materials used for the production of solar cells are mainly based on organic materials. And in this chapter we have tried to provide a basis for saying that conduction of light is indeed possible using a dye-based polymer solar cell, though the efficiency obtained is minimal. But with the passage of time the research going on in this field is sure to increase. In this chapter it has been shown how polymers and dyes (especially natural dyes) can be used for generating electricity, thus making the production of electricity a green method.

In summation, it should be stated that this century has seen a solar cell boom in the market and we have learned of their significance due to decreasing resources. Now these solar cells are classified into three generations and the one reported on herein is the third generation: organic photovoltaic solar cells. An organic solar cell consists of a substrate, two transport layers, two electrodes and an active layer (which is responsible for dissociation into charged particles). By incorporating the functioning of a PSC, life-cycle analysis of organic solar cells can be performed.

In this chapter, problems related to PSCs which are yet to be solved are listed as the current condition of polymer solar cells and after that material synthesis techniques generally used for PSC production are provided. Out of the listed techniques, direct arylation polymerization is of greater importance as it does not involve any toxic substance emitted as by-product. The degradation and stability nuances of a PSC are broadly discussed; degradation may be possible due to physical or chemical causes and can be reduced to a significant level. Compared to silicon solar cells, these cells are insensitive to impurities in the fabrication process, which accelerates their transition from research laboratory to mass production line. Many other benefits of polymer/organic cells, such as flexibility and portability, have attracted considerable company investment and government funding.

References

1. European Wind Energy Association, *Wind in Power: 2011 European Statistics*, pp. 1–11, Technical Report, EWEA, 2012, 1–11.
2. Plasticphotovoltaics.org-a hub for OPV research, http://www.plasticphotovoltaics.org/, 2016.
3. Zhang, Y.D., *et al.*, How to design dye-sensitized solar cell modules. *Sol. Energ. Mat. Sol. Cells* 95, 2564–2569, 2011.
4. Grätzel, M., Photoelectrochemical cells. *Nature* 414, 338, 2001.
5. Gupta, A., and Sharma, S., Organic solar cell – A renewable energy resource. *IJESDM* 4, 3, 2013.
6. Hagfeldt, A., Didrikrisson, B., Parmqvist, T., Lindström, H., Södergren, S., Rensmo, H., and Lindquist, S., Verification of high efficiencies for the Grätzel-cell. A 7% efficient solar cell based on dye-sensitized colloidal TiO_2 films. *Sol. Energ. Mat. Sol. Cells* 31, 481–488, 1994.
7. Mohamed, S.H., El-Hossary, F.M., Gamal, G.A., and Kahlid, M.M., Properties of indium tin oxide thin films deposited on polymer substrates. *Acta. Phys. Pol. A* 115, 704–708, 2009.
8. Smalley, R.E., Future global energy prosperity: The terawatt challenge. *MRS Bull.* 30, 412–417, 2015.
9. Heo, J.H., and Sung, Y.M., Synthesis of nanoporous TiO_2 materials using sol-gel combustion method and its photovoltaic characteristics. *J. Electr. Eng. Technol.* 58, 322–326, 2009.
10. Goto, K., Kawashima, T., and Tanabe, N., Heat-resisting TCO films for PV cells. *Sol. Energ. Mat. Sol. Cells* 90, 3251–3260, 2006.
11. Aziz, F., and Ismail, A.F., Spray coating methods for polymer solar cells fabrication: A review. *Mater. Sci. Semicond. Process.* 39, 416–425, 2015.
12. Søndergaard, R., Helgesen, M., Jørgensen, M., and Krebs, F.C., Fabrication of polymer solar cells using aqueous processing for all layers including the metal back electrode. *Adv. Energ. Mater.* 1, 68–71, 2011.
13. Stille, J.K., The palladium-catalyzed cross-coupling reactions of organotin reagents with organic electrophiles. *Angew. Chem. Int. Ed.* 25, 508–524, 1986.

14. Espinet, P., and Echavarren, A.M., The mechanisms of the Stille reaction. *Angew. Chem. Int. Ed.* 43, 4704–4734, 2004.
15. Suzuki, A., Recent advances in the cross-coupling reactions of organoboron derivatives with organic electrophiles, 1995–1998. *J. Organomet. Chem.* 576, 147–168, 1999.
16. So, C.M., and Kwong, F.Y., Palladium-catalyzed cross-coupling reactions of aryl mesylates. *Chem. Soc. Rev.* 40, 4963–4972, 2011.
17. Rudenko, A.E., Direct arylation polymerisation (DARP) for the synthesis of conjugated polymer, Doctoral Dissertation, Retrieved from USC Digital Library, 2014.
18. Stevens, M.P., *Polymer Chemistry: An Introduction*, Longman Higher Education, 1975.
19. Pron, A., and Rannou, P., Processible conjugated polymers: From organic semiconductors to organic metals and superconductors. *Prog. Polym. Sci.* 27, 135–190, 2002.
20. Nunzi, J.-M., Organic photovoltaic materials and devices. *C. R. Physique* 3, 523–542, 2002.
21. Sariciftci, N.S., Smilowitz, L., Heeger, A.J., and Wudi, F., Photoinduced electron transfer from a conducting polymer to buckminsterfullerene. *Science* 258, 1474–1476, 1992.
22. Lugenschmied, C., Dennler, G., Czeremuzskin, G., Latreche, M., Neugebauer, H., and Sariciftcia, N.S., Flexible encapsulation for organic solar cells. *Photonics for Solar Energy Systems, Proc. SPIE* 6197, 619712, 2006.
23. Uy, R.L., Price, S.C., and You, W., Structure-property optimizations in donor polymers via electronics, substituents, and side chains toward high efficiency solar cells. *Macromol. Rapid Commun.* 33, 1162–1177, 2012.
24. Biniek, L., Fall, S., Chochos, C.L., Anokhin, D.V., Ivanov, D.A., Leclerc, N., Lévêque, P., and Heiser, T., Impact of the alkyl side chains on the optoelectronic properties of a series of photovoltaic low-band-gap copolymers. *Macromolecules* 43, 9779–9786, 2010.
25. Carlé, J.E., Andreasen, J.W., Jørgensen, M., and Krebs, F.C., Low band gap polymers based on 1,4-dialkoxybenzene, thiophene, bithiophene donors and the benzothiadiazole acceptor. *Sol. Energ. Mat. Sol. Cells* 94, 774–780, 2010.
26. Li, Y., Chen, Y., Liu, X., Wang, Z., Yang, X., Tu, Y., and Zhu, X., Controlling blend film morphology by varying alkyl side chain in highly coplanar donor–acceptor copolymers for photovoltaic application. *Macromolecules* 44, 6370–6381, 2011.
27. Liang, Y., Feng, D., Wu, Y., Tsai, S.-T., Li, G., Ray, C., and Yu, L., Highly efficient solar cell polymers developed via fine-tuning of structural and electronic properties. *J. Am. Chem. Soc.* 131, 7792–7799, 2009.
28. Liu, F., Gu, Y., Shen, X., Ferdous, S., Wang, H.-W., and Russell, T.P., Characterization of the morphology of solution-processed bulk heterojunction organic photovoltaics. *Prog. Polym. Sci.* 38, 1990–2052, 2013.

29. Shi, Q., Fan, H., Liu, Y., Chen, J., Ma, L., Hu, W., Shuai, Z., Li, Y., and Zhan, X., Side chain engineering of copolymers based on bithiazole and benzodithiophene for enhanced photovoltaic performance. *Macromolecules* 44, 4230–4240, 2011.
30. Zhang, Z.-G., Min, J., Zhang, S., Zhang, J., Zhang, M., and Li, Y., Alkyl chain engineering on a dithieno[3,2-*b*:2′,3′-*d*]silole-alt-dithienylthiazolo[5,4-*d*] thiazole copolymer toward high performance bulk heterojunction solar cells. *Chem. Commun.* 47, 9474–9476, 2011.
31. Zhang, G., Xu, H., Liu, K., Li, Y., Yang, L., and Yang, M., Synthesis, characterization and comparative photovoltaic behavior of diketopyrrolopyrrole (DPP)-containing poly(phenylenevinylene)s varying at N-alkyl or 3,6-aryl substituents in DPP units. *Synth. Met.* 160, 1945–1952, 2010.
32. Yu, G., Gao, J., Hummelen, J.C., Wudl, F., and Heeger, A.J., Polymer photovoltaic cells: Enhanced efficiencies via a network of internal donor-acceptor heterojunctions. *Science*, 270, 1789–1791, 1995.
33. Bundgaard, E., Helgesen, M., Carlé, J.E., Krebs, F.C., and Jørgensen, M., Advanced functional polymers for increasing the stability of organic photovoltaics. *Macromol. Chem. Phys.* 214, 1546–1558, 2013.
34. Liu, F., Gu, Y., Shen, X., Ferdous, S., Wang, H.-W., and Russell, T.P., Characterization of the morphology of solution-processed bulk heterojunction organic photovoltaics. *Prog. Polym. Sci.* 38, 1990–2052, 2013.
35. Chen, L.-M., Hong, Z., Li, G., and Yang, Y., Recent progress in polymer solar cells: Manipulation of polymer: fullerene morphology and the formation of efficient inverted polymer solar cells. *Adv. Mater.* 21, 1434–1449, 2009.
36. Rivaton, A., Chambon, S., Manceau, M., Gardette, J.-L., Lemaître, N., and Guillerez, S., Light-induced degradation of the active layer of polymer-based solar cells. *Polym. Degrad. Stab.* 95, 278–284, 2010.
37. Manceau, M., Rivaton, A., and Gardette, J.-L., Involvement of singlet oxygen in the solid-state photochemistry of P3HT. *Macromol. Rapid Commun.* 29, 1823–1827, 2008.
38. Koch, M., Nicolaescu, R., and Kamat, P.V., Photodegradation of polythiophene-based polymers: Excited state properties and radical intermediates. *J. Phys. Chem. C* 113, 11507–11513, 2009.
39. Manceau, M., Rivaton, A., Gardette, J.-L., Guillerez, S., and Lemaître, N., The mechanism of photo- and thermooxidation of poly(3-hexylthiophene) (P3HT) reconsidered. *Polym. Degrad. Stab.* 94, 898–907, 2009.
40. Racal, N., Vaishaly, A.G., Sharma, H., and Mathew, B.B., Dye sensitized solar cells: The emerging technology. *EPES* 2, 46–52, 2015.
41. Calogero, G., Bartolotta, A., Di Marco, G., Di Carlo, A., and Bonaccorso, F., Vegetable-based dye-sensitized solar cells. *Chem. Soc. Rev.* 44, 3244–3294, 2015.
42. Senthil, T.S., Muthukumarasamy, N., Velauthapillai, D., Agilan, S., Thambidurai, M., and Balasundaraprabhu, R., Natural dye (cyaniding 3-O-glucoside) sensitized nanocrystalline TiO_2 solar cell fabricated using

liquid electrolyte/quasi-solid-state polymer electrolyte. *Renew. Energy* 36, 2484–2488, 2011.
43. Jusof Khadidi, M.H., and Hamid, E.A., A new flocculant-coagulant with potential use for industrial wastewater treatment. *2nd Int. Conf. Environ., Energ. Biotechnol.* 51, 139–142, 2013.

5

A Simple Route to Synthesize Nanostructures of Bismuth Oxyiodide and Bismuth Oxychloride (BiOI/BiOCl) Composite for Solar Energy Harvesting

I. D. Sharma[1,2]*, Chander Kant,[1] A. K. Sharma,[1] Ravi Ranjan Pandey[1] and K. K. Saini[1]*

[1]CSIR-National Physical Laboratory, New Delhi, India
[2]Vikram Seva Sadan Inter College, Pisawa, Aligarh, India

Abstract
In this chapter, bismuth oxychloride, bismuth oxyiodide and a composite of these two compounds have been synthesized by wet chemical route. Synthesized material was characterized by SEM, XRD and FTIR. The composite of bismuth oxychloride and bismuth oxyiodide exhibits strong absorption ~ 2 eV in the visible range, which can support photodegradation reactions by absorption of visible energy abundantly available in the solar spectrum with high throughput comparable to that of bismuth oxychloride with absorption of UV energy.

***Keywords*:** Photocatalyst, visible light photoctalyst, composite material, layered transition metal oxides, bismuth oxyhalides

5.1 Introduction

Rapid development of global industry and steadily worsening environmental pollution is a very challenging combination and has the potential of a global crisis. For sustainable development of our society, the development of both pollution-free technologies for environmental remediation and alternative clean energy supplies is the urgent agenda. Among the wide

**Corresponding authors*: krishanksaini@gmail.com; idsharma1974@gmail.com

Sanjay K. Nayak *et al.* (eds.) Trends and Applications in Advanced Polymeric Materials, (83–100) 2018 variety of green and renewable energy projects underway, semiconductor photocatalysis has emerged as one of the most promising technologies because it represents an easy way to utilize either natural sunlight energy or artificial indoor illumination, which is abundantly available everywhere in the world. The potential applications of photocatalysis are found mainly in the fields of photolysis of water to yield hydrogen fuel, photodecomposition or photooxidization of hazardous substances; artificial photosynthesis; photoinduced superhydrophilicity; photoelectrochemical conversion, decontamination of water, decomposition of crude oil, self-sterilizing photocatalytic coatings, and so forth [1].

The fundamental principles of semiconductor photocatalysis are to initiate or accelerate specific reduction and oxidation (redox) reactions on the surface of irradiated semiconductor particles. Light absorption and the consequent photoexcitation of electron-hole pairs take place when the energy of the incident photons matches or exceeds the band gap energy [1–3].

The conduction-band (CB) electrons ($e_{cb}-$) in semiconductors have a chemical potential of –0.5 to +1.5 V versus the normal hydrogen electrode (NHE), hence they can act as reductants. The valence-band (VB) holes ($h_{vb}+$) exhibit a strong oxidative potential of +1.0 to +3.5 V versus NHE [1]. Initially, the energy of the incident photons is stored in the semiconductor by photoexcitation; it is then converted into chemical form by a series of electronic processes and surface/interface reactions. In contrast to the thermodynamics of conventional catalysis, not only spontaneous reactions ($\Delta G < 0$) but also non-spontaneous reactions ($\Delta G > 0$) can be promoted by photocatalysis. In the former case, the input energy is used to overcome the activation barrier so as to facilitate photocatalysis at an increased rate or under milder conditions. In the latter case, part of the input energy is converted into chemical energy that is accumulated in the reaction products.

In general, a semiconductor photocatalytic cycle comprises three steps, as shown in Figure 5.1: first, illumination induces a transition of electrons from the VB to the CB, leaving an equal number of vacant sites (holes); second, the excited electrons and holes migrate to the surface; third, they react with absorbed electron donors (D) and electron acceptors (A), respectively. In the second step, a large proportion of electron-hole pairs recombine, dissipating the input energy in the form of heat or emitted light. To prevent the recombination of electron-hole pairs, the approach that has generally been applied is to load cocatalysts, such as platinum (Pt), palladium (Pd), nickel(II) oxide (NiO) and ruthenium(IV) oxide (RuO_2), on the semiconductor surface [4–6]. The heterojunctions that are formed between the host semiconductor and the cocatalyst provide an internal electric field

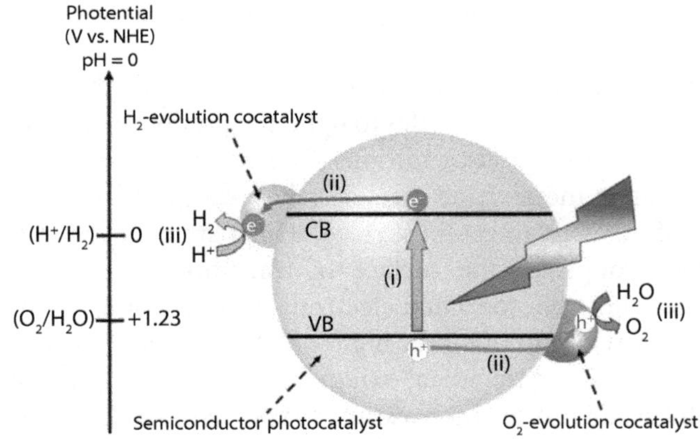

Figure 5.1 Redox processes at the semiconductor surface by photogenerated electrons and holes.

that facilitates separation of the electron-hole pairs and induces faster carrier migration.

The probability of electronic transition, excitation of valence-band electron to the conduction band by absorption of a photon, is proportional to the square of the amplitude of the radiation field, E_0, and the square of the transition dipole moment $|\mu_{if}|$.

$$P \propto E_o^2 \left|\mu_{if}\right|^2 \tag{5.1}$$

The quantum yield for an ideal system undergoing photocatatalytic events is measured as the number of events occurring per photon absorbed. The typical assumption in determining the quantum yield is that all photons are absorbed on the semiconductor surface, and the quantum yield is referred to as the apparent quantum yield. This assumption is necessary due to the difficulty in measuring the actual photons absorbed by the solid surface. The relation between the quantum yield (φ), the rate of charge transfer, k_{CT}, and the electron-hole recombination rate, k_R, is given by the equation:

$$\varphi \propto \frac{k_{CT}}{\left(k_{CT} + k_R\right)} \tag{5.2}$$

Photoinduced molecular transformations at transition metal oxide surfaces can be organized into two general classes. Photoexcitation of the adsorbate which then reacts with the catalyst substrate is classified as a catalyzed photoreaction. Photoexcitation of the catalyst followed by interaction of the catalyst with a ground state reactant is classified as a sensitized photoreaction.

Photoassisted adsorption and desorption can both be promoted by exposure of transition metal oxides to light, the predominant process being controlled by experimental conditions. The process by which adsorption and desorption on metal oxide surfaces takes place is related to the photogeneration of holes on the solid surface, which are believed to be trapped by hydroxyl groups on the surface of transition metal oxides. These trapped holes allow photoexcited electrons to be available for chemisorption. Doping of a cation of either higher or lower valence can change the electronic properties of the metal oxide. Doping with a higher valence cation typically results in an increase in *n*-type semiconductivity, or raises its Fermi energy, while doping with a lower valence cation should lower the Fermi energy level and reduce the metal oxides *n*-type semiconductivity. The process of doping indicates that a cation other than the transition metal cation, experienced in the majority of the bulk is incorporated into the crystal structure of the semiconductor, either by replacing the cation or interstitially adding to the matrix.

Research on photocatalytic materials and applications has experienced continuous exponential growth but titanium dioxide (TiO_2) is still by far the most used photocatalyst. The more recent burst of alternative photocatalysts to TiO_2 has been mainly fostered by the renewed interest in artificial photosynthesis processes, the advances in materials science, and acknowledgment that the results which could be obtained by introduction of dopants into titania are strongly limited. Extending the spectral sensitivity of TiO_2 to the visible range by doping inherently decreases the reactivity of the active sites, and therefore seems to prove that with these modifications the photoactivity obtained under visible light is only a fraction of that obtained under UV radiation. Moreover, it is becoming increasingly clear that visible-light active non-doped photocatalytic materials require the combination of multiple cations, and the search for the so-called "third generation photocatalysts" with improved characteristics is underway. Many visible-light active photocatalysts are tested using only visible radiation ($\lambda > 400$ nm), despite the fact that the sunlight spectrum contains a small fraction of UV light. Figure 5.2 shows some of the investigated photocatalytic materials with their respective energy band edge positions.

Despite the significant progress achieved by photocatalysis in the last decade, there are still important challenges ahead for the full development of this technology. The obvious interest in the implementation of more sustainable processes allows us to hope for a possibly long, but surely brilliant trajectory of photocatalysis towards maturity as a widespread technology.

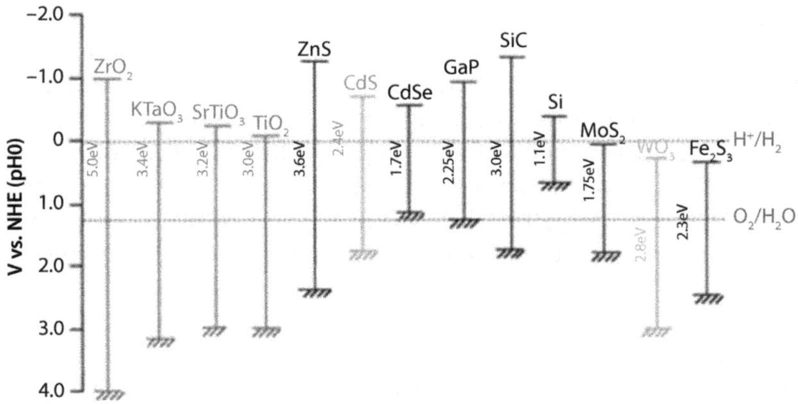

Figure 5.2 Energy diagram (positions of conduction and valence bands) for typical semiconductors.

One of the principal challenges in this field is the understanding of the mechanism responsible for the enhancement of the photocatalytic activity. The dependence of the photocatalytic activity on the exposed facets has been explored in several materials, viz., $BiVO_4$, Ag_3PO_4, BiOCl and layered niobate photocatalysts [7]. In addition to facet engineering, the increasing control of preparation procedures allows the synthesis of complex heterostructures at the nanoscale. The interesting UV-photocatalyst BiOCl and other visible-light active members of this series, viz., BiOBr and BiOI, have been studied to gain an in-depth understanding of the photoactivity mechanisms. A facet-dependent photocatalytic activity and nanosheet synthesis with preferentially exposed facets are being investigated. These are related to the formation of oxygen vacancies under irradiation in the oxygen-rich planes.

5.1.1 Bismuth Oxyhalide [BiOX (X = Cl, Br, I)]: General Remarks

Bismuth oxyhalides, BiOX (X = Cl, Br, and I), belong to a new class of promising layered materials for photocatalytic energy conversion and environment remediation which have unique layered-structure mediated fascinating physicochemical properties and suitable band structure [8, 9]. High chemical and optical stability, nontoxicity, low cost, and corrosion resistance are favorable properties of these materials. As an important class of V–VI–VII ternary oxide semiconductors, they have attracted much attention as a new family of promising photocatalysts [9–11]. The bismuth oxyhalides crystallize with a tetragonal matlockite structure (PbFCl-type) having space group P4/nmm. These layer-structured compounds consist of [X–Bi–O–Bi–X] slices stacked together by the nonbonding (van der Waals) interaction through the halogen atoms along the c-axis [12]. A bismuth center is surrounded by four oxygen and four halogen atoms in each [X–Bi–O–Bi–X] layer, which generate an asymmetric decahedral geometry. The optimized structures have been described with an ordered packing of five-atom layers in the sequence of X–Bi–O–Bi–X along the c-axis direction [13]. The layered structure of BiOCl unit cell contains six atoms (2O, 2Cl, 2Bi), in which one O atom is coordinated to four Bi atoms, forming a tetragonal pyramid with the O atom at the center.

The special layered structures of BiOX are characterized by $[Bi_2O_2]^{2+}$ slabs interleaved by double slabs of halogen atoms with an internal static electric field perpendicular to each layer, which are effective in improving the separation of photoinduced electron-hole pairs, and thus have demonstrated excellent photocatalytic and photoelectrochemical performances [13–15].

The valance band maxima in these materials is mainly composed of O 2p and X np states (n = 3, 4 and 5 for X = Cl, Br and I, respectively) where the Bi 6p states dominate the conduction band minimum [12, 16–19]. As the atomic number of X increases, the contribution of X ns states increases remarkably, and the dispersive characteristic of band energy level becomes more and more striking, thereby narrowing the band gap. Obviously, the band gap values and the redox potentials of BiOX are highly associated with the atomic number of X, namely, the composition of the layered structure. Theoretical and experimental studies have shown that the band gaps of BiOCl, BiOBr and BiOI are ca. 3.4, 2.8 and 1.8 eV, respectively. It is evident that the tunable layered structures endowed bismuth oxyhalide materials with excellent optical absorption ranging from UV to visible light, suggesting that they have potential for photocatalytic applications. It is found that BiOCl exhibits the best performance in the degradation of air or water pollutants, but under UV exposure. We have synthesized composites of BiOCl and BiOI to achieve equivalent degradation performance under visible light exposure. Results of our investigations are reported in Section 5.3 of this chapter.

The BiOX single crystalline material has a layered two-dimensional (2D) structure derived from the calcium fluorite (CaF_2) structure as shown in figure 5.3. The Bi atom is coordinated with a square antiprism with four O atoms in one base and four X (halogen) atoms in another. The O atom is tetrahedrally coordinated with four Bi atoms. The X atom is bonded with four Bi atoms in a planar square to form a pyramid and with its nonbonding (lone pair) electrons pointing to the other side of the square. These nonbonding electrons convert the three-dimensional (3D) fluorite-like structure into a 2D layered structure. In the 3D Bi_2O_3, the Bi atoms are coordinated with six O atoms. This structural difference between them is the major reason for BiOCl having a wider optical band gap. The BiOX layers are stacked together by the van der Waals interaction through the X atoms along the **c**-axis. Therefore, the structure is not closely packed in this direction. When one photon excites one electron from Cl 3p states to Bi 6p states in BiOCl, one electron-hole pair (e^-–h^+ pair) is generated. The layered structure of BiOX can provide a space large enough to polarize the related atoms and orbitals. The induced dipole can separate the hole-electron pair efficiently, enhancing photocatalytic activities. BiOX has an indirect transition band gap, so the excited electron has to travel a certain k space distance to be emitted to VB. Therefore, the BiOX single crystals can be described as 2D structured materials. This k space varies with X (X = Cl, Br, I), and therefore different BiOX exhibit varying activities in radiation exposure. This reduces the recombination probability of the excited electron and the hole, especially in BiOCl. Both the open structure, i.e., loose packed, and indirect transition may benefit the hole-electron separation and the charge transport. Thus, these features are both favorable to the photocatalytic reactions.

5.1.2 Synthesis of Bismuth Oxyhalide

Many solution routes have been applied to synthesize different members of the BiOX family in nanocrystalline form, viz., nanowires [20], nanobelts [21], nanosheets [22], nanoparticles [23], nanoflowers [24, 25], and hierarchical nanostructures [26–32].

Zhang *et al.* [26] synthesized BiOX microspheres with hierarchical superstructures by solvothermal process by employing ethylene glycol as solvent. An *et al.* [33] synthesized sheet-shaped BiOX (X = Cl, Br, I) samples by reaction between Bi_2O_3 and concentrated halogen acid. Wang *et al.* [34] and Sharma *et al.* [35] synthesized BiOX nanosheets by a simple wet chemical route using one-pot synthesis process. Shang *et al.* [36] synthesized lamellar BiOBr by a hydrothermal route using cetyltrimethyl ammonium bromide (CTAB), which acted not only as the template but also as the Br source. They reported photoactivity of this compound for the photodegradation of MO nearly four times higher than that obtained with BiOBr sample prepared using KBr instead of CTAB. Zhang *et al.* [10] synthesized 3D flower-like BiOBr architectures by ethylene glycol mediated self-assembly process. These BiOBr materials were composed of many radially grown

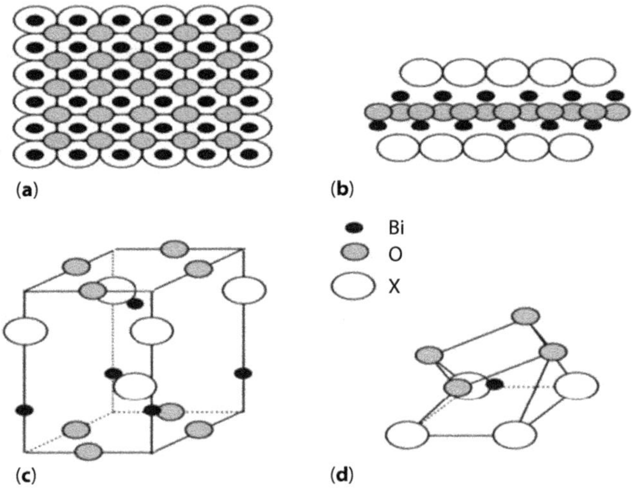

Figure 5.3 Crystal structure of bismuth oxyhalides.

nanosheets as petals. These structures exhibited much greater activity than BiOBr bulk plates. Hierarchical BiOBr nanoplate microspheres prepared by a nonaqueous sol-gel method can efficiently remove NO in indoor air under visible light irradiation conditions [37, 38].

To choose a route for the synthesis of a particular sample, a judicious decision should be made, depending upon the quality of the desired product and complexity of the approach. Solvothermal processes have been most commonly used to synthesize high quality samples. Simple wet methods are also used to synthesize BiOX, but rarely, as most of the researchers are not confidant about the control parameters and often end up with poor quality of the samples with reference to crystal structure and stoichiometry. Here we report the synthesis of high quality BiOX samples by a simple wet chemical route, obtained by tight control over the synthesis parameters.

We have synthesized three compositions; BiOCl, BiOI and composite (BiOCl/BiOI) with bismuth nitrate [$Bi(NO_3)_3.5H_2O$], potassium chloride (KCl) and potassium iodide (KI) precursor materials. Our solvent is mixture of ethanol and deionized water (S ~ 18 MΩ) in equivolume (50:50) ratio. All the chemicals were of analytical grade procured from Merck India and used as procured, i.e., without further purification. To synthesize BiOCl, 0.25 moles of bismuth nitrate and 0.25 moles of potassium chloride were dissolved separately in the solvent to yield clear solution. Potassium chloride solution was added at a controlled rate to the bismuth nitrate solution with continuous stirring. The pH value of the mixture was adjusted to 2 by slowly adding ammonia solution with continuous stirring.

Precipitates so obtained were filtered out, washed with deionized water and dried at 80 °C under ambient conditions. This gives us BiOCl nanostructured powder. BiOI was synthesized in a similar way by simply replacing KCl by KI. To synthesize BiOCl/BiOI composite, we dissolved 0.5 moles of bismuth nitrate in deionized water and 0.25 moles of KCl and 0.25 moles of KI were dissolved in another beaker and the rest of the procedure was as described above.

All three synthesized powders were characterized by XRD, SEM and Raman studies. X-ray data were taken on a Bruker D8 AXS diffractometer which employs copper Kα radiations as probing radiations. The FTIR spectra were recorded on a Nicolet 5700 FTIR spectrometer and surface morphology studies have been carried out on an EVO50 scanning electron microscope (SEM).

5.2 Photocatalytic Activity Measurements

Photocatalytic activities of the synthesized samples were evaluated by studying the degradation kinetics of methylene blue (MB) in aqueous solution (0.02 mM) under natural sunlight (Figure 5.6). First, 50 mg of the synthesized samples (BiOCl, BiOI, BiOCl/BiOI) was suspended in 200 ml of aqueous MB solution taken in a quartz container and exposed to the light source from bottom. Solution was constantly stirred by bubbling dry nitrogen. About 2 ml of MB solution was taken out from the quartz reactor after one hour, two hours, three hours and four hours of exposure durations. Amount of methylene blue (MB) present in the aqueous solution was estimated by recording the absorption spectra of the extracted solution in visible range (400–800 nm) and then calculating area under the peak centered at 664 nm. Kinetic constant for the dye degradation was calculated by assuming pseudo-first-order mechanism for the degradation reaction as per relation $C = C_0 \exp(-kt)$, where C_0 and C are concentrations of MB in the solution before exposure and after exposure respectively for time duration 't' and 'k' is degradation reaction rate constant. Methylene blue exhibits strong absorption peak at 664 nm; the area under this peak has been used to evaluate methylene blue content present in the solution.

5.3 Results and Discussion

Figure 5.4a–f shows SEM micrographs of the synthesized BiOCl, BiOI and BiOCl/BiOI composite samples recorded at two different magnifications. All the samples exhibited unevenly shaped platelet-like structure of

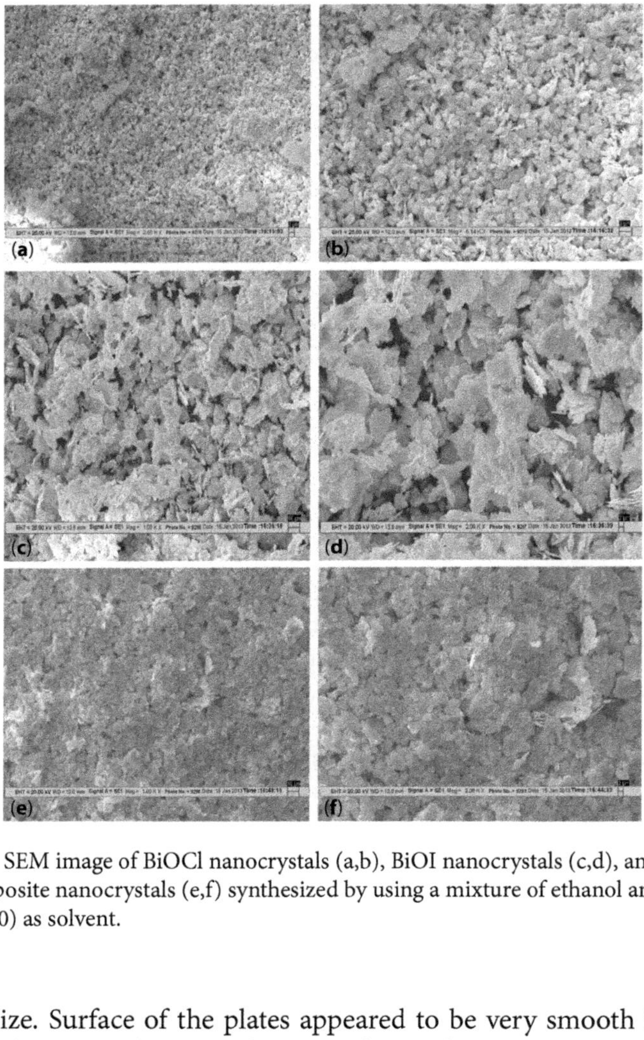

Figure 5.4 SEM image of BiOCl nanocrystals (a,b), BiOI nanocrystals (c,d), and (BiOCl/BiOI) composite nanocrystals (e,f) synthesized by using a mixture of ethanol and distilled water (50:50) as solvent.

varying size. Surface of the plates appeared to be very smooth in BiOCl samples whereas in the case of BiOI and BiOCl/BiOI composite, a rough surface morphology is evident from the SEM image.

The FTIR spectra of BiOCl, BiOI and BiOCl/BiOI composite recorded on a Nicolet 5700 FTIR spectrometer in 450–4500 cm^{-1} range are shown in Figure 5.5. FTIR curves obtained from BiOI and BiOCl samples are in agreement with the reported values. Different nature of the FTIR spectra recorded for the BiOCl/BiOI composite supports the formation of composite as expected.

Figure 5.6a–c shows X-ray diffraction patterns for the three samples, viz., BiOCl, BiOI and BiOCl/BiOI recorded for 2θ range 10–80° on a Bruker D8 AXS diffractometer using copper Kα radiations. Figure 5.6a,b confirms

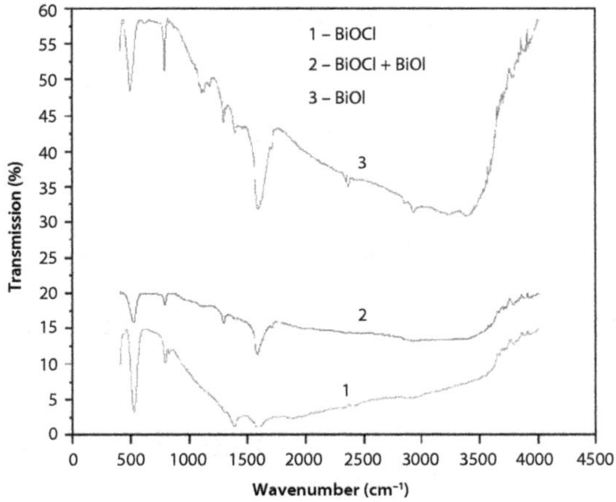

Figure 5.5 FTIR spectra of BiOCl, BiOI and BiOCl/BiOI composite in 450 to 4500 cm^{-1} wavenumber range.

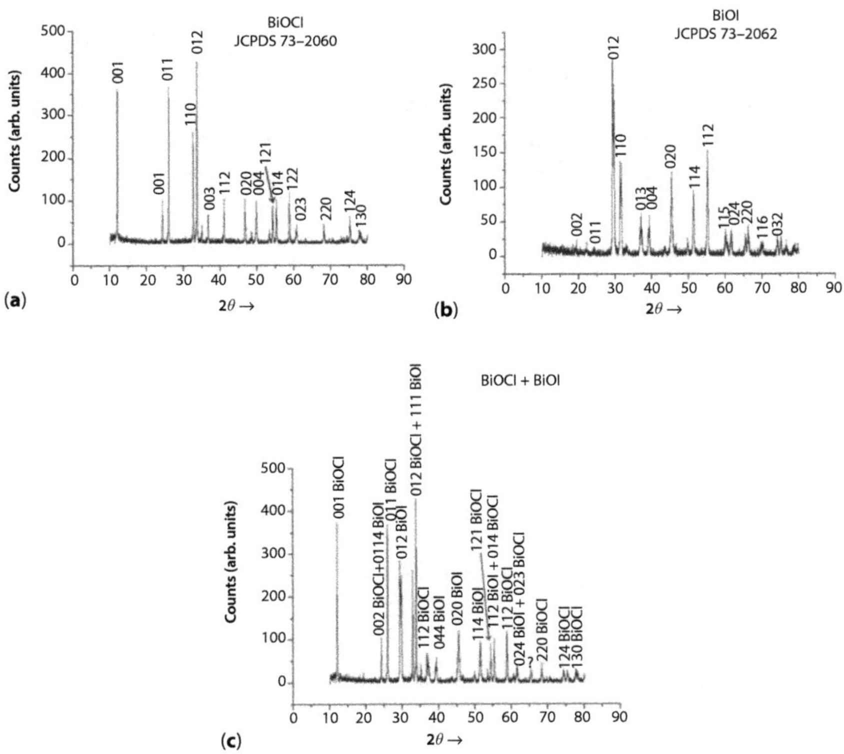

Figure 5.6 (a) X-ray diffraction of BiOCl, (b) BiOI, and (c) BiOCl/BiOI composite sample synthesized in a mixture of ethanol and distilled water (50:50) for 2θ value between 10–80°.

the formation of pure BiOCl and BiOI compounds respectively. All the observed XRD peaks can be well assigned miller indices as per JCPDS data card 73-2060 for BiOCl and 73-2062 for BiOI. Figure 5.3c shows x-ray diffraction pattern of the composite (BiOCl + BiOI). All the major observed peaks can be assigned BiOCl or BiOI, thereby confirming the formation of composite material of both the compounds.

Spectrophotometery results show sharp absorption at ~600 nm in the absorption spectra of the composite (Figure 5.7), which corresponds to ~2.0 eV. Therefore, a strong absorption in visible range is confirmed by the synthesized composite. This suggests the excitation of valence band electron to the conduction band and thus the creation of electron-hole pairs in the material by absorption of visible energy. These photogenerated electrons and holes subsequently carry out chemical reactions leading to the degradation of organic pollutants.

Figure 5.8a shows typical absoption curves of the dye solution after different durations of exposure of (BiOCl + BiOI) composite material under visible light from the Sun. Pure BiOCl sample synthesized here was exposed to UV radiations from the mercury vapor lamp in a similar solution under identical conditions. Its absorption curves are shown in figure 5.8b. We have observed almost identical degradation from the two samples (fig. 5.8 a & b) but one (composite of BiOCl+ BiOI) under natural visible light from the Sun and other (BiOCl) under UV exposure.

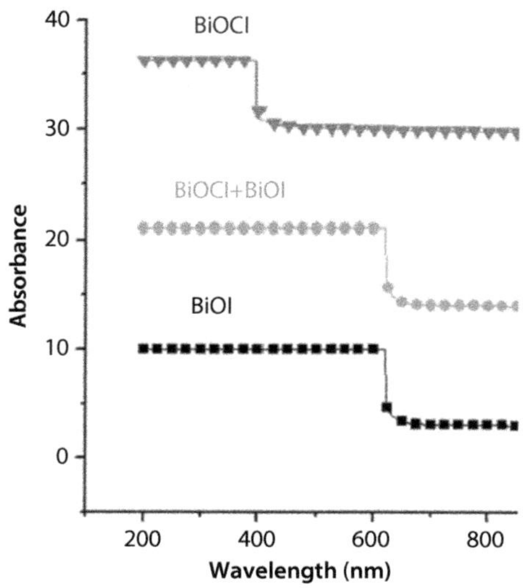

Figure 5.7 Absorption spectra of BiOCl, BiOI and composite (BiOCl/BiOI).

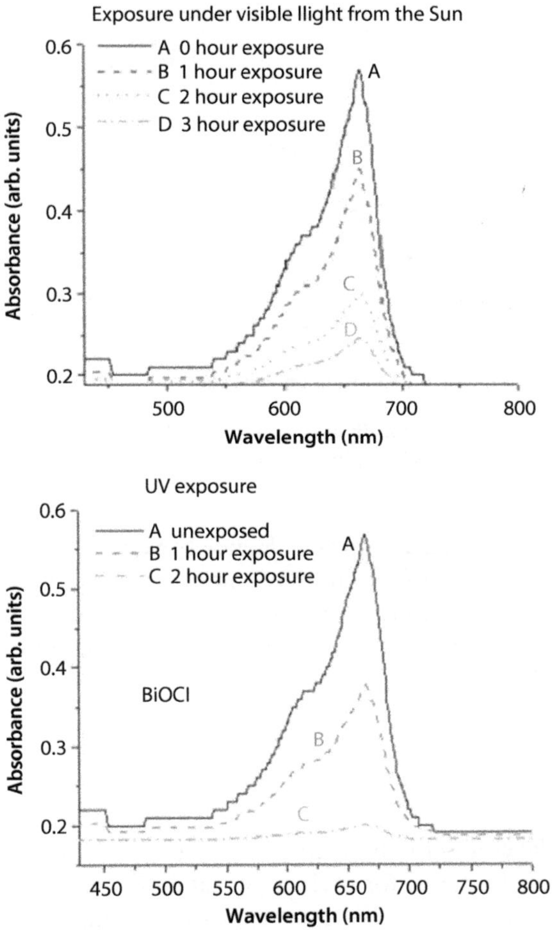

Figure 5.8 Typical absorption spectra of the dye solution after different durations of exposure to visible light from the Sun in the presence of BiOCl/BiOI composite and UV light exposure in the presence of BiOCl photocatalyst.

To further confirm the above observation, the degradation reaction kinetic curves from these absorption spectra are drawn in Figure 5.9. One can see the importance of composite material from these figures.

The composite material exhibits excellent photodegradation of the dye under natural visible light from the Sun whereas BiOCl exhibits about an order smaller photoactivity under identical exposure conditions. However, this material (BiOCl) shows excellent photoactivity under UV exposure due to the large energy band gap (~ 3.4 eV) which is not available in the

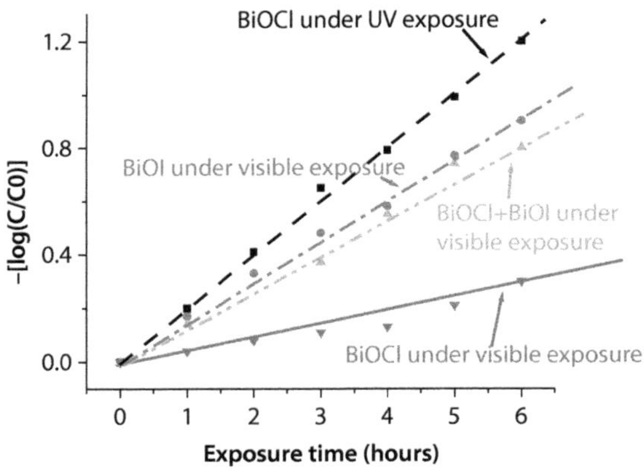

Figure 5.9 Degradation kinetics for BiOCl in UV and visible range and for BiOI and BiOCl/BiOI composite under visible light exposure.

solar spectrum on the Earth's surface. Hence the environmental remediation reactions are not feasible with BiOCl alone under visible exposure but the composite material supports these reactions, thereby proving its utility to combat environmental pollutants with the help of natural visible light available on the Earth's surface.

5.4 Conclusion

Bismuth-based layered semiconductors, BiOX (X = Cl, Br, I) have recently been of exceptional interest for their photocatalytic behavior. Pure BiOCl exhibits photoactivity under UV light exposure owing to the large band gap energy whereas BiOBr and BiOI are visible-light active photocatalysts. With regard to photoactive performance for the degradation of organic molecules which are also environmental pollutants, the performance of BiOCl is unmatchable as compared to the other two members of this family. Therefore there is a need to induce visible-light-driven photocatalytic performances in BiOCl to effectively fight environmental problems by using freely available energy from the Sun on Earth's surface. There have been attempts to achieve this goal mostly by synthesizing nanostructures, viz., multi-length scale microstructure modulation in the form of 1D templated nanofibers, 2D intrinsic nanosheets, 3D assembled hierarchical superstructure and supported thin films; favorable hybridization with semiconductors, metals, and cocatalysts/sensitizers; rational structural design by doping, solid solution formation and composition tailoring.

Despite much progress, the study of BiOX (X = Cl, Br, I) photocatalysts is still going on, and additional challenges need to be addressed. Confined by their CB levels, BiOX (X = Cl, Br, I) fail to fulfill the requirements to split water into H_2 to produce hydrogen fuel. To date, most of the reports concerning BiOX photocatalysts are focused on the decomposition of organic pollutants. Other applications, such as CO_2 reduction and selective organic transformation, should be extended utilizing BiOX photocatalysts. Since the photocatalytic performance of a semiconductor is dependent on its light-responsive range and carrier-separation capacity, rational design of the BiOX-based complex architecture systems (e.g., 3D ordered inverse opals, non-symmetric Janus metal/semiconductor structure, metal/insulator/semiconductor) that meet the requirements are crucial for constituting highly efficient and robust heterogeneous photocatalysts. For example, the periodic structures in inverse opals could enhance their light harvesting by multiple scattering, and the unintermitted pore tunnels would accelerate the separation of carriers, thereby making it a promising candidate for next generation photocatalysis. Moreover, with the aid of voltage bias, it is fascinating to study the photoelectrocatalysis of BiOX, which facilitates the charge separation and offers new opportunities to enhance their catalytic properties. In the development of novel photocatalytic nanostructures, chemical transformation of the existing material into another demonstrates an effective and versatile synthetic strategy. As the crystal structure of BiOX (X = Cl, Br, I) is built by the interlacing of $[Bi_2O_2]$ slabs with double halogen slabs to form a layered structure, such a strong ionic feature could allow ion exchange reactions or lattice-directed topotactic transformations between BiOX precursors and incoming species. By adopting this method, a variety of well-defined nanostructures, such as Bi_2WO_6 hollow microspheres, Bi_2S_3 superstructured networks and Bi_2E_3 (E = S, Se, Te) core-shell microspheres, can be obtained. The advent of nanoscience and nanotechnology has brought great guidance and protocols to develop highly efficient semiconductor photocatalysts. With ever-growing in-depth investigations, it is anticipated that BiOX (X = Cl, Br, I)-based photocatalysts will be more efficient and important applications in practical environmental purifications.

Bismuth oxyhalide composites can be synthesized by simple wet chemical route in large quantities, an industrially viable process. These composites can be utilized to carry out photodegradation reactions with visible light energy abundant in the solar spectrum, with high throughput, as the present investigations have established. These reactions were otherwise possible only with UV(B) energy, which is not available in the solar spectrum.

Acknowledgments

The authors are thankful to the director of the CSIR-National Physical Laboratory, New Delhi, for allowing this work and his keen interest in the work.

References

1. Geim, A.K., and Grigorieva, I.V., van der Waals heterostructures. *Nature* 499, 419, 2013.
2. Chhowalla, M., Shin, H.S., Eda, G., Li, L.J., Loh, K.P., and Zhang, H., The chemistry of two-dimensional layered transition metal dichalcogenide nanosheets. *Nat. Chem.* 5, 263, 2013.
3. Nicolosi, V., Chhowalla, M., Kanatzidis, M.G., Strano, M.S., and Coleman, J.N., Liquid exfoliation of layered materials. *Science* 340, 6139, 2013.
4. Zhang, T., Wang, S., and Chen, F., Pt–Ru bimetal alloy loaded TiO_2 photocatalyst and its enhanced photocatalytic performance for CO oxidation. *J. Phys. Chem. C* 120, 9732–9739, 2016.
5. Xie, S., Zhang, Q., Liu, G., and Wang, Y., Photocatalytic and photoelectrocatalytic reduction of CO_2 using heterogeneous catalysts with controlled nanostructures. *Chem. Commun.* 52, 35–59, 2016.
6. Liu, G., Wang, L., Yang, H.G., Cheng, H.-M., and Lu, G.Q. (Max), Titania-based photocatalysts—crystal growth, doping and heterostructuring. *J. Mater. Chem.* 20, 831–843, 2010.
7. Liu, X., Su, Y., Zhao, Q., Du, C., and Liu, Z., Constructing $Bi_{24}O_{31}Cl_{10}$/BiOCl heterojunction via a simple thermal annealing route for achieving enhanced photocatalytic activity and selectivity, *Sci. Rep.* 6, article number 28689, 2016.
8. Rotmensch, J., Whitlock, J.L., Dietz, M.L., Hines, J.J., Reba, R.C., Horwitz, E.P., and Harper, P.V., *Abstr. Pap. Am. Chem. Soc.* 216, U926, 1998.
9. Xiao, X., and Zhang, W.D., Facile synthesis of nanostructured BiOI microspheres with high visible light-induced photocatalytic activity. *J. Mater. Chem.* 20, 5866–5870, 2010.
10. Zhang, J., Shi, F., Lin, J., Chen, D., Gao, J., Huang, Z., Ding, X., and Tang, C., Self-assembled 3-D architectures of BiOBr as a visible light-driven photocatalyst. *Chem. Mater.* 20, 2937–2941, 2008.
11. Zhang, K.L., Liu, C.M., Huang, F.Q., Zheng, C., and Wang, W.D., Study of the electronic structure and photocatalytic activity of the BiOCl photocatalyst. *Appl. Catal. B* 68, 125–129, 2006.
12. Wang, W.C., Yang, W.J., Chen, R., Duan, X.B., Tian, Y.L., Zeng, D.W., and Shan, B., Investigation of band offsets of interface BiOCl:Bi_2WO_6: A first-principles study. *Phys. Chem. Chem. Phys.* 14, 2450, 2012.

13. Zhao, L.J., Zhang, X.C., Fan, C.M., Liang, Z.H., and Han, P.D., First-principles study on the structural, electronic and optical properties of BiOX (X = Cl, Br, I) crystals. *Phys. B* 407, 3364, 2012.
14. Xu, J., Li, L., Guo, C., and Zhang, Y., Photocatalytic degradation of Rhodamine B by BiOX (X=Cl, Br, I) microspheres. 978-1-4244-5089-3/11/IEEE, 2011.
15. Jiang, J., Zhao, K., Xiao, X., and Zhang, L., Synthesis and facet-dependent photoreactivity of BiOCl single-crystalline nanosheets. *J. Am. Chem. Soc.* 134, 4473–4476, 2012.
16. Zhang, X., Ai, Z.H., Jia, F.L., and Zhang, L.Z., Generalized one-pot synthesis, characterization, and photocatalytic activity of hierarchical BiOX (X = Cl, Br, I) nanoplate microspheres. *J. Phys. Chem. C* 112, 747–753, 2008.
17. Zhang, H.J., Liu, L., and Zhou, Z., Towards better photocatalysts: First-principles studies of the alloying effects on the photocatalytic activities of bismuth oxyhalides under visible light. *Phys. Chem. Chem. Phys.* 14, 1286, 2012.
18. Zhang, X.C., Zhao, L.J., Fan, C.M., Liang, Z.H., and Han, P.D., Effects of oxygen vacancy on the electronic structure and absorption spectra of bismuth oxychloride. *Comput. Mater. Sci.* 61, 180, 2012.
19. Zhang, H.J., Liu, L., and Zhou, Z., First-principles studies on facet-dependent photocatalytic properties of bismuth oxyhalides (BiOXs). *RSC Adv.* 2, 9224, 2012.
20. Wu, S., Wang, C., Cui, Y., *et al.*, Synthesis and photocatalytic properties of BiOCl nanowire arrays. *Mater. Lett.* 64, 115–118, 2010.
21. Deng, H., Wang, J., Peng, Q., Wang, X., and Li, Y., Controlled hydrothermal synthesis of bismuth oxyhalide nanobelts and nanotubes. *Chem. Eur. J.* 11, 6519–6524, 2005.
22. Xiong, J., Cheng, G., Li, G., Qin, F., and Chen, R., Well-crystallized square-like 2D BiOCl nanoplates: Mannitol-assisted hydrothermal synthesis and improved visible-light-driven photocatalytic performance. *RSC Adv.* 1, 1542–1553, 2011.
23. Pare, B., Sarwan, B., and Jonnalagadda, S.B., The characteristics and photocatalytic activities of BiOCl as highly efficient photocatalyst. *J. Mol. Struct.* 1007, 196–202, 2012.
24. Song, J., Mao, C., Niu, H., Shen, Y., and Zhang, S., Hierarchical structured bismuth oxychlorides: Self-assembly from nanoplates to nanoflowers via a solvothermal route and their photocatalytic properties. *CrystEngCommun* 12, 3875–3881, 2010.
25. Wang, D.H., Gao, G.Q., Zhang, Y.W., *et al.*, Nanosheet-constructed porous BiOCl with dominant {001} facets for superior photosensitized degradation. *Nanoscale* 4, 7780–7785, 2012.
26. Zhang, X., Ai, Z., Jia, F., and Zhang, L., Generalized one-pot synthesis, characterization, and photocatalytic activity of hierarchical BiOX (X = Cl, Br, I) nanoplate microspheres. *J. Phys. Chem. C* 112, 747–753, 2008.

27. Xiong, J., Cheng, G., Qin, F., *et al.*, Tunable BiOCl hierarchical nanostructures for high-efficient photocatalysis under visible light irradiation. *Chem. Eng. J.* 220, 228–236, 2013.
28. Cui, Z.K., Mi, L.W., and Zeng, D.W., Oriented attachment growth of BiOCl nanosheets with exposed {110} facets and photocatalytic activity of the hierarchical nanostructures. *J. Alloys Compd.* 549, 70–76, 2013.
29. Zhu, L.P., Liao, G.H., Bing, N.C., *et al.*, Self-assembled 3D BiOCl hierarchitectures: Tunable synthesis and characterization. *CrystEngComm* 12, 3791–3796, 2010.
30. Lei, Y., Wang, G., Song, S., Fan, W., and Zhang, H., Synthesis, characterization and assembly of BiOCl nanostructure and their photocatalytic properties. *CrystEngComm* 11, 1857–1862, 2009.
31. Xiong, J., Jiao, Z., Lu, G., *et al.*, Facile and rapid oxidation fabrication of BiOCl hierarchical nanostructures with enhanced photocatalytic properties. *Chem. Eur. J.* 19, 9472–9475, 2013.
32. Liu, J., Hu, J., Ruan, L., and Wu, Y., Facile and environment friendly synthesis of hierarchical BiOCl flowery microspheres with remarkable photocatalytic properties. Chin. Sci. Bull. 59, 802–809, 2014.
33. An, H., Du, Y., Wang, T., Wang, C., Hao, W., and Zhang, J., Photocatalytic properties of BiOX (X = Cl, Br, and I). *Rare Metals* 27, 243–250, 2008.
34. Wang, C., Zhang, X., Zhang, Y., Jia, Y., Yang, J., Sun, P., and Liu, Y., Hydrothermal growth of layered titanate nanosheet arrays on titanium foil and their topotactic transformation to heterostructured TiO_2 photocatalysts. *J. Phys. Chem. C* 115, 22276–22285, 2011.
35. Sharma, I.D., Tripathi, G.K., Sharma, V.K., Tripathi, S.N., Kurchania, R., Kant, C., Sharma, A.K., and Saini, K.K., *Cogent Chemistry* 1, 1076371, 2015. http://dx.doi.org/10.1080/23312009.2015.1076371
36. Shang, M., Wang, W., and Zhang, L., Preparation of BiOBr lamellar structure with high photocatalytic activity by CTAB as Br source and template. *J. Hazard. Mater.* 167(1–3), 803–809, 2009.
37. Ai, Z., Ho, W., Lee S., and Zhang, L., Efficient photocatalytic removal of NO in indoor air with hierarchical bismuth oxybromide nanoplate microspheres under visible light. *Environ. Sci. Technol.* 43, 4143–4150, 2009.
38. Zhang, H., Liu, L., and Zhou, Z., Towards better photocatalysts: First-principles studies of the alloying effects on the photocatalytic activities of bismuth oxyhalides under visible light. *Phys. Chem. Chem. Phys.* 14, 1286, 2012.

6

Investigation of DC Conductivity, Conduction Mechanism and CH_4 Gas Sensor of Chemically Synthesized Polyaniline Nanofiber Deposited on DL-PLA Substrate

Muktikanta Panigrahi[1], Debabrat Pradhan[2], Subhasis Basu Majumdar[2] and Basudam Adhikari[2]*

[1]Department of Metallurgical Engineering, Gandhi Institute of Engineering and Technology (GIET), Gunupur, Odisha, India
[2]Materials Science Centre, Indian Institute of Technology (IIT) Kharagpur, West Bengal, India

Abstract

In this chapter, the *in-situ* synthesis of DL-polylactic acid/polyaniline-emeraldine salt (DL-PLA/PANI-ES) composites, particularly their electrical properties, are presented. The attenuated total reflection–Fourier transform infrared (ATR-FTIR) spectra indicated the presence of different functional groups in the as-prepared DL-PLA/PANI-ES composites. The ultraviolet-visible (UV-Vis) absorption spectroscopic analysis showed the presence of polaron band, suggesting PANI-ES form. The room temperature dc conductivity, temperature variation dc conductivity (in the presence and absence of magnetic field), and magnetoresistance (MR) were analyzed. The highest room temperature dc conductivity value was obtained from H_2SO_4 doped DL-PLA/PANI-ES composites and all prepared conductive composites followed ohms law. The low temperature dc conductivity was carried out in order to study the semiconducting nature of prepared materials. The Mott-type variable range hopping (VRH) model was found to be well fitted with the conductivity data and described the density of states at the Fermi level, which is constant in this temperature range. From MR plots, a negative MR was observed, which

**Corresponding author*: muktikanta2@gmail.com

Sanjay K. Nayak *et al.* (eds.) Trends and Applications in Advanced Polymeric Materials, (101–126) 2018 © Scrivener Publishing LLC

described the quantum interference effect on hopping conduction. Presented in this chapter is a study on the methane gas response of as-prepared DL-PLA/PANI-ES composite film, which at 500 ppm was found to be 9%.

***Keywords*:** Conducting polymer, nanofibers, dc conductivity, VRH model, activation energy, magnetoresistance, methane gas sensor

6.1 Introduction

Conductivity is a prime characteristic of polymer for application in light-emitting diodes, supercapacitors and sensors [1–3]. The synthesis parameters can be varied in a controlled manner to obtain polymer with a desired conductivity range. Doping and structural modification are two common processes to improve the conductivity of a polymer. However, it depends upon the polymer and its properties as to which doping and structural modifications are performed. Therefore, suitable dopants and polymers need to be chosen. For example, PANI is a very common intrinsically conducting polymer (ICP), which is not stable under ambient conditions. The stability of PANI can be improved by making it in composite form with thermoplastic polymer as stabilizer [4]. A variety of PANI composites have been prepared using different types of synthetic polymers, such as poly(methyl methacrylate), polyvinyl chloride, polystyrene and polyurethane [5–8], and their electrical properties were studied in detail in the presence and absence of magnetic field [9–12].

To understand the electrical properties of various doped polymers, several models are proposed. Variable range hopping is especially employed to understand the transport mechanism. In particular, Li and coworkers proposed a VRH model with the transport mechanism of hydrochloric acid (HCl) doped and dodecylbenzene sulfonic acid (DBSA) doped PANI [13]. Kapil *et al.* [14] analyzed the transport mechanism of p-toluene sulphonic acid (PTSA) doped PANI by Arrhenius model, VRH model and Kivelson model over a wide temperature range of 30–300 K. The conduction mechanism of PANI organic film and metal particles embedded in insulating materials and PANI were also explained by using VRH and charging energy limited tunneling (CELT) models. The conductivity measurements are also studied in variations of magnetic field at constant temperature to measure the magnetoresistance (MR). This helps to understand the depth of conduction mechanism.

A few conducting polymers have shown positive MR such as PANI [15–18], polypyrrole [19–21], PEDOT films [22], and PANI composites at low temperature ($0 < 10$ K). The positive MR of conducting polymer is ascribed to the shrinkage of localized wave functions of electron in the presence of magnetic field [23] or electron-electron interactions [24, 25]. Moreover, there are also a few polymers which show negative MR. Lee and Ramakrishnan [26] reported the negative MR from highly conducting polyacetylene at low temperature (1.5 K), which was attributed to the weak localization effects.

Methane is the chief constituent of natural gas which is hazardous to both humans and the environment. Due to its highly flammable and explosive characteristics, even at a low level (5–14%) concentration it poses a serious threat. With the global population boom, more and more human lives are being endangered by the effect of CH_4 gas exposure [27]. That creates a demand for monitoring CH_4 gas to provide safe environments in homes, industries and mines. There is a great need to develop sensors for detecting methane gas quickly and accurately in coal mines at ambient conditions. Generally, metal oxide-based gas sensors are used, such as ZnO, MoO_3, SnO_2, TiO_2, ZrO_2, spinel compounds, etc. [28–33].

All the above sensors require high temperature for sensor operation, which is expensive. Researchers have put effort into making sensors based on room temperature. In spite of these considerable efforts, room temperature-based sensors for CH_4 gas sensing have not yet been realized, the problem being vital to industry as well as the general public. To meet this demand, a new class of sensor materials is underway, including efforts to prepare the sensor, such as conducting polymer-based sensor, which can work at ambient conditions. In the conducting polymer family, PANI is one of the members with good environmental and thermal stability coupled with ease of processability [34] and doping-dedoping features [35]. In spite of the above advantages, PANI film is brittle and highly moisture sensitive. This is due to the ionic electroactive nature of PANI polymer, which limits its use for commercial applications. The stability of PANI polymer can be improved by making a composite with a thermoplastic polymer such as poly(methyl methacrylate) (PMMA), polyvinyl chloride (PVC), polystyrene (PS), and polyurethane (PU) as stabilizer [36–39]. The stability of PANI polymer could be enhanced by covalently grafting with polymers [40, 41]. We selected DL-PLA polymer in the composite preparation because of its strong interaction with ionic electroactive polymer such as PANI polymer. This interaction can potentially change the surface properties, such as surface charge and wettability, to induce conformational and dimensional changes on bioactive materials.

Despite all the above-mentioned properties, one of the most important issues related to the applications of biodegradable polymer in electronic engineering is to reduce the degradation of conducting polymers. There are some reports on the use of PLA with multiwalled carbon nanotubes (MWCNTs) [42], conducting polymers such as PANI [43], and polypyrrole [44] to form composites, which showed improved surface resistance/conductivity properties compared to neat PLA. Therefore, the use of biodegradable polymer with conductive polymers is a very important and challenging task.

Our goal was to prepare an environmentally stable PANI-ES composite that would show better electrical properties and, therefore, could be potentially used for gas sensor and other electronic applications. We selected DL-PLA polymer in the composite preparation because of its high mechanical properties compared to other biodegradable polymers and its strong interaction with ionic electroactive polymer such as PANI. Inorganic acids (HCl or H_2SO_4 or H_3PO_4) were doped into DL-PLA/PANI-ES to improve the electrical properties of the as-prepared composites by *in-situ* polymerization technique. The low and room temperature conductivity of HCl, H_2SO_4 and H_3PO_4 doped DL-PLA/PANI-ES composite was investigated by dc conductivity measurement and possible conduction mechanism was proposed. Furthermore, the MR of H_2SO_4 and H_3PO_4 doped DL-PLA/PANI-ES composites have been studied. Also, in this chapter, we have reported the methane gas sensing characteristics of H_3PO_4 doped DL-PLA/PANI-ES composite. For the first time, the characteristics of H_3PO_4 doped DL-PLA/PANI-ES composite have been demonstrated by us to achieve CH_4 gas response at room temperature.

6.2 Experimental Details

6.2.1 Preparation of Desired Materials

The DL-PLA films were first prepared by solution casting technique using DL-PLA and $CHCl_3$ as polymer and solvent, respectively. DL-PLA (2 g) was added to 20 mL of $CHCl_3$ in a 100 mL beaker and stirred for 3 h at room temperature. The soluble product was poured into a Petri dish (diameter 10 cm) and left until the solvent evaporated. A DL-PLA film formed after solvent evaporation. Such films were removed from the Petri dish and cut into small pieces for characterizations.

The DL-PLA/PANI-ES composites were prepared by *in-situ* oxidation polymerization process at room temperature using aniline and DL-PLA film (1 cm^2), which acts as monomer and base material, respectively. To prepare H_2SO_4 doped DL-PLA/PANI-ES composite, 3 mL of aniline was added to 105 mL of aqueous 1 M H_2SO_4 and stirred for 0.5 h until slightly yellowish solution (solution 1) was obtained. Then a DL-PLA film (1 cm^2) was put into solution 1 and stirred for 12 h. Solution 2 was prepared by adding 7.47 g of ammonium perdisulphate (APS) in 60 mL of 1 M H_2SO_4. Then solution 2 was added dropwise into the DL-PLA mixed solution 1 for 1 h for the polymerization to occur. The color of DL-PLA film was observed to change from white to light green and finally to dark green as polymerization continued. The reaction mixture was maintained under constant stirring for 10 h to complete the polymerization. The resulting H_2SO_4 doped DL-PLA/PANI-ES composite was finally washed with deionized water several times and dried in the ambient atmosphere [45]. The same procedure was followed to prepare HCl, HNO_3 and H_3PO_4 doped DL-PLA/PANI-ES composites using 1 M HCl, HNO_3 and H_3PO_4 in place of 1 M H_2SO_4.

6.2.2 Characterization of DL-PLA Films and DL-PLA/PANI-ES Composites

The ATR-FTIR spectra were recorded on a Thermo Nicolet Nexus 870 spectrophotometer in the range of 400–4000 cm^{-1}. The instrument settings were kept constant (50 scan at 4 cm^{-1} resolution, absorbance mode). Ultraviolet-visible (UV-Vis) spectra of HCl, HNO_3, H_2SO_4 and H_3PO_4 doped DL-PLA/PANI-ES composites were recorded by using a Micropack UV-Vis-NIR, DH-2000 in the wave length region of 200–1000 cm^{-1}. Baseline was corrected before recording the spectra. Surface morphology of HCl, HNO_3, H_2SO_4 and H_3PO_4 doped DL-PLA/PANI-ES composites was analyzed by scanning electron microscopy (SEM) using a Carl Zeiss Supra-40 scanning electron microscope. Gold coating was done before performing SEM experiment. Operating voltage was maintained at 4 kV. Room temperature (~30 °C) dc conductivity of the prepared HCl, HNO_3, H_2SO_4 and H_3PO_4 doped DL-PLA/PANI-ES composites was measured using the linear four-point probe technique. Contact was made with nonconducting silver paste.

A constant current (I) from a current source (Keithley 2400 programmable current source) was allowed to pass through two terminal leads of the four-point probe and the voltage (V) across the other two leads was measured using a multimeter (Keithley 2000 digital multimeter). Also, resistivity with and without magnetic field as a function of temperature was measured using linear four-point probe technique for an understanding of the transport phenomenon of HCl, HNO_3, H_2SO_4 and H_3PO_4 doped DL-PLA/PANI-ES composites. Resistivity measurement with magnetic field was investigated using a helium compressor (HC) (Sumitomo model

HC-4E1) in combination with a GAINS model cryostat (Janis Research Company) equipped with a 0.8 T superconducting magnet (Lake Shore electromagnet). A Lake Shore 331 temperature controller was used. Measurements were performed at 5000 gauss and at various temperatures using a computer controlling measuring system. A constant current (*I*) from a current source (Keithley 220 programmable current source) was allowed to pass through two terminal leads of a four-point probe and the voltage (*V*) across the other two leads was measured using a multimeter (Keithley model 2182 Nanovoltameter). For gas sensing studies, the prepared composite film (~25 °C) was used. For the sensing measurements, the surface of the sensing material was electroded with silver paste strips of 8 mm length, 1.5 mm width, which were separated from each other by 4 mm. The methane gas sensing behavior was estimated by measuring the resistance change before and after gas exposure. This measurement was done using an electrometer (model 6517A, Keithley Instruments, USA) under dynamic flow of methane gas. The pre-diluted (~10,000 ppm) reducing gases (methane) in nitrogen were mixed with carrier gas (air) using a mixing unit before flowing into the reactor unit. The mass flow controller and electrometer were all interfaced with a PC equipped with a GPIB card (National Instruments) and LabVIEW 8.5 software (National Instruments). To evaluate the gas sensing characteristics of the polymer membrane, its planar resistance transients were measured as a function of temperature and gas concentration.

6.3 Results and Discussion

6.3.1 Scanning Electron Microscopic (SEM) Analysis

Figure 6.1a–c and 7.1d–f shows the SEM images of pure DL-PLA film, H_2SO_4 (low and higher magnification) and H_3PO_4, HCl and HNO_3 doped DL-PLA/PANI-ES composites, respectively. The pure DL-PLA film appears quiet smooth (Figure 6.1a) whereas doped DL-PLA/PANI-ES composites have fibrous morphology (Figure 6.1b–f). Figure 7.1c,d shows the SEM images of H_2SO_4 doped DL-PLA/PANI-ES composite at two different types of magnification with a network-type surface morphology (Figure 6.1c) on the DL-PLA film. This network-type morphology is formed (Figure 6.1c) after a DL-PLA film is put into the solution 1 (i.e., before polymerization). The change in the surface morphology is believed to be due to the reaction of hydrogenium ion (from dopant) and anilinium salt with ester group present in DL-PLA chain.

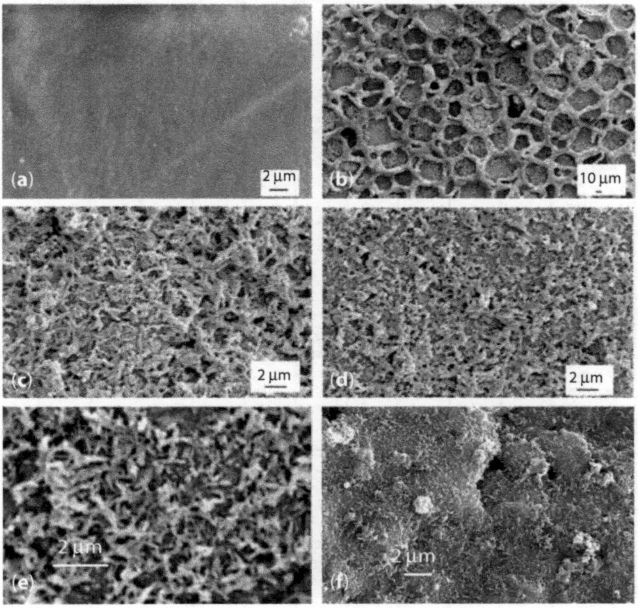

Figure 6.1 SEM images of (a) DL-PLA film, H_2SO_4 [(b) low and (c) higher magnification], H_3PO_4 (d), HNO_3 (e), and HCl (f) doped DL-PLA/PANI-ES composites.

After polymerization (slow addition of solution 2 into solution 1), the network-type morphology remains with the formation of additional fibrous morphology all over the surface of DL-PLA film at lower magnification. The magnified image in Figure 6.1c shows the fibrous morphology of H_2SO_4 doped DL-PLA/PANI-ES composite. The average diameter of as-grown fibers was measured to be 153 nm. A similar type of fibrous morphology is also obtained for H_3PO_4, HCl and HNO_3 doped DL-PLA/PANI-ES film (Figure 6.1d–f) with average fiber diameter of 163 nm, 178 nm and 162 nm, respectively. The different sized diameter formed may be due to the presence of various sized counterions in the as-prepared composites. The formation of fiber morphology is favorable for the sensor application due to the inherent increase in the surface area of composite film.

6.3.2 Attenuated Total Reflectance Fourier Transformation Infrared (ATR-FTIR) Spectroscopic Analysis

All the characteristic bands and their corresponding assignments for preparation of materials are presented in Table 6.1. The absorption

Table 6.1 FTIR peak positions and their assignments of DL-PLA, PANI-ES, HCl, HNO_3, H_2SO_4, and H_3PO_4 doped DL-PLA/PANI-ES composites.

Peak assignments	Peak positions (cm^{-1})					
	DL-PLA	PANI-ES	HCl doped DL-PLA/ PANI-ES	HNO_3 doped DL-PLA/ PANI-ES	H_2SO_4 doped DL-PLA/ PANI-ES	H_3PO_4 doped DL-PLA/PANI-ES
N–H (stretch.)	---	3217	3289	3291	3291	3280
Quinoid (stretch.)	---	1554	1533	1537	1530	1531
Benzoid (stretch.)	---	1475	1453	1450	1451	1454
C=N (stretch.)	---	1108	1067	1085	1077	1067
C–H (stretch.)	2995	2926	2965	2968	2965	2968
C=O	1759	---	1728	1720	1725	1724
C-O	1616	---	1652	1650	1656	1653
CH (def.)	1363	---	1375	1363	1374	1374
CH (bending)	1268	---	1262	1267	1259	1260
C-O-C	1216	---	1226	1223	1229	1226

bands of DL-PLA at 2995, 1759, 1616 and 1216 cm^{-1} have been attributed to C-H stretching, C=O stretching, C-O stretching of ester and C-O-C stretching vibration, respectively, whereas the bands at 1453, 1361 and 1363 cm^{-1} represent the stretching vibration of C-H deformation of DL-PLA polymer [46]. This indicates that the characteristic absorption features of DL-PLA polymer are retained in the prepared DL-PLA film. The main FTIR bands of PANI-ES are found at 1554, 1475 and 1108 cm^{-1}, corresponding to quinoid, benzenoid and C=N stretching, respectively [47].

From our observation, the presence of quinoid and benzenoid ring vibrations are exhibited at 1475 and 1554 cm^{-1} respectively, indicating the presence of oxidation state of PANI-ES. The characteristic band obtained in the ATR-FTIR spectrum of DL-PLA/PANI-ES composite film indicates the formation of conducting DL-PLA/PANI-ES composite films. From Table 6.1 it is observed that N-H bands of HCl doped DL-PLA/PANI-ES, H_2SO_4 doped DL-PLA/PANI-ES, and H_3PO_4 doped DL-PLA/PANI-ES appeared at 3289, 3291 and 3280 cm^{-1} respectively. The different N-H stretching peak positions represent the strong dependency on acid strength. The C=N, C=O, C-O and C-O-C stretching bands do not change significantly. The entire bands indicate that both PANI-ES and DL-PLA are retained in the composites.

6.3.3 Ultraviolet Visible (UV-Vis) Absorption Spectroscopic Analysis

It is found from the literature that PANI-ES shows π-π^* of benzene ring, polaron to π^*, benzenoid to quinoid ring and polaron transition respectively [48]. It is noticed from Figure 6.2a that there is no transition in DL-PLA film [49]. Various transitions are observed in HCl, HNO_3, H_2SO_4 and H_3PO_4 doped DL-PLA/PANI-ES composites, which are shown in Figure 6.2. Its peak positions with their assignments are mentioned in Table 6.2. From Figure 6.2 and Table 6.2 we found two types of bands, such as π to localized polaron band and π-π^* band of benzenoid ring, for HCl, HNO_3, H_2SO_4 and H_3PO_4 doped DL-PLA/PANI-ES composite. Both the π-π^* band of benzenoid ring and polaron band are suggested to be due to the presence of anilinic unit and oxidation unit in emeraldine salt form of composite films [48, 50].

The different peak positions and also peak areas may be happening due to the presence of a variety of nano-regime PANI-ES chains on DL-PLA polymer film. This indicates that the conjugation lengths affect the band gap energy. Hence, electrons are delocalized in the excitation band [50].

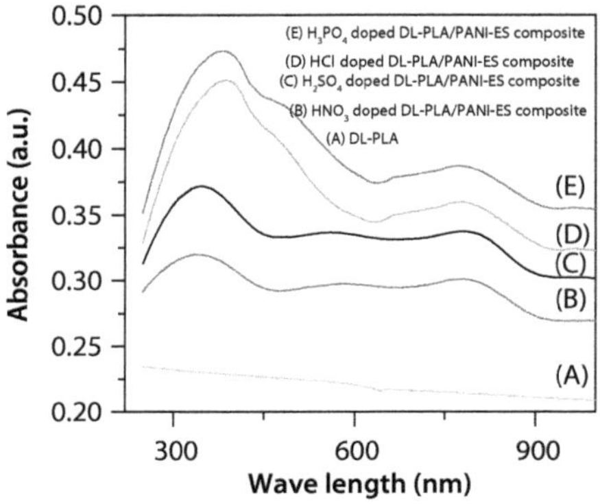

Figure 6.2 UV-Vis spectra of neat DL-PLA (A), HNO_3 (B), H_2SO_4 (C), HCl (D), and H_3PO_4 (E) doped DL-PLA/PANI-ES composites.

Table 6.2 UV-Vis peak positions and their assignments and direct band gap of DL-PLA, HCl, HNO_3, H_2SO_4, H_3PO_4 doped DL-PLA/PANI-ES composites.

Materials	Polaron band	π–π*	Band gap (eV)
DL-PLA	---	---	---
HCl doped DL-PLA/PANI-ES composite	780	383	1.60
H_2SO_4 doped DL-PLA/PANI-ES composite	797	344	1.37
H_3PO_4 doped DL-PLA/PANI-ES composite	787	381	1.46
HNO_3 doped DL-PLA/PANI-ES composite	791	344	1.69

Photon absorption in the UV-Vis region of semiconducting materials is observed by the Tauc expression [51]:

$$ahv = A\left(hv - E_g\right)^n \tag{6.1}$$

where α = optical absorption coefficient; hv = photon energy; E_g = energy gap calculated from graph; A = absorption constant; and n = type of transition occuring. $n = 2$ indicated allowed indirect transitions and $n = 1/2$ indicated allowed direct transitions. We have plotted $(\alpha h\upsilon)^2$ vs. hν for direct band energy analysis of prepared samples, which is shown in Figure 6.3. The linear portion of it was extrapolated to $\alpha = 0$ value to obtain direct band gap energy.

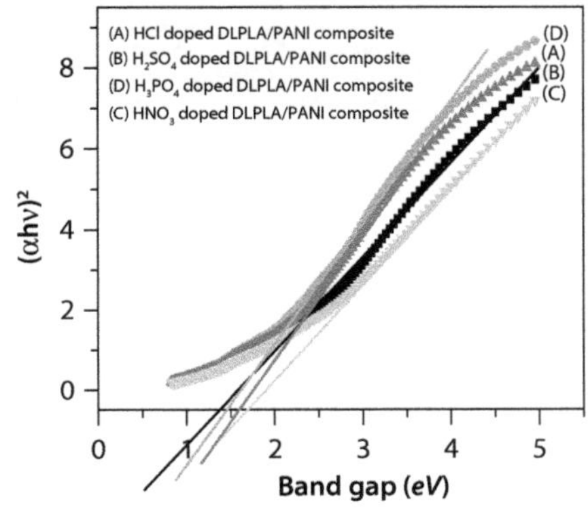

Figure 6.3 Optical (direct) band gap of HCl (A), H_2SO_4 (B), HNO_3 (C), H_3PO_4 (D) doped DL-PLA/PANI-ES composites.

The band gap energy value is found to be 1.60 eV for HCl doped DL-PLA/PANI-ES composite, 1.69 eV for HNO_3 doped DL-PLA/PANI-ES composite, 1.46 eV for H_3PO_4 doped DL-PLA/PANI-ES composite and 1.37 eV for H_2SO_4 doped DL-PLA/PANI-ES composite, respectively. The difference in band energy occurred due to the effect of acid strength during the formation of PANI-ES on DL-PLA polymer film. It is observed from Table 6.2 that the H_2SO_4 doped DL-PLA/PANI-ES composite showed the highest band gap value compared to the other three prepared composites. This may have happened due to the higher strength of H_2SO_4, which forms higher conjugation of PANI-ES chains and affects the band gap energy [48, 51].

6.3.4 DC Electrical Analysis

The dc conductivity of HCl, HNO_3, H_2SO_4 and H_3PO_4 doped DL-PLA/PANI-ES composites along with DL-PLA polymer film was measured at room temperature using linear four-point probe technique. The expression used is $\sigma = 1/\rho$, where ρ is resistivity [52]. The resistivity was measured from the relation $\rho = 2\pi S$ (V/I), where S is the probe spacing (0.15 cm), I is the supplied current (in nA) and V is the corresponding voltage (in mV). Figure 6.4 shows the I-V characteristics of HCl, HNO_3, H_2SO_4 and H_3PO_4 doped DL-PLA/PANI-ES composites at room temperature. The measured I-V characteristics show the linear behavior. The linear fit straight line passes through the origin, indicating the ohmic behavior. The conductivity measured from I-V data for the HCl, HNO_3, H_2SO_4 and H_3PO_4 doped DL-PLA/PANI-ES composites is presented in Table 6.3. The conductivity

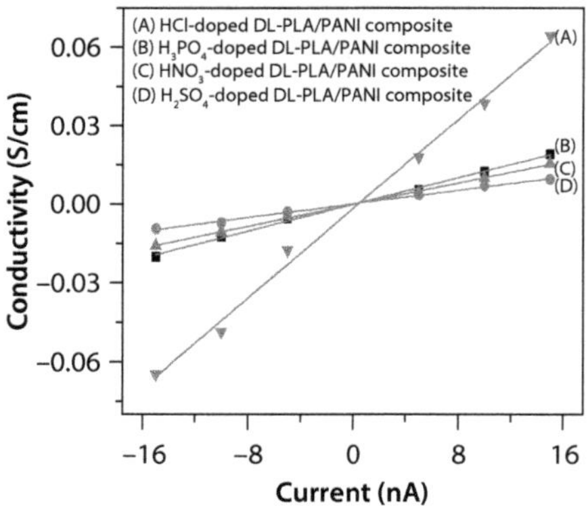

Figure 6.4 DC conductivity of HCl (A), H_2SO_4 (B), HNO_3 (C) and H_3PO_4 (D) doped DL-PLA/PANI-ES composites measured at room temperature.

Table 6.3 DC conductivity of HCl, H_2SO_4, HNO_3 and H_3PO_4 doped DL-PLA/PANI-ES composites at room temperature.

Materials	**DC conductivity (S/cm)**
DL-PLA	6.05×10^{-15}
HCl doped DL-PLA/PANI-ES composite	0.285×10^{-4}
H_2SO_4 doped DL-PLA/PANI-ES composite	0.162×10^{-2}
H_3PO_4 doped DL-PLA/PANI-ES composite	0.1097×10^{-3}
HNO_3 doped DL-PLA/PANI-ES composite	0.774×10^{-3}

of H_2SO_4 doped DL-PLA/PANI-ES composite (0.15×10^{-2} S/cm) is found to be higher than that of HCl, HNO_3 and H_3PO_4 doped DL-PLA/PANI-ES composite (0.285×10^{-4}, 0.774×10^{-3} and 0.309×10^{-3} S/cm). This could be due to that stronger H_2SO_4 dopant, which exerts more force towards better ordering of PANI polymer chains, leading to higher conjugation. The conjugation is a favorable factor for the intermolecular mobility of charged species along the chains and to some extent on the intermolecular hopping because of the proper arrangement of polymeric chains [53]. Additionally, H_2SO_4 doped composite produces more semiquinone conducting group and PANI-ES as compared to HCl, HNO_3 and H_3PO_4 doped composite because of the highly reactive nature of H_2SO_4 dopant [54, 55].

The temperature-dependent dc conductivity is shown in Figure 6.5. It shows the conductivity of HCl, H_2SO_4 and H_3PO_4 doped DL-PLA/PANI-ES

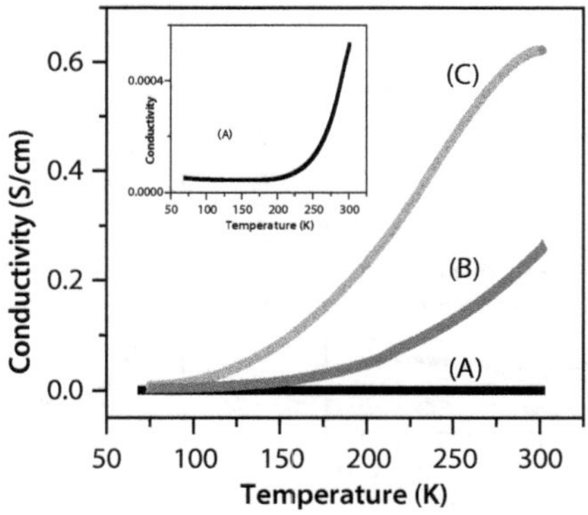

Figure 6.5 DC conductivity vs. temperature (77–300 K) of HCl (A), H_2SO_4 (B) and H_3PO_4 (C) doped DL-PLA/PANI-ES composite.

composites as a function of temperature. It is evident from Figure 6.5 that the composites showed increase in conductivity with an increase in the temperature from 77 to 300 K, which is similar to an inorganic semiconductor. Hence, it can be called an organic semiconductor [56].

Various models, including variable range hopping (VRH), were used to understand the probable mechanism of current transport in the organic semiconductor [57–60]. According to VRH, the temperature (T) dependence of dc conductivity follows Mott's expression [57–60]:

$$\sigma = \sigma_0 \exp\left(-\frac{T_0}{T}\right)^t \tag{6.2}$$

where T_0 is the Mott characteristic temperature and σ_0 the limiting value of conductivity at infinite temperature and the exponent 'r' is related to the dimensionality of the transport process via the expression $r = 1/(1+d)$; here, d = 1, 2 and 3 for one-, two-, and three-dimensional (1D, 2D and 3D) conduction process, respectively. Out of the three processes, 3D processes are more fitted.

Figure 6.6a,b shows the 3D-VRH plots and Arrhenius plots on measured conductivity in logarithmic scale for HCl, H_2SO_4 and H_3PO_4 doped DL-PLA/PANI-ES composites in the temperature range of 77–300 K, respectively. The conductivity data were carefully fitted linearly to both the 1D and 3D VRH as well as Arrhenius plots for conduction processes. The obtained regression values are presented in Table 6.4. It is clear from Table 6.4 that Mott's 3D-VRH model of as-prepared composite fits better into the experimental data (Regression value) than that of 1D model

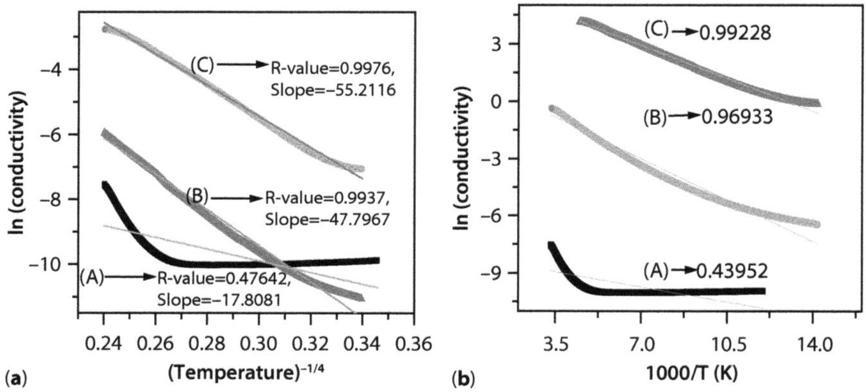

Figure 6.6 (a) 3D VRH plot and (b) Arrhenius plot of HCl (A), H_2SO_4 (B) and H_3PO_4 (C) doped DL-PLA/PANI-ES composites.

Table 6.4 Regression values (3D-, 1D-VRH model and Arrhenius model) of HCl, H_2SO_4, and H_3PO_4 doped DL-PLA/PANI-ES composites.

Regression value (R-value)		**HCl doped DL-PLA/ PANI-ES**	**H_2SO_4 doped DL-PLA/ PANI-ES**	**H_3PO_4 doped DL-PLA/ PANI-ES**
R-value (3D)	VRH	0.47642	0.9976	0.9937
R-value (1D)			0.974	0.993
R-value	Arrhenius	0.43952	0.99228	0.96933

and Arrhenius model. This suggests that charge carrier can hop both in between the chains, i.e., interchain hopping, and along the chain, i.e., intra-chain hopping, as PANI has the chain structure [57–60].

In the 3D-VRH model, the temperature dependence of DC conductivity can be written as:

$$\sigma = \sigma_0 \exp\left(-\frac{T_0}{T}\right)^{\frac{1}{4}} \tag{6.3}$$

Figure 6.6a shows that the linear dependence of $\ln\sigma$ vs $T^{-\frac{1}{4}}$ is better than that of $\ln\sigma$ vs $T^{-\frac{1}{2}}$. The estimated linear factor, i.e., regression value (R-value), for the 3D-VRH model (0.9976) was better than that of the 1D-VRH model (0.974), supporting the reported literature [61–65].

T_0 and σ_0 values were also calculated from the VRH plot of H_2SO_4 doped DL-PLA/PANI-ES composite and are presented in Table 6.5. Unlike the H_2SO_4 doped composite, the regression value for H_3PO_4 doped DL–PLA/PANI-ES composite was measured to be almost the same, i.e., 0.9937 and 0.993 from the 3D and 1D fitted VRH model, respectively. This indicates that the 3D-VRH model can be applied to describe the conduction mechanism in the H_3PO_4 doped DL-PLA/PANI-ES composite. In the H_2SO_4 doped DL-PLA/PANI-ES composite, sulphuric acid (H_2SO_4) is a strong dopant which exerts stronger force, leading to an alignment of chains and increase in the compactness [53]. Therefore, charge carriers could hop easily from one chain to another chain, obeying the 3D-VRH conduction mechanism model.

Other than the regression values, other parameters like density of states, hopping distance and hopping energy are supported to understand the conduction mechanism. Estimation of localization length (L_{loc}) is an

Table 6.5 VRH conduction parameters obtained by analyzing the low temperature resistivity (with and without magnetic field) data of HCl, H_2SO_4, H_3PO_4 doped DL-PLA/PANI-ES composites.

Conducting parameters	**L_H (Magnetic length in °A)**	**HCl doped DL-PLA/ PANI-ES composite**	**H_2SO_4 doped DL-PLA/ PANI-ES composite**	**H_3PO_4doped DL-PLA/ PANI-ES composite**
Slope at 0 T (3D VRH)	49.39	−17.8081	−55.21167	−47.7966
Slope at 0.4T		187.399	386.08	291.918
T_{Mott} (K)		1.005×10^5	9.29×10^6	5.21×10^6
L_{loc}(°A)		95.11	48.48	50.28
N (E_F) (no. states/ eV/cm^3)		2.14×10^{18}	1.755×10^{17}	2.08×10^{17}
$R_{Hop, Mott}$ (°A) at 125 K		189.74	300.18	269.47
$\Delta_{Hop, Mott}$ (meV at 125 K)		19.34	44.48	38.49
Temp. at M-I (K)		125	125	125

important parameter for the calculation of density of states, hopping distance and hopping energy. Resistivity of HCl, H_2SO_4 and H_3PO_4 doped DL-PLA/PANI-ES composites was measured as a function of temperature in the presence of magnetic field (0.4 T). We analyzed the resistivity data at 0.4 T (Figure 6.7) along with the 3D-VRH model described by Mott's law [13] and given below, and calculated the localization length (L_{loc}) and density of states. Using these important parameters, we calculated hopping distance ($R_{hop, Mott}$) and hopping energy ($\Delta_{hop, Mott}$). The above estimated parameters help to understand the transport phenomenon.

$$\rho(T) = \rho_0 e^{\left(\frac{T_{Mott}}{T}\right)^{\frac{1}{4}}} \tag{6.4}$$

$$T_{Mott} = \frac{16}{\left[K_B N(E)_F L_{loc}^3\right]} \tag{6.5}$$

where K_B is the Boltzmann constant, $N(E_F)$ is the density of states at the Fermi level, and L_{loc} is the localization length. The plot of $\ln\rho$ vs $T^{-\frac{1}{4}}$ exhibits a straight line for HCl, H_2SO_4 and H_3PO_4 doped DL-PLA/PANI-ES composites (Figure 6.7). T_{Mott} can be evaluated from the slope of the straight line and is listed in Table 6.5. The localization length L_{loc} can also be calculated from magnetoresistivity data, as shown in Figure 6.7. From the VRH

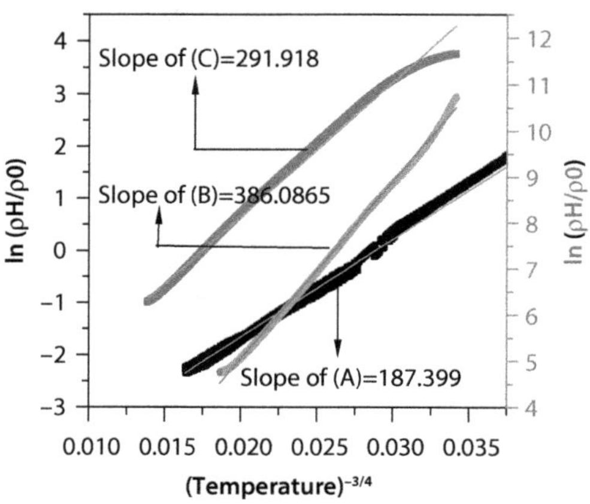

Figure 6.7 Plots of [lnρ(H)/ρ(0)] vs. $T^{-4/3}$ for HCl (A), H_2SO_4 (B), H_3PO_4 (C) doped DL-PLA/PANI-ES composites at 0.4 T (77–300 K).

model, the resistivity at different temperatures at a particular magnetic field can be written as [66, 67]:

$$\ln\left[\frac{\rho(H)}{\rho_0}\right] = t\left(L_{loc} / L_H\right)^4 \left(\frac{T_{Mott}}{T}\right)^{\frac{3}{4}} \tag{6.6}$$

where $t = \frac{5}{2016}$ and $L_H = \left[\frac{hc}{2\pi eH}\right]^{\frac{1}{2}}$ is magnetic length, c = velocity of light (3×10^{10} cm/s), h = Planck's constant (6.62×10^{-27}erg.sec), e = electronic charge (1.6×10^{-19} C) and H = 0.4 T is the applied magnetic field. From the slope of the curves in Figure 6.7, we determined the L_{loc} values and listed them in Table 6.5. Using the $T_{0,Mott}$ and L_{loc} of HCl, H_2SO_4 and H_3PO_4 doped DL-PLA/PANI-ES composites in Equation 7.5, the density of states ($N(E_F)$) can be calculated.

The $T_{0,Mott}$, L_{loc} and ($N(E_F)$) values were used to calculate the mean hopping distance R_{hop} and the energy difference between sites (Δ_{hop}) by using the following expression [13, 67] and are presented in Table 6.5.

$$R_{hop,Mott} = \left(\frac{3}{8}\right)\left(\frac{T_{Mott}}{T}\right)^{\frac{1}{4}} L_{loc} \tag{6.7}$$

$$\Delta_{hop,Mott} = \left(\frac{1}{4}\right)\left(k_B T\right)\left(\frac{T_{Mott}}{T}\right)^{\frac{1}{4}} \tag{6.8}$$

By putting the required parameters into the above expressions, the $R_{hop,Mott}$ and $\Delta_{hop,Mott}$ values were calculated at M-I transition temperature (125 K), which is presented in Table 6.5. In addition, the obtained hopping parameters were found to satisfy the 3D-VRH criteria.

Recently, the negative MR of hopping systems has been explored and the various possible ways of hopping path were interpreted by quantum interference effect of spins in the presence of magnetic field [68–71]. This effect is ascribed to the statistics of self-crossed trajectories in conductors, i.e., phase coherence of the electron's wave function between different conduction paths is destroyed in the presence of magnetic field [72]. In the present work, the low temperature conductivity has indicated the hoping conduction mechanism but not the weak localization system. The concept of weak localization effect is valid for diffusive motion of current carriers, and can be extended to hopping conduction [46, 47]. In the weak field region, the negative MR can be expressed as [46, 47]:

$$MR = \left[\frac{R(H) - R(0)}{R(0)}\right] a - H^x T^{-y} \tag{6.9}$$

where the exponent $x = 1$ [46] or $x = 2$ [47]; $y = \frac{3}{4}$ for 3D VRH and $y = \frac{3}{2}$ for 1D VRH. The interesting observation was found by Sivan *et al.* [46] and predicted quadratic field dependence over most of the weak field range. Also, Nguyen *et al.* [47] reported the negative MR, which is linear in the field. Using Equation 7.9, MR values of H_2SO_4 and H_3PO_4 doped DL-PLA/PANI-ES composites were calculated at room temperature. Figure 6.8 shows the plot of negative MR as a function of magnetic field (H). In the case of H_2SO_4 doped DL-PLA/PANI-ES composites, MR were found to linearly decrease with increase in the magnetic field whereas in the case of H_3PO_4 doped DL-PLA/PANI-ES composites, the MR value was found to slightly increase above 0.4 T. The above negative MR results are in qualitative agreement with the predicated features [46, 47].

The H_3PO_4 doped DL-PLA/PANI-ES composite was prepared for testing the CH_4 gas in different concentrations and balanced with synthetic air. During this gas sensing testing, the different gas concentrations (50 ppm, 100 ppm, 200 ppm and 500 ppm) were used. The sensing measurement was operated at room temperature. Some have reported that PANI polymeric nanofibers are a well-known sensing material for hydrocarbon gases such as CH_4 and H_2 [73, 74]. H_3PO_4 doped DL-PLA/

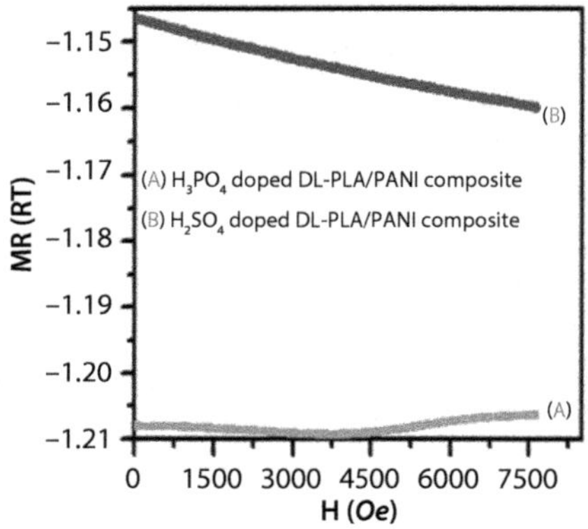

Figure 6.8 Magnetoresistance (MR) of H_3PO_4 (A) and H_2SO_4 (B) doped DL-PLA/PANI-ES composites at room temperature.

PANI-ES composite is employed for sensor performance against a well-known polymer system.

Figure 6.9 shows the dynamic responses of the prepared sensor. Recorded typical resistance transients of the tested materials are shown for CH_4 gas. As shown in the Figure 6.9, when the test gas is on, the resistance of sensor materials is reduced, whereas the same increased during recovery, i.e., when the synthetic air is on, the base resistance is regained. The response (S) percentage at room temperature is estimated through the relation:

$$\text{Response}(\%) = \frac{R_a - R_g}{R_a} \times 100 \tag{6.10}$$

where (R_a) and (R_g) are the sensor resistance measured in air and test gas respectively. The response (%) was calculated to be 0.00288%, 0.4255%, 3.5% and 9% for 50 ppm, 100 ppm, 250 ppm and 500 ppm concentration of methane, respectively. Minimal baseline drift is observed when the sensing element is switched back and forth between air and test gas environment. In addition, the baseline is found to be shifted downward, as indicated in Figure 6.9. For better understanding, we have plotted % response vs. concentration and the results are shown in Figure 6.9. It shows that the % response increases with the increase in CH_4 concentration which is found to be nonlinear in nature. Similar behavior has been reported for a variety of semiconducting materials-based sensors [75, 76]. Several physical phenomena could be invoked to explain such behavior [75, 76]. The change in resistance is due to the interaction between the interface of PANI and CH_4 gas molecules [77]. The resistance of 1.174 KΩ at 500 ppm CH_4 is measured

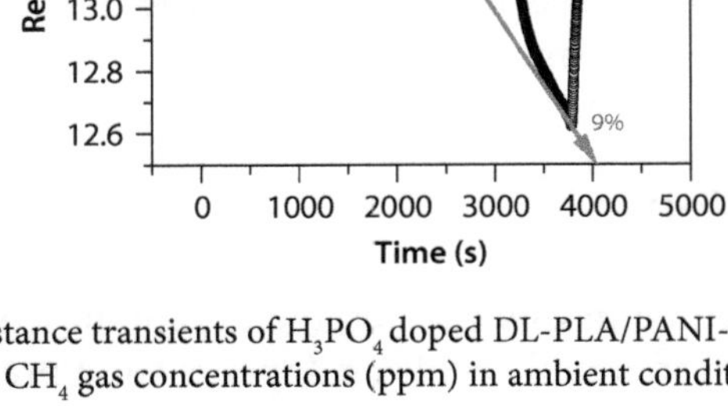

Figure 6.9 The resistance transients of H_3PO_4 doped DL-PLA/PANI-ES composite film measured at various CH_4 gas concentrations (ppm) in ambient conditions.

to be much lower than that of 0.0384 KΩ at 50 ppm CH_4. At lower concentration (50 ppm), a smaller number of gas molecules are expected to interact with PANI surface, giving poor response, i.e., small change in the resistance [78].

During the exposure to 50 ppm of CH_4 gas balanced in synthetic air, the prepared sensor responded negligibly (0.00288%). However, for higher and higher concentration of CH_4 gas, i.e., 50 ppm to 100 ppm, 250 ppm and 500 ppm, the prepared sensor responded differently. Fowler *et al.* [78] suggested that the chemisorbed H_2 gas can react with oxygen from the environment to produce H_2O. They also showed that the presence of H_2O ruins the ability of PANI to sense CH_4. Similar effects might also deteriorate the response of our PANI sensors at high CH_4 concentrations (Figure 6.9) due to the presence of 21% oxygen (from synthetic air). Our observation is supported by published results in an earlier paper where it was found that the diameter of the PANI nanofibers has a direct effect on the CH_4 gas-sensing performance [78].

Because CH_4 gas is reducing in nature, the mechanisms of its interaction with the sensitive materials considered in this report are largely governed by the chemical properties of each material. The CH_4 sensing mechanism for PANI is not fully understood. Chemisorption of CH_4 gas molecules at the charged amine nitrogen sites may occur. The dissociation of the CH_4 bond then leads to the formation of new N–H bonds with the amine nitrogen on the PANI chain. Subsequent charge transfer between adjacent amine nitrogen returns the PANI back to its polaronic form, i.e., doped state. Removal of the gas source and exposure to synthetic air releases the CH_4 from the polymer chains, resulting in a completely reversible reaction [74]. The polaronic form of PANI is the state of high conductivity. Therefore, the PANI film resistance is expected to decrease during the interaction with CH_4 gas, as observed from Figure 6.9. The test gas molecules diffuse more in H_3PO_4 doped DL-PLA/PANI-ES composite. This is because the PANI nanofibers provide more surface area for gas diffusion. PANI nanofibers have been observed in SEM images (Figure 6.1). Therefore, the charged amine nitrogen sites on the PANI backbone are largely accessible to CH_4 gas molecules, resulting in the dominant PANI CH_4 gas-sensing mechanism.

6.4 Conclusion

In the study presented in this chapter, membrane-based PANI composites using different inorganic acids as dopants were prepared and their spectral, electrical and thermal properties were investigated. Methane gas was sensed at room temperature using H_3PO_4 doped DL-PLA/PANI-ES composite. Such an effective technique and alternative material is an initial approach which encourages the improvement of conducting PANI materials and can possibly find application in other electronic industries.

Acknowledgments

We thank the Indian Institute of Technology, Kharagpur, for the financial support in the frame of the Institute Research Fellowship. I also acknowledge the Central Research Facility, IIT Kharagpur, India, for electrical characterizations (Hall measurement lab).

References

1. Burroughes, J.H., Bradley, D.D.C., Brown, A.R., Marks, R.N., Mackay, K., Friend, R.H., Burns, P.L., and Homes, A.B., Light-emitting diodes based on conjugated polymers. *Nature* 347, 539, 1990.
2. Frackowiak, E., Khomenko, V., Jurewicz, K., Lota, K., and Beguin F., Supercapacitors based on conducting polymers/nanotubes composites. *J. Power Sourc.* 153, 413, 2006.
3. Conn, C., Sestak, S., Baker, A.T., amd Unsworth. J., A polyaniline-based selective hydrogen senso. *Electroanalysis* 10, 1137, 1998.
4. Abdelkader, R., Amine, H., and Mohammed, B., Thermally stable forms of pure polyaniline catalyzed by acid-exchanged montmorillonite clay called maghnite-H^+ as an effective catalyst. *Int. J. Polym. Sci.* 2012, 1, 2012.
5. Yang, S., and Ruckenstein, E., Processable conductive composites of polyaniline/poly (alkyl methacrylate) prepared via an emulsion method. *Synth. Met.* 59, 1, 1993.
6. Singh, V., Mohan, S., Singh, G., Pandey, P.C., and Prakash, R., Synthesis and characterization of polyaniline-carboxylated pvc composites: Application in development of ammonia sensor. *Sens. and Actuators B* 132, 99, 2008.
7. Okubo, M., Fujii, S., and Minami, H., Production of electrically conductive, core-shell polystyrene/polyaniline composite particles by chemically oxidative seeded dispersion polymerization. *Colloid and Polym. Sci.* 279, 139, 2001.
8. Spirkova, M., Stejskal, J., and Quadrat, O., Electrically anisotropic polyaniline-polyurethane composites. *Synth. Met.* 102, 1264, 1999.
9. Aleshin, A., Kiebooms, R., Menon, R., and Heeger, A.J., Electronic transport in doped poly (3,4-ethylenedioxythiophene) near the metal-insulator transition. *Synth. Met.* 90, 61, 1997.

10. Shklovskii, B.I., and Efros, A.L., *Electronic Properties of Doped Semiconductors*, Springer: Berlin, 1984.
11. Long, Y.Z., Yin, Z.H., and Chen, Z.J., Low-temperature magnetoresistance studies on composite films of conducting polymer and multiwall carbon nanotubes. *J. Phys. Chem. C* 112, 11507, 2008.
12. Zuppiroli, L., Bussac, M.N., Paschen, S., Chauvet, O., and Forro, L., Hopping in disordered conducting polymers. *Phys. Rev. B* 50, 5196, 1994.
13. Li, J., Fang, K., Qiu, H., Li, S., and Mao, W., Micromorphology and electrical property of the HCl-doped and DBSA-doped polyaniline. *Synth. Met.* 142, 107, 2004.
14. Kapil, A., Taunk, M., and Chand, S., Preparation and charge transport studies of chemically synthesized polyaniline. *J. Mater. Sci.: Mater. Electron.* 21, 399, 2010.
15. Reghu, M., Yoon, C.O., Moses, D., Smith, P., Heeger A.J., and Cao,Y., Magnetoresistance in polyaniline networks near the percolation threshold. *Synth. Met.* 69, 271, 1995.
16. Clark, J.C., Ihas, G.G., Rafanello, A.J., Meisel, M.W., Reghu M., Yoon, C.O., Cao, Y., and Heeger, A.J., Resistivity and magnetoresistance of metallic polyaniline and polypyrrole at millikelvin temperatures. *Synth. Met.* 69, 215, 1995.
17. Long, Y., Chen, Z., Zhang, X., Zhang, J., and Liu, Z., Synthesis and electrical properties of carbon nanotube polyaniline composites. *Appl. Phys. Lett.* 85, 1796, 2004.
18. Sangeeth, C.S.S., Jiménez, P., Benito, A.M., Maser, W.K., Menon, R., Charge transport properties of water dispersible multiwall carbon nanotube-polyaniline composites. *J. Appl. Phys.* 107, 103719, 2010.
19. Abou-Elazab, T.F., Migahed, M.D., Park, H., Park, Y.W., MacNeillis, P., Rabenau, T., Magnetoresistance of polypyrrole and polyacetylene. *Synth. Met.* 76, 281, 1996.
20. Mativetsky, J.M., and Datars, W.R., Morphology and electrical properties of template-synthesized polypyrrole nanocylinder. *Physica B* 324, 191, 2002.
21. Park, J.G., Lee, S.H., Kim, B., and Park, Y.W., Electrical resistivity of polypyrrole nanotube measured by conductive scanning probe microscope: the role of contact force. *Appl. Phys. Lett.* 81, 4625, 2002.
22. Aleshin, A., Kiebooms, R., Menon, R., and Heeger, A.J., Electronic transport in doped poly (3,4-ethylenedioxythiophene) near the metal-insulator transition. *Synth. Met.* 90, 61, 1997.
23. Nath, C., and Kumar, A., Effect of temperature and magnetic field on the electrical transport of polyaniline nanofibers. *J. Appl. Phys.* 113, 093707, 2013.
24. Long, Y., Chen, Z., Shen, J., Zhang, Z., Zhang, L., Huang, K., Wan, M., Jin, A., Gu, C., and Duvail, J.L., Magnetoresistance studies of polymer nanotube/wire pellets and single polymer nanotubes/wires. *Nanotechnology* 17, 5903, 2006.
25. Long, Y., Huang, K., Yuan, J., Han, D., Niu, L., Chen, Z., Gu, C., Jin, A., and Duvail, J.L., Electrical conductivity of a single au/polyaniline microfiber. *Appl. Phys. Lett.* 88, 162113, 2006.

26. Lee, P.A., and Ramakrishnan, T.V., Disordered electronic systems. *Rev. Modern Phys.* 57, 287, 1985.
27. Panigrahi, M., Singh, N.K., Gautam, R.K., Banik, R.M., and Maiti, P., Improved biodegradation and thermal properties of poly(lactic acid)/layered silicate nanocomposites. *Compos. Interface.* 17, 143, 2010.
28. Mathai, C.J., Saravanan, S., Anantharaman, M.R., Venkitachalam, S., and Jayalekshmi, S., Effect of iodine doping on the band gap of plasma polymerized aniline thin films. J. *Phys. D: Appl. Phys.* 35, 2206, 2002.
29. do Nascimento, G.M., Constantino, V.R.L., and Temperini, M.L.A., Spectroscopic characterization of new type of conducting polymer-clay nanocomposite. *Macromolecules* 35, 7535, 2002.
30. Bocchini, S., and Frache, A., Comparative study of filler influence on polylactide photooxidation. *Express Polym. Lett.* 7, 431, 2013.
31. Shi, J., Wu, Q., Li, R., Zhu, Y., Qin, Y., and Qiao, C., The pH-controlled morphology transition of polyaniline from nanofibers to nanospheres. *Nanotechnology* 24, 175602, 2013.
32. Ansari, A.A., Khan, M.A.M., Khan, M.N., Alrokayan, S., Alhoshan, A.M., and Alsalhi, M.S., Optical and electrical properties of electrochemically deposited polyaniline/CeO_2 hybrid nanocomposite film. *J. Semiconduct.* 32, 043001, 2011.
33. Lux, F., Properties of electronically conductive polyaniline: A comparison between well-known literature data and some recent experimental findings. *Polym.* 35, 2915, 1994.
34. Langer, J.J., Krzyminiewski, R., Krucznski, Z., Gibinski, T., Czajkowski, I., and Framski, G., EPR and electrical conductivity in microporous polyaniline. *Synth. Met.* 122, 359, 2001.
35. Li, J., Fang, K., Qiu, H., Li, S., and Mao, W.J., Micromorphology and electrical property of the HCl-doped and DBSA-doped polyaniline. *Synth. Met.* 142, 107, 2004.
36. Chakraborty, G., Gupta, K., Meikap, A.K., and Jana, P.C., Direct current electrical transport and magneto transport properties of polyaniline nanocomposites. *J. Phys. Sci.* 14, 207, 2010.
37. Long, Y., Chen, Z., Shen, J., Zhang, Z., Zhang, L., Huang, K., Wan, M., Jin, A., Gu, C., and Duvail, J.L., Magnetoresistance studies of polymer nanotube/wire pellets and single polymer nanotubes/wires. *Nanotechnology* 17, 5903, 2006.
38. Wang, Z.H., Javadi, H.H.S., Ray, A., MacDiarmid, A.G., and Epstein, A.J., Electron localization in polyaniline derivatives. *Phys. Rev. B* 42, 5411, 1990.
39. Li, W., and Wan, M., Porous polyaniline films with high conductivity. *Synth. Met.* 92, 121, 1998.
40. Chakraborty, G., Gupta, K., Rana, D., and Meikap, A.K., Effect of multiwalled carbon nanotubes on electrical conductivity and magnetoconductivity of polyaniline. *Adv. Nature Sci.: Nanosci. Nanotechnol.* 3, 035015, 2012.
41. Ghosh, M., Barman, A., Meikap, A.K., De, S.K., Chatterjee, S., and Chattopadhyay, S.K., Electrical resistivity and magnetoresistivity of protonic

acid (H_2SO_4 and HCl)-doped polyaniline at low temperature. *JAPS* 75, 1480, 2000.
42. Faran, O., and Ovadyaho, Z., Magnetoconductance in the variable-range-hopping regime due to a quantum-interference mechanism. *Phys. Rev. B* 38, 5457, 1998.
43. Baumgartner, G., Carrard, M., Zuppiroli, L., Bacsa, W., Heer, W.A., and Forro, L., Hall effect and magnetoresistance of carbon nanotube films. *Phys. Rev. B* 55, 6704, 1997.
44. Yosida, Y., and Oguro, I., Variable range hopping conduction in bulk samples composed of single-walled carbon nanotubes. *J. Appl. Phys.* 86, 999, 1999.
45. Fuhrer, M.S., Holmes, W., Richards, P.L., Delany, P., Louie, S.G., and Zettl, A., Nonlinear transport and localization in single-walled carbon nanotubes. *Synth. Met.* 103, 2529, 1999.
46. Sivan, U., Entin-Wohlman, O., and Imry, Y., Orbital magnetoconductance in the variable-range-hopping regime. *Phys. Rev. Lett.* 60, 1566, 1988.
47. Nguyen, V.L., Spivak, B.Z., and Shklovskii, B.I., Aharonov-Bohm oscillation withn and superconductivity flux-quanta in hopping conduction. *Sov. Phys.-J. Exp. Theor. Phys.* 62, 1021, 1985.
48. Borowski, W.S., A review of methane and gas hydrates in the dynamic, stratified system of the Blake Ridge region, offshore southeastern North America. *Chem. Geology* 205, 311, 2004.
49. Jun, J.H., Yun, J., Cho, K., Hwang, I.-S., Lee, J.-H., and Kim, S., Necked ZnO nanoparticle-based NO_2 sensors with high and fast response. *Sens. Actuator. B* 140, 412, 2009.
50. Barazzouk, S., Tandon, R.P., and Hotchandani, S., MoO_3-based sensor for NO, NO_2 and CH_4 detection. *Sens. Actuator. B* 119, 691, 2006.
51. Xi, L., Qian, D., Tang, X., and Chen, C., High surface area SnO_2 nanoparticles: Synthesis and gas sensing properties. *Mater. Chem. Phys.* 108, 232, 2008.
52. Hu, P., Du, G., Zhou, W., Cui, J., Lin, J., Liu, H., Liu, D., Wang, J., and Chen, S., Enhancement of ethanol vapour sensing of TiO_2 nanobelts by surface engineering. *Appl. Mater. Interfaces* 2, 3263, 2010.
53. Fidelus, J.D., Lojkowski, W., Millers, D., Grigorjeva, L., Smits, K., and Piticescu, R.R., Zirconia based nanomaterials for oxygen sensors-generation, characterisation and optical properties. *Solid State Phenom.* 128, 141, 2007.
54. Satyanarayana, L., Reddy, C.V.G., Manorama, S.V., and Rao, V.J., Liquid-petroleum-gas sensor based on a spinel semiconductor, $ZnGa_2O_4$. *Sens. Actuator. B* 46, 1, 1998.
55. MacDiarmid, A.G., Chiang, J.C., Richter, A.F., and Epstein, A.J., Polyaniline: A new concept in conducting polymers. *Synth. Met.* 18, 285, 1987.
56. Focke, W.W., Wnek, G.E., and Wei, Y., Influence of oxidation state, pH, and counter ion on the conductivity of polyaniline. *J. Phys. Chem.* 91, 5813, 1987.
57. Omastova, M., Pavlinec, J., Pionteck, J., Simon, F., and Kosina, S., Chemical preparation and characterization of conductive poly (methyl methacrylate)/polypyrrole composites. *Polym.* 39, 6559, 1998.

58. Singh, V., Mohan, S., Singh, G., Pandey, P.C., and Prakash, R., Synthesis and characterization of polyaniline-carboxylated PVC composites: Application in development of ammonia sensor. *Sens. Actuator. B* 132, 99, 2008.
59. Cairns, D.B., Armes, S.P., Chehimi, M.M., Perruchot, C., and Delamar, M., X-ray photoelectron spectroscopy characterization of submicrometer-sized polypyrrole-polystyrene composites. *Langmuir* 15, 8059, 1999.
60. Njuguna, J., and Pielichowski, K., Review: Recent developments in polyurethane-based conducting composites. *J. Mater. Sci.* 39, 4081, 2004.
61. Peng Wang, P., Tan, K.L., Zhang, F., Kang, E.T., and Neoh, K.G., Synthesis and characterization of poly (ethylene glycol)-grafted polyaniline. *Chem. Mater.* 13, 581, 2001.
62. Chen, Y., Kang, E.T., Neoh, K.G., and Tan, K.L., Chemical modification of polyaniline powders by surface graft co-polymerization. *Polym.* 41, 3279, 2000.
63. Yoon, J.T., Lee, S.C., and Jeong, Y.G., Effects of grafted chain length on mechanical and electrical properties of nanocomposites containing polylactide-grafted carbon nanotubes. *Compos. Sci. Technol.* 70, 776, 2010.
64. Yang, J., Wang, X., Wang, X., Jia, R., and Huang, J., Preparation of highly conductive cnts/polyaniline composites through plasma pretreating and *in situ* polymerization. *J. Phys. Chem. Solids* 71, 448, 2010.
65. Lee, Y.K., Lee, K.J., Kim, D.S., Lee, D.J., and Kim, J.Y., Polypyrrole-carbon nanotube composite films synthesized through gas-phase polymerization. *Synth. Met.* 160, 814, 2010.
66. Long, Y.Z., Yin, Z. H., and Chen, Z.J., Low-temperature magnetoresistance studies on composite films of conducting polymer and multi-walled carbon nanotubes. *J. Phys. Chem. C* 112, 11507, 2008.
67. Choi, E.S., Kim, G.T., Suh, D.S., Kim, D.C., Park, J.G., and Park, Y.W., Magnetoresistance of the metallic polyacetylene. *Synth. Met.* 100, 3, 1999.
68. Reghu, M., Vakiparta, K., Cao. Y., and Moses, D., Pressure dependence of the conductivity and magnetoconductance in oriented iodine-doped polyacetylene. *Phys. Rev. B* 49, 16162, 1994.
69. Kozub, V.I., Aleshin, A.N., Suh, D.S., and Park, Y.W., Evidence of magnetoresistance for nanojunction-controlled transport in heavily doped polyacetylene. *Phys. Rev. B* 65 224204, 2002.
70. Kim, G.T., Choi, E.S., Suh, D.C., Park, Y.W., Liu, K., Duesberg, G., and Roth, S., Magnetoresistance of an entangled single-wall carbon-nanotube network. *Phys. Rev. B* 58, 16064, 1998.
71. Kang., N., Hu, J.S., Kong, W.J., Lu, L., Zhang, D.L., Pan, Z.W., and Xie, S.S., Consistent picture of strong electron correlation from magnetoresistance and tunneling conductance measurements in multiwall carbon nanotubes. *Phys. Rev. B* 66, 241403, 2002.
72. Lee, P.A., and Ramakrishnan, T.V., Disordered electronic systems. *Rev. Mod. Phys.* 57, 287, 1985

73. Wu, Z., Chen, X., Zhu, S., Zhou, Z., Yao, Y., Quan, W., and Liu, B., Room temperature methane gas sensor based on graphene nanosheets/polyaniline nanocomposite thin film. *IEEE Sensors J.* 13, 777, 2013.
74. Al-Mashat, L., Shin, K., Kalantar-Zadeh, K., Plessis, J.D., Han, S.H., Kojima, R.W., Kaner, R.B., Li, D., Gou, X., Ippolito, S.J., and Wlodarski, W., Graphene/polyaniline nanocomposite for hydrogen sensing. *J. Phys. Chem. C* 114, 16168, 2010.
75. Lu, G., Ocola, L.E., and Chen, J., Reduced graphene oxide for room-temperature gas sensors. *Nanotechnology* 20, 445502, 2009.
76. Fowler, J.D., Allen, M.J., Tung, V.C., Yang, Y., Kaner, R.B., and Weiller, B.H., Practical chemical sensors from chemically derived graphene. *ACS Nano* 3, 301, 2009.
77. Virji, S., Kaner, R.B., and Weiller, B.H., Hydrogen sensors based on conductivity changes in polyaniline nanofibers. *J. Phys. Chem. B* 110, 22266, 2006.
78. Fowler, J.D., Virji, S., Kaner, R.B., and Weiller, B.H., Hydrogen detection by polyaniline nanofibers on gold and platinum electrodes. *J. Phys. Chem. C* 113, 6444, 2009.
79. Sadek, A.Z., Baker, C.O., Powell, D.A., Wlodarski, W., Kaner, R.B., and Kalantar-zadeh, K., Polyaniline nanofiber based conductometri acoustic wave gas sensors-effect of nanofiber diameter on H_2 response. *J. IEEE Sens.* 7, 213, 2007.

7

Electrical Properties of Conducting Polymer-MWCNT Binary and Hybrid Nanocomposites

B.T.S. Ramanujam[1a,b*] and S. Radhakrishnan[2]

[1.a]Present Address: Department of Sciences, Amrita School of Engineering, Coimbatore, Amrita Vishwa Vidyapeetham, Amrita University, Tamil Nadu, India.
[1.b]Past Address: School of Science and Engineering, Navrachana University, Vadodara, Gujarat, India.
[2]Department of Polymer and Petrochemical Engineering, Maharashtra Institute of Technology, Pune, Maharastra

Abstract
The demand for lightweight, cost-effective electrically conducting polymers and conducting nanocomposites for various applications, such as sensors, actuators, etc., has been skyrocketing in recent years. Electrically conducting polymer composites (CPCs) can be developed by dispersing conducting particles in inherently conducting polymers or insulating polymer matrices. Due to the recent advancements in nanotechnology, it is possible to tailor make composite materials with exceptional electrical, mechanical and thermal properties. The properties of conducting polymer binary composites can be altered by careful selection and dispersion of fillers in the polymer matrix. In a similar way, conducting hybrid polymer composites for a particular application can be synthesized by exploiting the synergism in properties of two or more dispersed phases. It is not merely cost reduction that drives the market of CPCs but also environmentally benign synthesis procedures wherever possible and mass production of the end product with better energy savings has an immense influence on the demand for these materials. The unprecedented growth of nanotechnology offers better control over manipulation of filler size as well as polymers, and hence the composites can be tailor-made. Though there are many nanofillers, such as nanographite, carbon nanofiber, etc., in this chapter, more attention will be given to the electrical properties of multiwalled

**Corresponding author*: ramsrag@gmail.com

Sanjay K. Nayak *et al.* (eds.) Trends and Applications in Advanced Polymeric Materials, (127–144) 2018 © Scrivener Publishing LLC

carbon nanotube (MWCNT) filled polyethersulfone (PES) conducting composites, along with that of PES-graphite-MWCNT hybrid nanocomposites.

***Keywords*:** Conducting polymer composites, multiwalled carbon nanotube, electrical properties, charge transport, scanning/transmission electron microscopy

7.1 Introduction

Lightweight, high performance, high energy density and cost-effective products have driven the rapid growth in the conducting polymers market. Biomemetic technology and artificial muscle applications based on conducting polymers have led to significant market growth of conducting polymers. The wide range of applications of conducting polymers, such as batteries, sensors, antistatic packaging, etc., have proven the importance of this class of materials. One market research report clearly predicts that the market for conducting polymers will reach around 6.75 billion USD by 2020 [1]. Hence, to say that there is an extensive ongoing research being carried out towards the development of conducting polymers is not a mere exaggeration. Conducting polymers can be broadly classified into three major types: inherently conducting polymers, charge transfer complexes and conducting polymer composites (Figure 7.1).

Existence of π electrons and conjugation makes inherently conducting polymers electrically conductive. Polyaniline (PANI), polypyrrole (PPy) and polythiophene (PTh) are examples of this class of conducting polymers. Addition of acceptor molecules in the insulating polymer results in polymer charge transfer complexes. The orbital overlap due to donor-acceptor interaction improves delocalization of electrons. Organometallic polymeric conductors also belong to this class of conducting polymers in which the polymer consists of transition metal ions and highly conjugated ligands in the backbone. The last class, namely CPCs, are obtained by dispersing conducting particles in either the insulating polymer matrix/polymer blends or inherently conducting polymers to improve electrical conductivity of the composites.

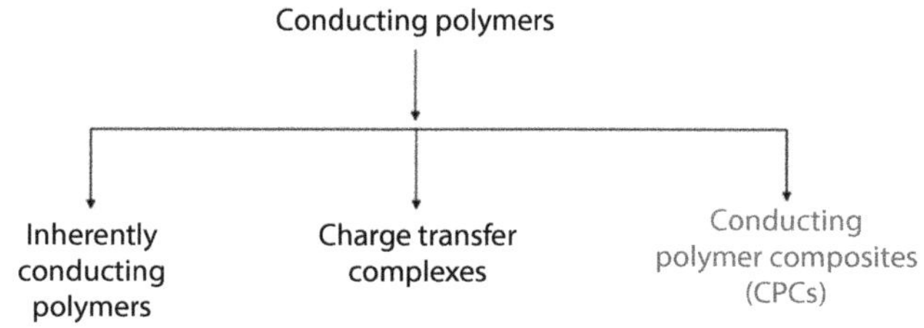

Figure 7.1 Classification of conducting polymers.

7.1.1 Theoretical Background of Electrical Conductivity in CPCs

The electrical conductivity varies nonlinearly for most of the binary and hybrid conducting composites [2] with respect to filler concentration (ϕ). The concentration of fillers at which insulator to semiconductor transition occurs is identified as electrical percolation threshold. Above a certain concentration of fillers the electrical conductivity saturates. The nonlinear variation of electrical conductivity can be explained in terms of percolation theory, which essentially takes into consideration the contacts between the filler particles. At electrical percolation threshold, contacts between fillers in the polymer matrix begin to form and saturation in electrical conductivity signifies saturation in contacts between the fillers. Classical percolation equations are given below:

$$\sigma_{eff} = \sigma_0\left(\varphi - \varphi_c/1 - \varphi_c\right)^t \qquad \varphi > \varphi c \tag{7.1}$$

$$\sigma_{eff} = \sigma_1\left(\varphi_c - \varphi/\varphi_c\right)^{-s} \qquad \varphi > \varphi c \tag{7.2}$$

where σ_{eff} is the effective conductivity of the composite consisting of a conductor with conductivity σ_0 and an insulator which is essentially the polymer with a conductivity σ_1, φ is the volume fraction of filler, and φ_c is the percolation threshold; '*t*' and '*s*' are critical exponents characterizing the conductivity in the conducting and the insulating phase, which depend on the dimensionality of the network formation. For three-dimensional network formation, '*t*' and '*s*' take the values 2 and 0.87 respectively [3].

7.1.2 Factors Affecting Electrical Percolation Threshold

The electrical percolation threshold is dependent on the dispersed filler particle size, shape, aspect ratio, orientation, processing routes, polymer characteristics, etc. It is very important to reduce the electrical percolation threshold as much as possible, which will help in easy processing of polymer composite materials. Very narrow particle size distribution, lower particle size and higher aspect ratios are preferred in order to reduce the percolation threshold. In this regard, not the micron-sized fillers but nanofillers, such as carbon nanofiber [4-5], graphene [6-7], etc., are being focused on. Out of the different carbonaceous fillers, multiwalled carbon nanotube (MWCNT) has been extensively focused in the literature [8–9]. This is due to the fact that it is cheaper than single-walled carbon nanotube and exhibits better conductivity than carbon nanofiber. Hence, in this chapter the electrical properties of MWCNT mixed polymer, especially PES, will be addressed.

7.1.3 Processing Methods of CPCs

There are four different ways of synthesizing CPCs:

a. *Powder mixing*: This method deals with physically mixing the filler particles with the polymer matrix which exists in powder form at room temperature. When nanofillers are to be incorporated, they have to be dispersed in a suitable solvent and the solvent must be removed from the system. There will be a chance of agglomeration of fillers in this method.
b. *Solution mixing*: In this method a suitable solvent is selected for dissolving the polymer. Well-dispersed fillers in the same solvent are then mechanically mixed with the dissolved polymer. Finally, the solvent should be evaporated from the composite system. This method results in better dispersion of filler particles in the matrix, as reported elsewhere [10].
c. *In-situ polymerization:* In this method conducting filler particles are swollen in the liquid monomer and then polymerization is started by adding initiators, followed by heating or irradiation. Many conducting polymer composites, such as polyphenylene sulfide (PPS)-expanded graphite [11], polymethyl methacrylate (PMMA)-expanded graphite [12], polystyrene (PS)-graphene [13], etc., have been synthesized and the electrical properties were investigated.
d. *Melt compounding:* This processing method of polymer composites is industrially viable as it involves only melting and solvents are not used.

7.1.4 Conduction Mechanism in CPCs

There are four different configurations which might exist in conducting polymer composites. These are best described as the uniform model, uniform channel model, non-tunneling barrier model and tunneling barrier model, as reported elsewhere [14]. Tunneling conduction is possible when a thin barrier of 50 A° or less is intercepted by the electron. The electron wave function under this condition can tunnel through the barrier and finite conductivity results. Just below percolation threshold, at a rather very low concentration of nanofiller, the hopping mechanism of charge transport may be dominant [15]. At and above percolation threshold the contact between filler particles usually dominate. It is possible to identify the hopping mechanism of charge transport through ac conductivity analysis of conducting polymer composites at a given temperature. The frequency-dependent conductivity at constant temperature obeys a power law, as given in Equation 7.3.

$$\sigma_{ac} = \sigma_{dc} + A \cdot \omega^{n} \tag{7.3}$$

where 'ω' is the angular frequency, σ_{dc} is the frequency-independent conductivity ($\omega \to 0$), A is the temperature-dependent constant and 'n' is the exponent which lies between $0 \leq n \leq 1$, depending on the temperature and frequency for hopping conduction [16]. The exponent has also been reported to be greater than one [16]. The electron can tunnel or hop through a barrier which depends on the shape of the barrier. In the variable range hopping (VRH) mechanism proposed by Mott [17, 18], the charge carriers can hop from one localized state to another localized state of same or variable energies.

$$\sigma(T) = \sigma_0 \exp\left(-T_0/T\right)^{\gamma} \tag{7.4}$$

where 'σ_0' is the limiting value of the conductivity at infinite temperature, T_0 is the temperature which determines thermally activated hopping which is a measure of disorder. The exponent 'γ' is a measure of dimensionality 'd' of the transport process through the relation:

$$\gamma = 1/\left(1+d\right) \tag{7.5}$$

When 'γ' = 0.5, 1D VRH hopping is operative in the system.

7.1.5 Multiwalled Carbon Nanotube (MWCNT) – Potential Conducting Filler

After the discovery of multiwalled carbon nanotube (MWCNT) by Ijima in 1991 [19], tremendous research has been carried out in CNT-based composites because of the excellent physical, chemical and mechanical properties exhibited by them. Due to its unusual structure, it exhibits very high electrical conductivity [20]. Also, both MWCNT and SWCNT exhibit amphoteric behavior by exchanging electrons with electron acceptors or donors to form positively or negatively charged counterion. Apart from the other two types, double-walled CNTs (DWCNTs) have also been synthesized. Due to the weak coupling between concentric cylinders of the MWCNTs, only the outer shell significantly contributes to electron transport. The diameters of SWCNT and MWCNT may lie in the range of 0.8–2 nm and 5–20 nm respectively. The elastic modulus and tensile strength of a single layer of MWCNT has been measured to be 1 TPa and 100 GPa [21]. MWCNTs exhibit a metallic range of electrical conductivity and can carry current up to 10^9 A/cm^2 [22]. Depending upon the orientation of graphene lattice with respect to the tube axis, i.e., chirality, CNTs can exhibit metallic or semiconducting properties. The influence of chirality on the mechanical properties of CNTs has been reported. Although chirality has a relatively small influence on the elastic stiffness, Stone-Wales transformation plays a major role in the plastic deformation of CNTs under tension. Stone-Wales transformation, which is nothing but a reversible diatomic interchange, introduces a new defect, namely the heptagon [23]. Many possible equilibrium shapes are possible due to the heptagonal defects.

The CNT-based research has grown many fold after the large-scale synthesis of MWCNT was industrially realized. Just a search for the term "polymer-CNT composites" in sciencedirect.com will yield the number of scientific papers published in different years, as shown in Figure 7.2. The number of publications after 2009 starts to steadily increase due to the extraordinary properties exhibited by this multifunctional material. The properties of CNTs can be altered by the synthesis method adopted. Hence, various synthesis routes of CNTs are outlined below.

7.1.5.1 Synthesis Methods of Carbon Nanotubes

The major routes to synthesize CNTs are: a) arc discharge method, b) laser ablation method, c) chemical vapor deposition and d) electrolysis.

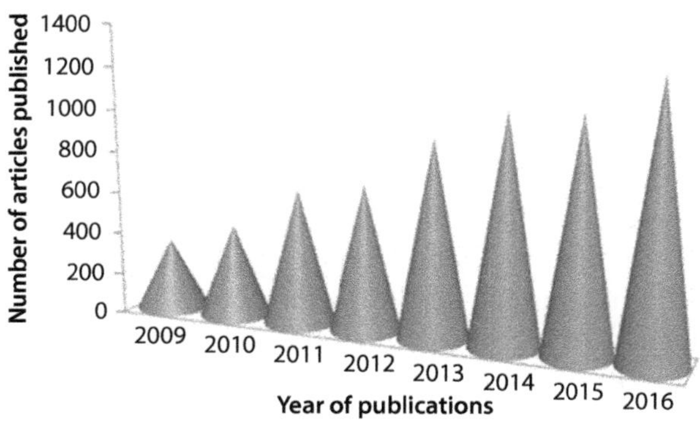

Figure 7.2 Growth in papers published in journals on the topic of polymer-CNT composites.

Chemical vapor depposition (CVD) is the most common technique to synthesize CNTs. However, large-scale synthesis of CNTs requires thermal annealing or chemical processing to remove contaminants which essentially affect the CNT properties. These processes can induce defects in the sidewalls of CNTs. Carbon arc discharge method is being used for the synthesis of MWCNTs. This method uses two graphite electrodes through which a direct current is passed under an inert atmosphere. The anode will be slowly consumed and deposition can be found at the cathode. SWCNTs can be produced in the same process provided metal catlysts have been inserted into the anode. Ando and Zhao [24] have shown that when pure iron catalyst is mixed with graphite rod after arc evaporation in the inert gas-hydrogen mixture atmosphere, SWCNT nets were produced.

Laser ablation technique is operated at similar conditions to the arc discharge method. In this method the carbon atoms from a graphite rod after evaporation are also condensed. Similar to arc discharge method, when graphite rod is mixed with metal catalyst and laser light is shed on the graphite electrode in inert gas atmosphere, SWCNT can be produced. Both arc discharge method and laser ablation techniques are not suitable for large-scale synthesis of CNTs. Further purification of CNTs is also an additional step required if CNTs are produced by those two methods. Hence, gas phase technique has gained more attention as far as CNT synthesis is concerned.

The catalytic decomposition of hydrocarbons has been well known to produce carbon fibers. Similar to that process, pyrolysis of hydrocarbons in the presence of metal catalyst generates both MWCNT and SWCNT. Lower temperatures (300–800 °C) are used to produce MWCNT, whereas SWCNT requires higher temperatures (600–1300 °C) in an inert gas atmosphere. The structure of MWCNT synthesized by both CVD and arc discharge methods is very different. The CVD process results in less crystalline MWCNTs with more defects than the arc discharge process. The CVD process is the most versatile technique when doped CNTs, especially those doped with boron or nitrogen or both, need to be synthesized [25, 26]. Further aligned CNTs can be synthesized through CVD technique.

When current is passed through two graphite electrodes immersed in molten ionic salts, MWCNTs can be produced. This method of production of CNTs, i.e., electrolysis, is the less common method. Thus, it is very clear that each process and processing parameter can result in different diameters of CNTs and defects can also be introduced which will affect the properties of the resultant carbon nanotube.

7.1.6 Electrical Properties of Polymer-MWCNT Composites

The electrical properties of polymer-MWCNT composites critically depend on the dispersion of MWCNTs, which will be affected due to different processing routes adopted. The processing route can also affect the aspect ratio of CNTs so that the electrical percolation threshold could be different for the same polymer-CNT systems. Very low electrical percolation threshold can be realized when MWCNTs are incorporated in different polymer matrices. Bauhofer and Kovacs [27] have reviewed the electrical percolation in CNT-polymer composites. The authors concluded that the production method and type of CNTs is less important than the polymer and dispersion method adopted. Many polymer composites with CNTs as the filler follow $\varphi_c = 1/\eta$, where 'η' is the aspect ratio of CNTs when the filler particles are homogeneously dispersed. Higher 'φ_c' is due to the non-homogeneous dispersion of CNTs and lower φ_c could be due to flocculation of CNTs as a result of kinetic percolation. Processing-induced anisotropic electrical percolation threshold has been reported for polypropylene-MWCNT composites after melt mixing followed by injection molding [28]. Two different electrical percolation thresholds, i.e., 2 wt% and 3 wt%, along longitudinal/transverse direction of mold filling flux have been observed for the PP-MWCNT system. Park *et al.* [29] have compared MWCNT synthesized by impregnation method and spray drying method. By controlling the catalyst supporter size, they have reported that it is possible to synthesize MWCNTs with different pore size, etc., and upon melt mixing with linear low density polyethylene (LDPE), less value of sheet resistance has been reported. Wang *et al.* [30] have investigated the formation of segregated structure in polyvinylidene fluoride (PVDF)-MWCNT composites and reported a percolation threshold lying between 1–2 wt% MWCNT. The PVDF-7 wt% MWCNT exhibited an EMI shielding efficiency of 30.89 dB in X-band frequency region. Ram *et al.* [31] have reported an electrical percolation threshold of 1 wt% MWCNT in PVDF when the composites were prepared by solution casting method. They have also reported that ac conductivity, dielectric constant and dielectric loss increase with the concentration of MWCNTs. The electrical properties of ethylene-1-octene copolymer-CB-CNT hybrid composites (both ac and dc) have been investigated by Burmistrov *et al.* [32].

Thus, various thermoplastic matrices, such as polypropylene, polyethylene, polystyrene, etc., have been employed to make conducting binary and hybrid composites. Pan and Li [33] have studied the electrical and rheological properties of PP-MWCNT composites. Kara *et al.* [34] have reported 1.8 wt% MWCNT as the percolation threshold along with optical percolation threshold in polystyrene (PS) matrix. Capriano *et al.* [35] have studied conductivity enhancement in melt processed polystyrene-MWCNT and polystyrene-CNF composites and reported conductivity of the composite was restored due to melt annealing above the glass transition temperature. In our group we have investigated the electrical properties of PES-graphite-CB hybrid composites [10]. Electrical and mechanical properties of polyethersulfone-MWCNT melt processed composites have been reported [36]. With the application of electric field of the order of 6 KV/m, very low electrical percolation threshold has been obtained. Though there are works reported related to electrical properties of MWCNT filled polymer composites, not much has been done related to polyethersulfone (PES)-MWCNT conducting composites synthesized by solution blending route. Hence, the dc and ac properties of solution blended PES-MWCNT and PES-graphite-MWCNT nanocomposites will be further discussed in the following section.

7.2 AC/DC Properties of Polyethersulfone (PES)-MWCNT, PES-Graphite-MWCNT Nanocomposites

Before discussing the ac/dc behavior of PES-MWCNT system, the material properties and composite preparation method are outlined below.

7.2.1 Material Properties

Polyethersulfone (PES) (3600 P), which is a high temperature thermoplastic with $T_g \sim 210$ °C exhibiting high resistance to heat and combustibity, was procured from Gharda Chemicals, Mumbai, India (currently known as Solvay Advanced Polymers). It is transparent and amorphous. Multiwalled carbon nanotube with tube diameter ranging from 110–170 nm and length from 5–9 microns was procured from Sigma-Aldrich. Graphite is a natural flake graphite powder (200 mesh) supplied by Carbon Enterprises, Pune, India.

7.2.2 Composite Preparation

According to weight fraction calculation, the required amount of MWCNT was added to 50 ml dichloromethane (DCM) and homogenized using a spindle connected to a motor with 1200 rpm (referred to as a homogenizer).

After 3 min of homogenization, MWCNT solution is immediately transferred to dissolved PES in DCM and homogenized for another 3 min and then poured into a Petri dish and dried. The resultant composite was crushed and pellets were made by applying a 3 ton load for three min using a hydraulic press. These pellets were used for dc/ac measurements. In a few cases, films were made especially at very low concentration of fillers and ac/dc properties were investigated. Similarly, hybrid composites were made after homogenizing the required amount of MWCNT and mixing it with PES-7 wt% graphite.

7.3 Discussion of Results

7.3.1 Electrical Behavior of Polyethersulfone (PES)-MWCNT Binary and PES-Graphite-MWCNT Hybrid Composites

Figure 7.3 depicts through-plane electrical conductivity variation in PES-MWCNT binary composites. The composites behave as insulators till 1 wt% MWCNT loading. Above that loading, the conductivity slowly increases. Above 2.5 wt% MWCNT in PES, the increase in electrical conductivity is reduced. The conductivity variation follows the percolation theory. As the contact between filler particles is established, there exists insulator to semiconductor transition. At higher loading of MWCNT, saturation in the contact between the filler particles results in saturation in the conductivity. Increase in MWCNT concentration decreases interparticulate distance in the polymer matrix and hence the effective dielectric constant increases at low frequency, i.e., at 0.01 Hz. The frequency-dependent effective dielectric constant variation is shown in Figure 7.4.

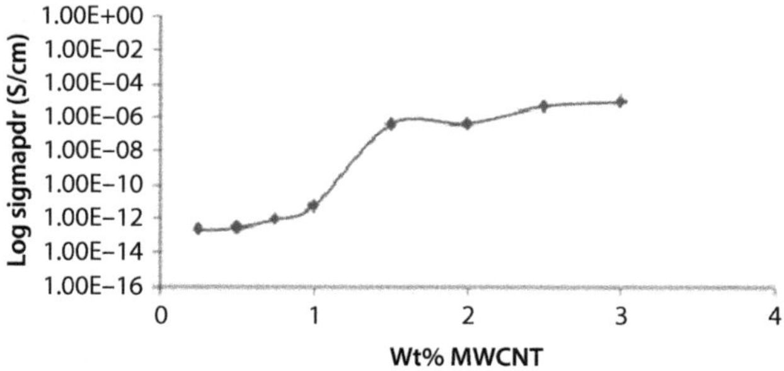

Figure 7.3 Through-plane conductivity of PES-MWCNT composites.

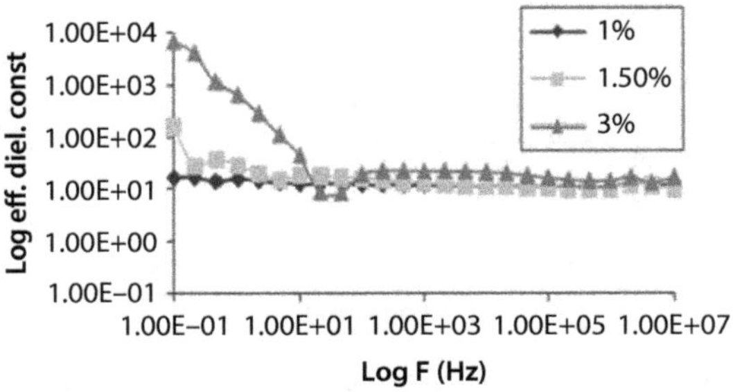

Figure 7.4 Frequency-dependent effective dielectric constant of PES-MWCNT composites.

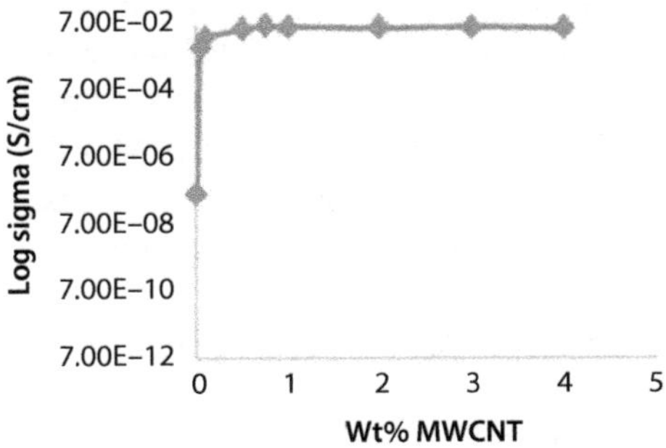

Figure 7.5 Through-plane conductivity of PES-7 wt% graphite–x wt% MWCNT hybrid composites.

The conducting fillers separated by insulating polymer matrix can be regarded as parallel plate capacitors and it is known that the capacitance is directly related to dielectric constant and inversely related to the distance between parallel plates. There exists almost a three orders of magnitude increase in effective dielectric constant at 0.01 Hz for 3 wt% MWCNT loading in PES compared to 1 wt% loading of MWCNT.

Figure 7.5 shows the variation of through-plane electrical conductivity in PES-7 wt% graphite–x wt% MWCNT homogenized samples. Due to high speed mixing without MWCNT, the PES-7 wt% graphite composite exhibits finite conductivity. This is due to the fact that graphite particle size is reduced and hence the number of particles is increased, though there can be reduction in the aspect ratio. With the addition of 0.05 wt% MWCNT, after dispersing in DCM the conductivity shoots up further to more than four orders of magnitude. Hence, the electrical percolation threshold is reduced in the case of hybrid composites. Hybrid composites are very effective in using nanofillers in small amounts along with cheap conventional fillers.

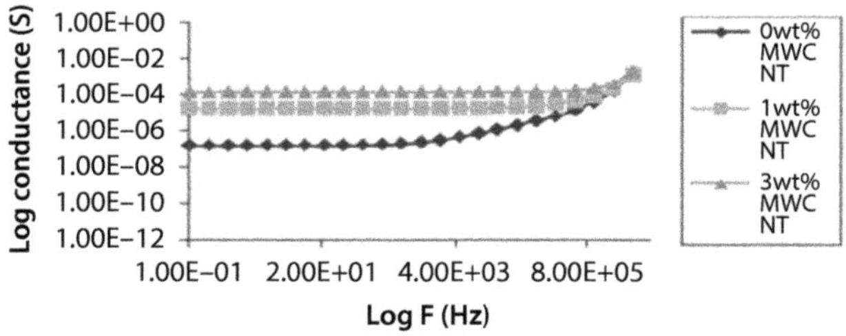

Figure 7.6 Frequency-dependent conductance of solution blended PES-7 wt% graphite–x wt% MWCNT (x = 0,1,3).

Figure 7.6 shows the ac conductance variation with the concentration of MWCNT (after sonication in DCM) in PES-7 wt% graphite. It can be clearly seen that even the conductance of 3 wt% MWCNT in PES-7 wt% graphite is less than that observed for 0.05 wt% MWCNT (homogenized) addition in the same. This clearly proves that in ultrasonicated MWCNT, agglomeration cannot be avoided and the dispersion can be improved by changing the solvent.

High speed mixing improves the degree of dispersion as well as the number of contacts. Even after 3 wt% MWCNT addition, the conductance increases by only two orders of magnitude when compared to 0 wt% addition of MWCNT in PES-7 wt% graphite. Though the frequency-dependent conductance is observed at higher frequency, the plateau region signifies dc conductance. Hence, a proper processing route has to be chosen for better dispersion of nanotubes. High speed mixing is a better method than sonication of MWCNT in DCM.

7.3.2 Transmission Electron Microscopy (TEM) Analysis

After dissolving PES in DCM, TEM analysis of PES-1 wt% MWCNT clearly shows that MWCNTs are twisted due to high speed mixing, as suggested in Figure 7.7. The high speed mixing reduces the tube length and increases the number of MWCNTs which are responsible for establishing contacts, and hence reducing the electrical percolation threshold.

Figure 7.8 clearly shows the interpenetrating MWCNTs and the broken edges due to high speed stirring. In order to show the difference in the dispersion state of MWCNTs in PES, PES-7 wt% graphite–x wt% MWCNT hybrid composites were prepared after dissolving PES in DCM and the required quantity of graphite was added, after which the stirring was continued for three hours. To this mixture, MWCNT sonicated for half an hour in DCM was added and then stirring continued for twelve more

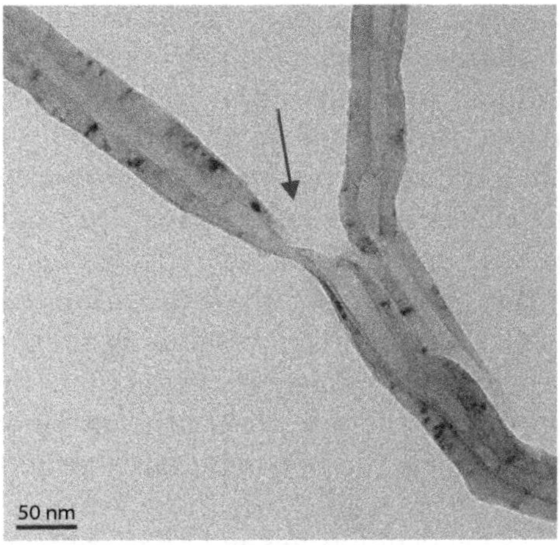

Figure 7.7 TEM image of PES-1 wt% MWCNT homogenized sample.

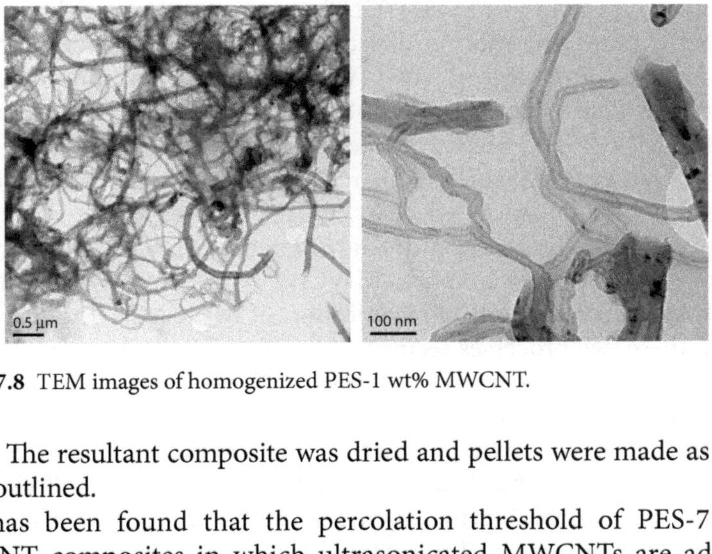

Figure 7.8 TEM images of homogenized PES-1 wt% MWCNT.

hours. The resultant composite was dried and pellets were made as previously outlined.

It has been found that the percolation threshold of PES-7 wt%-MWCNT composites in which ultrasonicated MWCNTs are added is higher than that of the same composition in which MWCNT is added after homogenization.

7.4 Conclusion and Future Perspectives

The development of CPCs for various applications, especially where electrical properties are harnessed, depends on the choice of conducting filler, aspect ratio, dispersion in the polymer matrix, processing routes adopted, polymer characteristics, etc. The nonlinear variation of electrical conductivity in both binary and hybrid composites/nanocomposites can be understood through percolation theory in most of the CPC systems. The reduction of electrical percolation threshold is critical, as the processing of CPCs becomes easier. Nanofillers such as multiwalled carbon nanotubes have proved to be very effective in reducing the electrical percolation threshold when mixed in various polymer matrices. The ac behavior of CPCs gives an understanding about the charge transport mechanism such as hopping-type conduction. The interparticular distance is vital in electrical conduction process. Hence, the improvement in the dispersion of filler in polymer matrix needs to be given much attention.

The PES-MWCNT nanocomposites and PES-7 wt% graphite–MWCNT hybrid composites were made by solution blending route after high speed stirring of MWCNT in DCM. It has been observed that the electrical percolation threshold in PES-MWCNT composites lies just above 1 wt% MWCNT loading. The effective dielectric constant increases to more than two orders of magnitude for 3 wt% MWCNT in PES. This is due to decrease in interparticulate distance. Analysis by TEM reveals that high speed mixing causes MWCNT to become twisted and acquire broken edges. Though the aspect ratio of MWCNT can be reduced, the number of tubes can be increased, which enhances effective contact between them, resulting in lower percolation threshold. A drastic difference exists in the percolation threshold of PES-7 wt% graphite–MWCNT hybrid composites prepared by high speed mixing and ultrasonication of MWCNTs. The dispersion states in these two processes seems to be very different and hence the percolation threshold is oberved to be different. Thus, it is very important to select a suitable processing route.

In this article, though the electrical behavior of polyethersulfone (PES)-MWCNT composites has been presented, plenty of room still exists for simultaneously improving the electrical, mechanical and thermal properties of CPCs in general. There should be strategies developed in order to improve dispersion of fillers in various polymer matrices. Applications which employ conducting polymers, such as solar cells, require improvement in order to have an energy efficiency closer to that of silicon-based solar cells. In a similar fashion, much work needs to be carried out in the area of theoretical modeling of electrical conductivity in multicomponent polymer composite systems because, as of now, there are no universal models which can predict the electrical conductivity on a range of polymer-filler systems.

Acknowledgment

One of the authors BTSR thanks Amrita University, Coimbatore, Tamil Nadu and Navrachana University, Vadodara, Gujarat for the support and encouragement.

References

1. Market Resource Store, http://www.marketresearchstore.com/news/global-conductive-polymers-market-214.
2. Ramanujam, B.T.S., and Radhakrishnan, S., Polyphenylene sulfide-graphite hybrid composites: Charge transport and impedance characteristics. *Int. J. Plastics Technol.* 14, 37, 2010.
3. Nan, C.W., Physics of inhomogeneous inorganic materials. *Prog. Mater. Sci.* 37(1), 1, 1993.
4. Chatterjee, A., Alam, K., and Klein, P., Electrically conductive carbon nanofiber composites with high-density polyethylene and glass fibers. *Mater. Manuf. Process.* 22, 62, 2007.
5. Higgins, B.A., and Brittain, W.J., Polycarbonate carbon nanofiber composites. *Eur. Polym. J.* 41, 889, 2010.
6. Stankovich, S., Dikin, D.A., Dommett, G.H.B., Kohlhaas, K.M., Zimney, E.J., Stach, E.A., *et al.*, Graphene based composite materials. *Nature* 442, 282, 2006.
7. Potts, J.R., Dreyer, D.R., Bielawski, C.W., and Ruoff, R.S., Graphene based polymer composites. *Polymer* 52, 5, 2011.
8. Muller, M.T., Krause, B., Kretzschmar, B., and Potschke, P., Influence of feeding conditions in twin-screw extrusion of PP/MWCNT composites on electrical and mechanical properties. *Compos. Sci. Technol.* 71, 1535, 2011.
9. Krause, B., Boldt, R., Häußler, L., and Potschke, P., Ultralow percolation threshold in polyamide 6.6/MWCNT composites. *Compos. Sci. Technol.* 114, 119, 2015.
10. Ramanujam, B.T.S., and Radhakrishnan, S., Solution-blended polyethersulfone–graphite hybrid composites: Formation of nanographite and electrical characterization. *J. Thermoplast. Compos. Mater.* 28, 835, 2015.
11. Ramanujam, B.T.S., Radhakrishnan, S., and Deshpande, S.D., Polyphenylene sulfide- expanded graphite nanocomposites: Processing route dependent electrical percolation. *J. Thermoplast. Compos. Mater.* Published online before print May 9, 2016.
12. Chen, G., Weng, W., Wu, D., and Wu, C., PMMA/graphite nanosheets composite and its conducting properties. *Eur. Polym. J.* 39, 2329, 2003.
13. Yu, J., Lu, K., Sourty, E., Grossiord, N., Koning, C.E., and Loos, J., Characterization of conductive multiwall carbon nanotube/polystyrene composites prepared by latex technology. *Carbon* 45, 2897, 2007.

14. Celzard, A., McRae, E., Furdin, G., and Mareche, J.F., Conduction mechanisms in some graphite-polymer composites: The effect of a direct-current electric field. *J. Phys. Condens. Matter* 9, 2225,1997.
15. Dyre, J.C., and Schrøder, T.B., Universality of ac conduction in disordered solids. *Rev. Mod. Phys.* 72, 873, 2000.
16. Ramanujam, B.T.S., Mahale, R.Y., and Radhakrishnan, S., Polyethersulfone-expanded graphite nanocomposites: Charge transport and impedance characteristics. *Compos. Sci. Technol.* 70, 2111, 2010.
17. Mott, N.F., Electrons in disordered structures. *Adv. Phys.* 16, 49, 1967.
18. Mott, N.F., Conduction in non-crystalline solids. *Philos. Mag.* 19(160), 835, 1969.
19. Iijima, S., Helical microtubules of graphitic carbon. *Nature* 354, 56, 1991.
20. Ebbesen, T.W., Lezec, H.J., Hiura, H., Bennett, J.W., Ghaem, H.F., and Thio, T., Electrical conductivity of individual carbon nanotubes. *Nature* 382, 54, 1996.
21. Peng, B., *et al.*, Measurements of near-ultimate strength for multiwalled carbon nanotubes and irradiation-induced cross-linking improvements. *Nat. Nanotechnol.* 3, 626, 2008.
22. Wei, B.Q., Vajtai, R., and Ajayan, P.M., Reliability and current carrying capacity of carbon nanotubes. *Appl. Phys. Lett.* 79, 1172, 2001.
23. Thostenson, E.T., Ren, Z., and Chou, T-W., Advances in the science and technology of carbon nanotubes and their composites: A review. *Compos. Sci. Technol.* 61, 1899, 2001.
24. Ando, Y., and Zhao, X.L., Synthesis of carbon nanotubes by arc-discharge method. *New Diam. Front. C. Tec.* 16, 123, 2006.
25. Burch, H.J., *et al.*, Electrical conductance and breakdown in individual CNx multiwalled nanotubes. *Appl. Phys. Lett.* 89, 143110, 2006.
26. Wang, W.L., *et al.*, Direct synthesis of B-C-N single-walled nanotubes by bias-assisted hot filament chemical vapor deposition. *J. Am. Chem. Soc.* 128, 6530, 2006.
27. Bauhofer, W., and Kovacs, J.Z., A review and analysis of electrical percolation in carbon nanotube polymer composites. *Compos. Sci. Technol.* 69, 1486, 2009.
28. Cesano, F., Zaccone, M., *et al.*, Relationship between morphology and electrical properties in PP/MWCNT composites: Processing-induced anisotropic percolation threshold. *Mater. Chem. Phys.* 180, 284, 2016.
29. Park, S., *et al.*, Electrical properties of MWCNT-based composites with size-controllable catalyst supporters. *Synth. Met.* 217, 1, 2016.
30. Wang, H., Zheng, K., *et al.*, Segregated poly(vinylidene fluoride)/MWCNTs composites for high-performance electromagnetic interference shielding. *Compos. Part A* 90, 606, 2016.
31. Ram, R., Rahaman, M., and Khastgir, D., Electrical properties of polyvinylidene fluoride (PVDF)/multi-walled carbon nanotube (MWCNT) semi-transparent composites: Modelling of DC conductivity. *Compos. Part A* 69, 30, 2015.

32. Burmistrov, I., Gorshkov, N., *et al.*, Improvement of carbon black based polymer composite electrical conductivity with additions of MWCNT. *Compos. Sci. Technol.* 129, 79, 2016.
33. Pan, Y., and Li, L., Percolation and gel-like behavior of multiwalled carbon nanotube/polypropylene composites influenced by nanotube aspect ratio. *Polymer* 54, 1218, 2013.
34. Kara, S., Arda, E., Dolastir, F., Pecken, O.J., Electrical and optical percolations of polystyrene latex–multiwalled carbon nanotube composites. *Colloid Interface Sci.* 344, 395, 2010.
35. Capriano, B.H., Kota, A.K., Gershon, A.L., Laskowski, C.J., Kashiwagi, T., Bruck, H.A., and Raghavan, S.R., Conductivity enhancement of carbon nanotube and nanofiber-based polymer nanocomposites by melt annealing. *Polymer* 49, 4846, 2008.
36. Bouchard, J., Cayla, A., Devaux, E., and Campagne, C., Electrical and thermal conductivities of multiwalled carbon nanotubes-reinforced high performance polymer nanocomposites. *Compos. Sci. Technol.* 86, 177, 2013.

8

Polyaniline-Based Sensors for Monitoring and Detection of Ammonia and Carbon Monoxide Gases

Neha Patni*, Neha Jain and Shibu G. Pillai

Department of Chemical Engineering, Institute of Technology, Nirma University, India

Abstract

Gas sensor devices normally use metal oxide semiconductors requiring high operating temperature, due to which the sensor often suffers from instability, response variation, poor selectivity and low sensitivity. Due to these limitations, intrinsically conducting polymers, such as polyaniline (PANI), polypyrrole, polythiophenes and their derivatives, are progressively being used as sensing layers operating at room temperature. PANI, in particular, has been extensively used due to its ease of fabrication, high stability, reversible doping/dedoping chemistry, stable electrical conduction mechanism, and high environmental stability. This chapter gives an overview of the detection of different kinds of gases, specifically ammonia and carbon monoxide, using polyaniline sensors.

Keywords: Sensor, gas, ammonia, carbon monoxide

8.1 Introduction

Over the past 20 years, extensive innovative research in the field of sensors has greatly expanded in terms of the number of financial investments, published writings, and dynamic specialists. An essential component of a sensor is to give data on our physical state, organic substances and the environment, and also for spill identification. Substance sensors play a vital part in observing the earth, measuring oxygen consumption, providing data on

**Corresponding author*: neha.patni@nirmauni.ac.in

Sanjay K. Nayak *et al.* (eds.) Trends and Applications in Advanced Polymeric Materials, (145–162) 2018 © Scrivener Publishing LLC

mechanical procedures/assemblies, release recognition, detecting burnable/combustible gases, locating structural damage, controlling food quality and detecting water pollutants from modern effluents and agribusiness/farm field runoffs and so on. A sensor is a gadget or instrument that changes synthetic data, running from the grouping of a particular example segment to aggregate organization investigation, into a systematic helpful sign [1].

Sensor contains two essential parts, a receptor and transducer. In the receptor part of a sensor the compound data is changed into a type of vitality which might be measured by the transducer. And this transducer part is equipped for changing the vitality conveying the synthetic data about the specimen into a helpful systematic sign. For manufacture of synthetic sensor, diverse materials are utilized, for example, semiconductors, separators, strong electrolytes, metals and synergist materials. Synthetic and physical properties of polymers might be customized by scientific experts for specific needs, who handpick significant properties in the development of sensor gadgets.

Gas sensors are essential for various industrial applications such as toxic gas leakage, oxygen depletion, environmental pollution detection, explosive detectors, combustible/flammable gases, petrochemical processing industries, automotive industry, mining, tunneling and offshore industries, etc. [2–5].

There are many different types of sensors, for example, solid state gas sensors [6], conducting polymer gas sensors using, e.g., polyaniline [7], mixed oxide gas sensors [8], ampere metric gas sensors [9], catalytic field-effect devices [10], or gas sensors. Here we are mainly focusing on polyaniline-based ammonia and carbon monoxide sensors.

8.2 Conducting Polymers

A conducting polymer is a polymer that is organic in nature and can act as a conductor or a semiconductor. It is a new kind of material with tremendous technological application in a growing amount of new technologies such as energy storage, molecular recognition, electromagnetic interference shielding and optoelectronic devices. All conducting polymers display highly reversible redox behavior with a unique chemical memory and hence are clearly noted as new materials for the manufacture of chemical sensors [11].

Conducting polymers (as shown in Figure 8.1) are generally used as sensitive layers in chemical microsensors. Polymers such as polypyrrole (PPy), polythiophene, polyindol, polyaniline (PANI), poly(3,4-ethylene-dioxythiophene), polyacetylene, polyfuran and their derivatives are

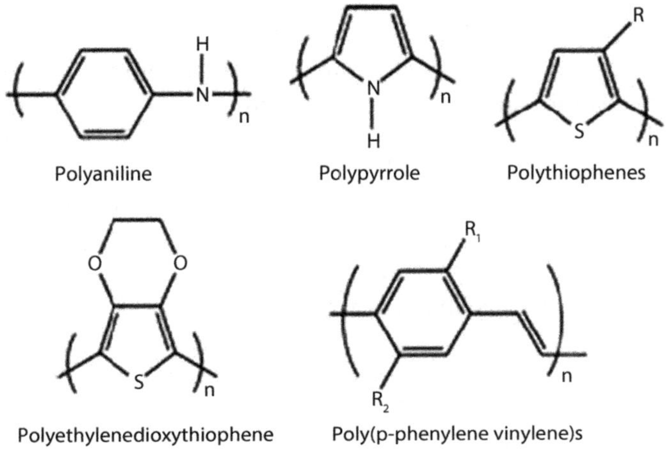

Figure 8.1 Molecular structures of conducting polymers.

extensively used as sensitive materials for conductometric polymer sensors [12].

There are two methods by which conducting polymers can be synthesized: oxidative polymerization or electrochemical polymerization. By passing electricity through a solution of constant current or varying potential, electrosynthesis of conducting polymers takes place. Conducting polymers possess a unique property in that they can be tempered by an electric voltage. Mostly alteration of the electrical property of polymer takes place due to change in color. For example, in the case of polyaniline, its non-conducting and conducting forms display different colors. The delocalized electronic states of the polymer molecules are responsible for this electrochromic behavior. A number of advantages of the conducting polymers are: simplicity of technology, nonmetallic temperature dependence of conductivity, high performance, light weight, low cost, flexible, fast response, simple fabrication process, and can be operated at room temperature, etc. [13–14]. There are many applications of conducting polymers in the various fields of chemical sensors. The majority of sensor devices utilize a number of conducting polymers with definite roles through immobilizing the species or in the sensing mechanism.

8.2.1 Polyaniline

The German chemist Friedlieb Ferdinand Runge was the first to discover polyaniline (PANI) in 1834. It is a mixed oxidation state polymer which contains two units: reduced benzoid unit and oxidized quinoid unit [15].

8.2.1.1 Structure of Polyaniline

As previously mentioned, PANI is a mixed oxidation state polymer consisting of benzenoid and oxidized quinonoid units. It has three dissimilar PANI oxidation states, as shown in Figure 8.2 [16]. PANI is present as fully reduced leucoemeraldine (LE) where $1-y = 0$, fully oxidized pernigraniline (PE) where$1-y = 1$ and half oxidized emeraldine base (EB) where $1-y = 0.5$. The EB is composed of two benzenoid units and one quinonoid unit that alternate and has been demonstrated to be the most beneficial form of PANI due to its high stability at room temperature. It is known to be a semiconductor [17].

8.2.1.2 Properties of Polyaniline

Polyaniline is especially attractive because of its properties. The converting properties of PANI (e.g., optical, conductive, and solubility) separate it from other conducting polymers.

8.2.1.2.1 Conductivity Properties of PANI

Polyaniline has many attractive properties that differentiate it from other conducting polymers. For example, the conductivity of PANI varies with the extent of oxidation (variation in the number of electrons) and the degree of protonation (variation in the number of protons), as shown in Figure 8.3.

8.2.1.2.2 Solubility Properties of PANI

The emeraldine base is a less ionic state of PANI. As there are no cationic charges present in the polymer backbone in this state, it is soluble in a number of organic solvents such as dimethyl sulfoxide, chloroform, tetrahydrofuran, dimethylformamide, and methylpyrrolidinone. However,

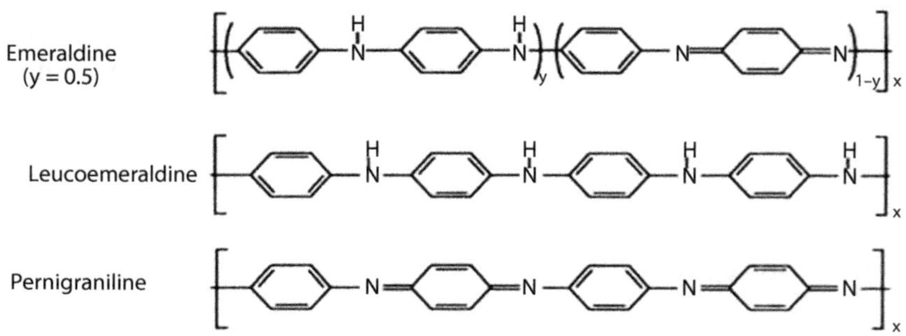

Figure 8.2 Three oxidation states of polyaniline.

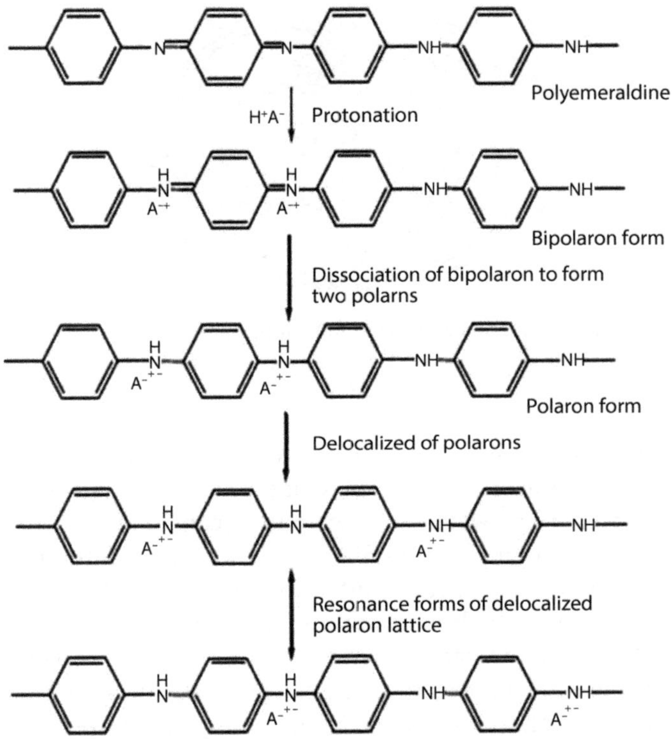

Figure 8.3 The doping of EB with protons to form the conducting emeraldine salt (PANI/HA) form of polyaniline [18].

solubility can also be improved by introducing functionalized protonic acid dopants such as sulfonic acids into the ES salt [19].

8.3 Ammonia Detection

8.3.1 Sources of Ammonia

Ammonia is a natural gas that is present everywhere in the environment from low ppb to sub-ppb level. Figure 8.4 shows the approximate yearly ammonia deposition rate worldwide, showing a maximum deposition in Central and Western Europe [20].

Nowadays, major emission of ammonia into our atmosphere is directly or indirectly caused by humans. The worldwide emission of ammonia per year was estimated in 1980 by the Commission of the European Communities for Environment and Quality of Life to be 20–30 Tg [21].

Figure 8.4 Annual ammonium deposition (100 mg/m^2) [20].

Ammonia is generally used in many industries like pulp and paper, petrochemical, fertilizer and oil. There are a number of small sources of ammonia (e.g., surface water, soil, sea and ocean water) which act as a huge source of ammonia. Ammonia is produced due to the presence of ammonium ions which are transmitted to the gaseous ammonia by alkaline rainwater [22]. Anhydrous ammonia (NH_3) is also very broadly used as a coolant in large industrial refrigeration systems. As a matter of fact, the ammonia used as a refrigerant has increased significantly over the last several years as an alternative to environmentally unfavorable chlorofluorocarbon refrigerants. Moreover, apart from being used as a standard in the food or beverage industry, ammonia refrigeration is now also found in pharmaceutical production, air-conditioning equipment for many public buildings and electric power generation plants [23].

Ammonia is correctly stated to be a highly hazardous chemical. Exposure to only moderate concentrations of ammonia (50–150 ppm) can cause eye, throat and skin irritation. It may affect the bloodstream and successive disturbances of oxygen uptake by hemoglobin after prolonged exposure. At high concentrations (>150 ppm) ammonia can scar lung tissue, causing pulmonary edema and inflammation of lung tissue. According to the European Environment Agency, the ammonia/ammonium nitrate explosion which occurred on September 21, 2001, at a fertilizer plant in Toulouse, France, is considered to have been the worst industrial accident of the last twenty years. It killed 30 people and more than 2,000 workers and nearby residents were injured. According to the government investigation, it would have been worse if intervening buildings had not broken the force of the explosion and prevented the possible detonation of 20 more railroad tank cars filled with anhydrous ammonia [24].

Table 8.1. Sensors used for detection of ammonia gas.

Type of sensor	Material used	Features	Ref.
Chemical sensor	Polyaniline (emeraldine base)	Nanocomposite ultrathin films of polyaniline and isopolymolybdic acid	[26]
Gas sensor	PANI–PMMA composite	Sensitivity of PANI-PMMA coatings are 10–4000 ppm and reversible response	[27]
Gas sensor	PPY–PVA composite	As NH_3 concentration increases, resistance increases but becomes irreversible beyond 10% NH_3 up to 100 ppm NH_3	[28]
Gas sensor	Acrylic acid doped polyaniline	Highly sensitive to even 1 ppm of NH_3 at room temperature and shows steady responses up to 120 day	[29]
Gas sensor	Polypyrrole	Response time 20 s; recovery time 60 s	[30]
Gas sensor	PANI–isopolymolybdic acid nanocomposite	Resistance increases with NH_3 concentration and it is reversible up to 100 ppm NH_3	[31]
Gas sensor	ZnO/SiO_2 bilayer nanofilms on ST-cut quartz (surface acoustic waves)	Exhibits a frequency of 2000 Hz in 300 ppm ammonia along with good stability and repeatability	[32]
Fiber optic ammonia gas sensor	Surface plasmon resonance using ITO (In_2O_3 + SnO_2) in the ratio 90:10 and polyaniline	As the concentration of ammonia gas increases, resonance wavelength increases up to a certain concentration; better sensitivity	[33]
Gas sensor	Tellurium thin film based	Operable at room temperature and linear response in the range of 0–100 ppm.	[34]

(*Continued*)

Table 8.1 Cont.

Type of sensor	Material used	Features	Ref.
Gas sensor	Multilayer graphene based	Highly sensitive, flexible and low cost; detects ammonia concentration as low as 430 ppb	[35]
Gas sensor	Carbon nanoflake (CNFL)/tin oxide composite	Having 10% of SnO_2 showed 3 times higher sensor response and better repeatability than the gas sensor using pristine SnO_2 nanoparticles at the optimal temperature of 350 °C.	[36]
Surface acoustic wave sensor	Based on Co_3O_4/SiO_2 composite sensing films	Displayed the best sensing property; highest RMS value (3.72); shows a positive frequency shift of 3500 Hz to 1 ppm ammonia as well as excellent selectivity, stability and reproducibility at room temperature	[37]
Gas sensor	Carbon nanotubes (CNTs) grown on silicon substrate were merged	Detects ammonia with a linear response over the entire range from 1–160 ppm ammonia in air	[38]

Figure 8.5 Polyaniline strips.

There are many different ways to detect ammonia. The human nose is very sensitive to ammonia and ammonia gas has a very powerful odor, so high concentrations are very easy to detect [25]. For very low concentrations of ammonia, we have different types of chemical sensors, as shown in Table 8.1.

Table 8.1 gives an insight into different materials that can be used as sensors for detection of ammonia gas based on different working principles. Conducting polymers displayed promising applications for sensing ammonia gas.

Conducting polymer composites with other polymers and doped with other materials, such as PVC, PMMA, PANI, etc., have very good sensitivity towards ammonia at very low levels. Here we are only focusing on polyaniline because of its attractive properties.

8.3.2 Experiment: Ammonia Sensor

In this experiment we have used polyaniline to sense ammonia solution. Because polyaniline (PANI) has high conductivity compared to other polymers, we used it as a conductor and introduced ammonia around the PANI film to measure the change in resistance (based on the fact that due to interaction of PANI with chemicals, its resistivity changes).

Procedure:

- Take a Whatman filter paper and make a strip from it.
- Make a film of polyaniline on the filter paper using a glass rod (Figure 8.5).
- Dry it using a dryer for 25 minutes.
- Measure the resistance using a multimeter.
- Bring ammonia solution near the PANI film and check the resistivity of the film.

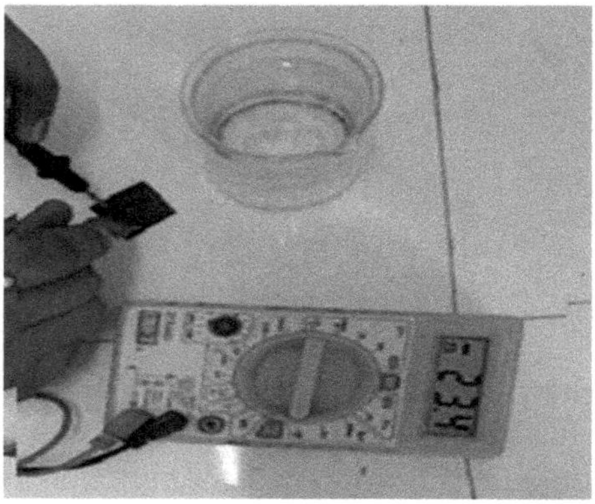

Figure 8.6 Varying voltage with ammonia concentration.

So, from this experiment we have concluded that PANI film can be used as sensing layer for ammonia. Potential changes as it senses the ammonia solution from 9 to 23.4 mV (Figure 8.6). So, from our tentative experiment we can conclude that PANI can be used as a sensing material for detection of ammonia.

8.4 Carbon Monoxide (CO) Detection

Carbon monoxide (CO) is an odorless, tasteless, highly toxic gas that is produced from the incomplete combustion of fossil fuels. It is frequently called "the silent killer," as it is almost impossible to detect without sensing technology [38].

8.4.1 Common Sources of CO

Carbon monoxide is produced from the incomplete combustion of fossil fuels. Improperly working machines and heating equipment, or limited venting during their operation, leads to the buildup of dangerous carbon monoxide in a given environment [39]. Primary sources of CO are heating systems, power tools, gas ranges/ovens, camp stoves/lanterns, charcoal grills or other charcoal sources. Other common sources of CO are gas water heaters, leaking, broken, corroded or disconnected flue/vent pipes, cracked heat exchangers, gas or wood burning fireplaces, etc.

8.4.2 Sensors Used for Detection of CO

Carbon monoxide is an extremely toxic gas that conglomerates with hemoglobin to produce carboxyhemoglobin, which affects the delivery of oxygen to bodily tissues. There are different types of sensors used for detection of CO that monitor the amount of CO in the air over a given time period. The three basic types of sensors used for CO detection are [40]:

a. Metal oxide semiconductor sensors;
b. Electrochemical sensors;
c. Biomimetic sensors.

Metal oxide semiconductor sensors: In metal oxide semiconductor (MOS) sensors, generally tin dioxide semiconductor is heated by an electric current at interrupted intervals. When its operating temperature is reached, the presence of carbon monoxide can change its resistance. Once the change in the resistance reaches its threshold, an alarm sounds. These types of sensors have a long lifespan and quick response but they are more expensive compare to electrochemical and biomimetic sensors [41].

Electrochemical sensors: In electrochemical sensors, which have platinum electrodes immersed in an electrolyte solution, an electric current is generated due to a reaction between carbon monoxide and the oxygen in the air. The alarm rings when carbon monoxide is present over some time and the current crosses the threshold limit. These sensors have a low current draw, quick response, high sensitivity, etc. [42].

Biomimetic sensors: Biomimetic sensors try to chemically mimic the effect that carbon monoxide has on hemoglobin. These sensors contain a gel-coated disc which will change color and darken in the presence of CO and then recognizes the color changes and sets off an alarm. These sensors are inexpensive, have a shorter life span, require very little electricity and have low reset capability, as shown in Table 8.2 [43].

A metal oxide-based sensor commonly consists of a semiconducting sensitive layer, an electrical connection and a heater [53].

A reactive molecule is chemisorbed on the metal-oxide surface and charge transfer takes place. When a metal-oxide crystal such as SnO_2 is heated at a certain high temperature in air, oxygen is adsorbed on the crystal surface, and a surface potential is made that hinders electron flow. As the surface is exposed to oxidizable gases, such as hydrogen, methane, and carbon monoxide, there is a decrease in the surface potential and increase in the conductivity [54]. Similarly, as the concentration of the target chemical increases, the magnitude of the change in resistance also

Table 8.2 Sensors used for detection of carbon monoxide.

Type of sensor	Material used	Features	Ref.
Semiconductor sensor	Nanostructured Au-doped CoOOH films	Good sensitivity to CO in the temperature range of 60–110 °C with the maximum CO response near 80 °C	[44]
Gas sensor	Dirhodium (II,II) based	The response–concentration relationship is linear in a wide range of CO concentrations (100–50,000 ppm)	[45]
Semiconductor Sensor	Tin oxide nanowire based	The nanowires act as sensitive, fast, stable, and reproducible gas sensors that can be easily integrated into a multicomponent array	[46]
Metal oxide gas sensor	In_2O_3 with the addition of alkali metal carbonates (Rb_2CO_3)	Detection of CO at level from 500–4000 ppm in wet air at 300 °C	[47]
Gas sensor	SnO_2/Ti thick film based	Ti-doped SnO_2 having higher sensitivity towards CO	[48]
Chemiresistive gas sensor	Template-free electrochemical polymerization of aniline on an interdigitated electrode	PANI nanosensor had an excellent responding ability on CO in hydrogen with a concentration as low as 1 ppm	[49]
Single-nanowire sensor array	Single-nanowire array consisting of four different materials, metal (palladium), conducting polymer (polypyrrole and polyaniline), and semiconductor (zinc oxide) on a chip via a site-specific electrochemical deposition process	Four different target gases, including H_2, CH_3OH, CO and NO_2, were selected for identification; with the help of PCA, the sensing patterns for all four target gases were successfully built up without overlap even at extremely low concentration levels (625 ppb)	[50]

Single nanowire-based nanosensor	Schottky barrier height tuned through the responsive variation of the surface chemisorbed gases and the amplification role played by the nanowire to Schottky barrier effect	Ultrahigh sensitivity of 32,000% was achieved using the Schottky contacted device operated in reverse bias mode at 275 °C for detection of 400 ppm CO, which is 4 orders of magnitude higher than that obtained using an Ohmic contact device under the same conditions. In addition, the response time and reset time were shortened by a factor of 7.	[51]
PANI nanocomposite thin film	Polyaniline (PANI), polyaniline/titanium dioxide (PANI/TiO_2), polyaniline/tin oxide (PANI/SnO_2) and polyaniline/indium oxide (PANI/In_2O_3) thin films were developed by using an *in-situ* self-assembly method at ~10 °C	Responses of PANI and its nanocomposite thin-film sensors to 23×10^{-6} CO in comparison with NH_3 gas sensitivity show less sensitivity to CO	[52]

increases. The relationship between the film's electrical resistance and a given oxidizable gas's concentration is described by the following empirical equation:

$$R_s = A[C]^{-\alpha} \tag{8.1}$$

where R_s is the sensor electrical resistance, A is a constant specific for a given film composition, C is the gas concentration, and α is the characteristic slope of R_s curve for that material and expected gas [55–56].

Carbon monoxide sensitivity depends upon the type of material, doping and de-doping chemistry, etc. So the improvement of CO sensitivity can be increased by:

a. Using nanostructure Au-doped COOH-SE. This can be operated at low temperature; the best range of CO was found at 80 °C;
b. Using a dirhodium (II,II) square complex molecular box as a semiconducting material for CO;
c. Using highly uniform SnO_2 nanowires, with bulk electronic properties directed by their surface chemistry. The nanowires act as sensitive, fast, stable, and reproducible gas sensors that can be easily integrated into a multicomponent array;
d. Using In_2O_3 for the detection of carbon monoxide gas. When In_2O_3 is added with alkali metal carbonates such as Rb_2CO, its sensitivity increases towards the CO. By using this combination CO can be detected in the range of 500–4000 ppm in wet air temperature at 300 °C;
e. Using SnO_2/Ti thick film-based sensor.

Due to their unique sensing mechanism and excellent sensing performance, the developed nanosensors are seen as very promising for application in a new type of sensor to monitor CO concentration.

8.5 Conclusion

Chemical sensors have received remarkable consideration in recent years. Ammonia and carbon monoxide gases are both hazardous in nature and represent the most harmful pollutants because of their interference with the mechanism of oxygen transport and storage in living systems. So the detection of concentration of these gases is very important. In this chapter we have reviewed the different types of sensors for detection of ammonia and carbon monoxide.

From our experiment, we concluded that polyaniline can be used for fabrication of ammonia sensors because of its tremendous technological potential. Mainly sensor properties depend upon the type of substance, doping and de-doping chemistry. Depending upon the type of material the sensing properties, such as response/recovery rate, long-term stability and selectivity, change. All the sensors have a different range of measurement of concentration of ammonia and carbon monoxide.

References

1. Adhikari, B., and Majumdar, S., Polymers in sensor applications. *Prog. Polym. Sci.* 29, 699–766, 2004.
2. Kohl, D., Function and applications of gas sensors. *J. Phys. D: Appl. Phys.* 34, R125, 2001.
3. Ampuero, S., and Bosset, J.O., The electronic nose applied to dairy products: A review. *Sens. Actuators B* 94, 1–12, 2003.
4. Nicolas, D., and Candel, S., Combustion control and sensors: A review. *Prog. Energy Combust. Sci.* 28, 107–150, 2002.
5. Riegel, J., Neumann, H., and Wiedenmann, H.M., Exhaust gas sensors for automotive emission control. *Solid State Ionics* 152–153, 783–800, 2002.
6. Dubbe, A., Fundamentals of solid state ionic micro gas sensors. *Sens. Actuators B* 88, 138–148, 2003.
7. Nicolas-Debarnot, D., and Poncin-Epaillard, F., Polyaniline as a new sensitive layer for gas sensors—a review. *Anal. Chim. Acta* 475, 1–15, 2003.
8. Zakrzewska, K., Mixed oxides as gas sensors. *Thin Solid Films* 391, 229–238, 2001.
9. Chang, S.C., Stetter, J.R., and Cha, C.S., Amperometric gas sensors—a review. *Talanta* 40, 461–477, 1993.
10. Lundstro, I., Sevensson, C., Spetz, A., Sundgren, H., and Winquist, F., From hydrogen sensors to olfactory images—twenty years with catalytic field-effect devices. *Sens. Actuators B* 13, 16–23, 1993.
11. Dhawan, S.K., Kumar, D., Ram, M.K., Chandra, S., and Trivedi, D.C., Application of conducting polyaniline as sensor material for ammonia. *Sens. Actuators B* 40, 99–103, 1997.
12. Ruckenstein, E., and Park, J.S., New method for the preparation of thick conducting polymer composites. *J. Appl. Poly. Sci.* 42, 925–934, 1991.
13. Albert, K.J., Lewis, N.S., Schauer, C.L., Sotzing, G.A., Stitzel, S.E., Vaid, T.P., and Walt, D.R., Cross-reactive chemical sensor arrays. *Chem. Rev.* 100, 2595–2626, 2000.
14. Harsanyi, G., Polymer films in sensor application: A review of present uses and future possibilities. *Sens. Rev.* 20, 98–105, 2000.
15. Molapo, K.M., Ndangili, P.M., Ajayi, R.F., Mbambisa, G., Mailu, S.M., Njomo, N., Masikini, M., Baker, P., and Iwuoha, E.I., Electronics of conjugated polymers (I): Polyaniline. *Int. J. Electrochem. Sci.* 7, 11859–11875, 2012.

16. Maddanimath, T., Mulla, I.S., Sainkar, S.R., Vijayamohanan, K., Shaikh, K.I., Patil, A.S., and Vernekar, S.P., Humidity sensing properties of surface functionalised polyethylene and polypropylene films. *Sens. Actuators B* 81, 141–151, 2002.
17. Masdarolomoor, F., Novel nanostructured conducting polymer systems based on sulfonated polyaniline, Doctral dissertation, Department of Chemistry, University of Wollongong, Australia, 2006.
18. Wallace, G.G., Spinks, G.M., Kane-Maguire, L.A.P., and Teasdale, P.R., *Conductive Electroactive Polymers: Intelligent Polymer Systems*, CRC Press: London, 2009.
19. Stafström, S., Brédas, J.L., Epstein, A.J., Woo, H.S., Tanner, D.B., Huangand, W.S., and MacDiarmid, A.G., Polaron lattice in highly conducting polyaniline: Theoretical and optical studies. *Phys. Rev. Lett.* 59, 1464–1467, 1987.
20. Warneck, P., *Chemistry of the Natural Atmosphere*, Academic Press Inc., 1998.
21. Istas, J.R., de Borger, R., de Temmerman, L., Guns, Meeus-Verdinne, K., Ronse, A., Scokart, P., and Termonia, M., Effect of ammonia on the acidification of the environment, European Communities Report No. EUR11857EN, 1988.
22. Lee, D.S., Halliwell, C., Garland, J.A., Dollard, G.J., and Kingdon, R.D., Exchange of ammonia at the sea surface—a preliminary study. *Atmos. Environ.* 32, 431–439, 1998.
23. Barrett, K., Oceanic ammonia emissions in Europe and their boundary fluxes. *Atmos. Environ.* 32, 381–391, 1998.
24. Sensidyne, www.sensidynegasdetection.com.
25. Erisman, J.W., Otjes, R., Hensen, A., Jongejan, P., Bulk, P.V.D., Khlystov, A., Mols, H., and Slanina, S., Instrument development and application in studies and monitoring of ambient ammonia. *Atmos. Environ.* 35, 1913–1922, 2001.
26. Li, D., Jiang, Y., Wu, Z., Chen, X., and Li, Y., Self assembly of polyaniline ultrathin films based on doping-induced deposition effect and applications for chemical sensors. *Sens. Actuators B.* 36, 125–127, 2000.
27. Nicho, M.E., Trejo, M., García-Valenzuela, A., and Saniger, J.M., Palacios J., Hu, H., Polyaniline composite coatings interrogated by nulling optical-transmittance bridge for sensing low concentrations of ammonia gas. *Sens. Actuators B* 76, 18–24, 2001.
28. Gangopadhyay, R., and De, A. Conducting polymer composites: Noble materials for gas sensing. *Sens. Actuators B* 77, 326–329, 2001.
29. Chabukswar, V.V., Pethkar, S., and Athawale, A.A., Acrylic acid doped polyaniline as an ammonia sensor. *Sens. Actuators B* 77, 657–663, 2001.
30. Yadong, J., Tao, W., Zhiming, W., Dan, L., Xiangdong, C., and Dan, X., Study on the NH3-gas sensitive properties and sensitive mechanism of polypyrrole. *Sens. Actuators B* 66, 280–282, 2000.
31. Tang, Y.-L., Li, Z.-J., Ma, J.-Y., Guo, Y.-J., Fu, Y.-Q., and Zu, X.-T., Ammonia gas sensor based on ZnO/SiO_2 bi-layer nanofilms on ST-cut quartz surface acoustic wave devices. *Sens. Actuators B* 201, 114–121, 2014

32. Mishra, S.K., Kumari, D., and Gupta, B.D., Surface plasmon resonance based fiber optic ammonia gas sensor using ITO and polyaniline. *Sens. Actuators B* 171–172, 976–983, 2012.
33. Sen, S., Muthe, K.P., Joshi, N., Gadkari, S.C., Gupta, S.K., Jagannath, Yakhmi, J.V., Roy, M., and Deshpande, S.K., Room temperature operating NH_3 based on tellurium thin film. *BARC Newsletter* issue no. 249.
34. Ghosh, R., Singh, A., Santra, S., Ray, S.K., Chandra, A., and Guha, P.K., Highly sensitive large-area multi-layered graphene-based flexible ammonia sensor. *Sens. Actuators B* 205, 67–73, 2014.
35. Lee, S.-K., Chang, D., and Kim, S.W., Gas sensor based on carbon nanoflake/tin oxide composities for ammonia detection. *J. Hazard. Mater.* 268, 110–114, 2014.
36. Tang, Y.-L., Li, Z.-J., Ma, J.-Y., Su, H.-Q., Guo, Y.-J., Wang, L., Du, B., Chen, J.-J., Zhou, W., Yu, Q.-K., and Zu, X.-T., Highly sensitive room-temperature surface acoustic wave (SAW) ammonia sensors based on Co_3O_4/SiO_2 composite films. *J. Hazard. Mater.* 280, 127–133, 2014.
37. Huang, J., Wang, J., Gu, C., Yu, K., Meng, F., and Liu, J., A novel highly sensitive gas ionization sensor for ammonia detection. *Sens. Actuators A* 150, 218–223, 2009.
38. Stewart, R.D., The effect of carbon monoxide on humans. *Ann. Rev. Pharmacol.* 15, 409–423, 1975.
39. Esfandyarpour, B., Mohajerzadeh, S., Famini, S., Khodadadi, A., and Asl Soleimani, E., High sensitivity Pt-doped SnO_2 gas sensors fabricated using sol–gel solution on micromachined (1 0 0) Si substrates. *Sens. Actuators B* 100, 190–194, 2004.
40. Gundal, L.A., Apte, M.G., and Nematollahi, A.R., Carbon Monoxide Detection Technology Comparision: Response to various gas, Berkeley Laboratory Report, 1998.
41. Wang, C., Yin, L., Zhang, L., Xiang, D., and Gao, R., Metal oxide gas sensor: Sensitivity and ınfluencing factors. *Sensors* 10, 2088–2106, 2010.
42. Warburton, P.R., Pagano, M.P., Hoover, R., Logman, M., Crytzer, K., and Warburton, Y.J., Amperometric gas sensor response times. *Anal. Chem.* 70, 998–1006, 1998.
43. Haupt, K., and Mosbach, K., Molecularly imprinted polymers and their use in biomimetic sensors. *Chem. Rev.* 100, 2495–2504, 2000.
44. Zhuiykov, S., Carbon monoxide detection at low temperatures by semiconductor sensor with nanostructured Au-doped CoOOH films. *Sensors Actuators B* 129, 431–441, 2008.
45. Lo Schiavo, S., Piraino, P., Bonavita, A., Micali, G., Rizzo, G., and Neri, G., A dirhodium(II,II) molecular species as a candidate material for resistive carbon monoxide gas sensors. *Sens. Actuators B* 129, 772–778, 2008.
46. Kolmakov, A., Zhang, Y., Cheng, G., and Moskovits, M., Detection of CO and O_2 using tin oxide nanowires. *Adv. Mater.* 15, 997–1000, 2003.

47. Yamaura, H., Jinkawa, T., Tamaki, J., Moriya, K., Miura, N., and Yamazoe, N., Indium oxide-based gas sensor for selective detection of CO. *Sens. Actuators B* 36, 325–332, 1996.
48. Zeng, W., Li, Y., Miao, B., Lin, L., and Wang, Z., Recognition of carbon monoxide with SnO_2/Ti thick-film sensor and its gas-sensing mechanism. *Sens. Actuators B* 191, 1–8, 2014.
49. Liu, C., Noda, Z., Sasaki, K., and Hayashi, K., Development of a polyaniline nanofiber-based carbon monoxide sensor for hydrogen fuel cell application. *Int. J. Hydrogen Energy* 37, 13529–13535, 2012.
50. Hu, Y., Lee, H., Kim, S., and Yun, M., A highly selective chemical sensor array based on nanowire/nanostructure for gas identification. *Sensors Actuators B* 18, 424–431, 2013.
51. Wei, T.-Y., Yeh, P.-H., Lu, S.-Y., and Wang, Z.L., Gigantic enhancement in sensitivity using Schottky contacted nanowire nanosensor. *J. Am. Chem. Soc.* 131, 17690–17695, 2009.
52. Tai, H., Jiang, Y., Xie, G., and Yu, J., Preparation, characterization and comparative NH3-sensing characteristic studies of PANI/inorganic oxides nanocomposite thin films. *J. Mater. Sci. Technol.* 26, 605–613, 2010.
53. Madou, M.J., and Morrison, S.R., *Chemical Sensing with Solid State Devices*, Academic Press: New York, 1989.
54. Wolfrum, E.J., Meglen, R.M., Peterson, D., and Sluiter, J., Metal oxide sensor arrays for the detection, differentiation, and quantification of volatile organic compounds at sub-parts per-million concentration levels. *Sens Actuators B* 115, 322–329, 2006.
55. Persaud, K., and Dodd, G., Analysis of discrimination mechanisms in the mammalian olfactory system using a model nose. Nature 299, 352–355, 1982.
56. Fraden, J., *Handbook of Modern Sensors*, pp. 569–605, Springer Science+ Business Media, LLC, 2010.

9

Synthesis and Characterization of Luminescent $La_2Zr_2O_7/Sm^{3+}$ Polymer Nanocomposites

Pramod Halappa* and C. Shivakumara*

Solid State and Structural Chemistry Unit, Indian Institute of Science, Bangalore, India

Abstract

Luminescence is one of the most important properties of nanoparticles. Applications are found largely in medical, biological or pharmaceutical areas. In general, luminescence is a property of aromatic organic moieties or single isolated semiconducting nanoparticles. Most of the earlier materials are toxic and carcinogenic and show limited thermodynamic stability against oxidation. These severe disadvantages make application difficult. Therefore, the search for nontoxic luminescent oxide nanoparticles or other rare earth doped oxides is underway. Looking at potential applications, there is a strong need for nontoxic oxide luminescent nanocomposites that are stable in ambient conditions. Luminescence of rare earth oxide nanoparticles is subject to rapid aging caused by formation of hydroxides at the surface of oxide nanoparticles, quenching luminescence. Coating the surface of the particles with a polymer protects the surface of the oxide nanoparticles against ambient air, and even water, thereby avoiding or at least reducing this problem. Additionally, polymer-coated oxide nanoparticles can be suspended in water, especially in biological applications. In this chapter, high surface area $La_2Zr_2O_7:Sm^{3+}$ phosphors were prepared by the nitrate - citrate gel combustion method and polymer composites were made by sonication methods. These compounds were characterized using powder X-ray diffraction, UV-visible spectroscopy, thermogravimetric analysis and photoluminescence technique. Both CIE chromaticity diagram and CCT values confirmed that these phosphors can be

**Corresponding authors*: pramochem@gmail.com; shiva@sscu.iisc.ernet.in

Sanjay K. Nayak *et al.* (eds.) Trends and Applications in Advanced Polymeric Materials, (163–190) 2018 © Scrivener Publishing LLC

useful in the fabrication of red component in white light-emitting diodes (WLEDs) for displays and other optical device applications.

***Keywords*:** Photoluminescence, phosphor, polymer nanocomposites, samarium, CIE chromaticity

9.1 Introduction

In the past few years, nanomaterials have become one of the thrust areas of current research in materials science and technology. It studies materials with morphological features on the nanoscale, and especially those that have special properties stemming from their nanoscale dimensions. Nanoscale is usually defined as smaller than one-tenth of a micrometer in at least one dimension, though this term is sometimes also used for materials smaller than one micrometer. These materials may be metals, semiconductors, metal oxides, organic materials or biomaterials. Thus, there is a tremendous scope for the design of new materials with unusual properties. Amongst the various types of nanomaterials, semiconductor nanoparticles have been widely investigated and exploited for various applications ranging from energy conversion to medicine. Semiconductors possess interesting and important optical and electronic properties useful for diverse technologies, including microelectronics, detectors, sensors, lasers and photovoltaics [1–5].

Nanostructures and nanomaterials possess a large fraction of surface atoms per unit volume, which makes new quantum mechanical effects possible. One of the most fascinating and useful aspects of nanomaterials is their optical properties. Applications based on optical properties of nanomaterials include optical detector, laser, sensor, imaging, phosphor, display, solar cell, photocatalysis, photoelectrochemistry and biomedicine. The optical properties of nanomaterials depend on parameters such as feature size, shape, surface characteristics, and other variables, including doping and interaction with the surrounding environment or other nanostructures. The simplest example is the well-known blue shift of absorption and photoluminescence spectra of semiconductor nanoparticles with decreasing particle size, particularly when the size is small enough. Many of the underlying principles are similar in these different technological applications that span a variety of traditional disciplines including chemistry, physics, biology, medicine, materials science and engineering, electrical and computer science and engineering.

9.1.1 Luminescence

Luminescence is an emission of light by a substance; all forms of light emission that derive their energy from other sources are termed luminescence. It can be caused by chemical reaction, electrical energy, subatomic motions or stress on a crystal. In luminescence, some energy source kicks an electron of an atom out of its lowest energy "ground" state into a higher energy "excited" state. Then the electron returns the energy in the form of light so it can fall back to its "ground" state. With few exceptions, the excitation energy is always greater than the energy (wavelength, color) of the emitted light.

To illustrate the diversity of luminescence emissions, Table 9.1 presents some of the more commonly observed types of luminescence, each named according to the source of energy or the trigger for the luminescence.

9.1.2 Photoluminescence

Photoluminescence is a result of absorption of photons or luminescence from any electromagnetic radiation. Today it is defined via the emission-based quantum mechanical mechanism for the orbital angular momentum multiplicity of the emitted electron (i.e., the singlet or triplet excited state). However, before the advent of quantum theory photoluminescence was defined solely on the basis of empirical evaluation of the duration of emissive lifetime. Photoluminescence can be classified into two types based on mechanism and time of decay.

9.1.2.1 Fluorescence

Fluorescence is a type of photoluminescence where no luminescence occurs after the cutoff of input energy source. It is the light emitted by an atom or molecule after a finite duration subsequent to the absorption of electromagnetic energy. Its lifetime is very small. Fluorescence is defined as a photoluminescent emission that arises from the singlet electronic state [7].

Absorption of an ultraviolet or visible photon promotes a valence electron from its ground state to an excited state with conservation of the electron's spin. For example, a pair of electrons occupying the same electronic ground state has opposite spins (Figure 9.1a) and are said to be in a singlet spin state. Absorbing a photon promotes one of the electrons to a singlet excited state (Figure 9.1b). This phenomenon is called "excitation." The excited states are not stable and will not stay indefinitely. If we observe

Table 9.1 Types of luminescence and their applications.

Excitation	Phosphors	Luminescence	Applications
Low energy photons (Ultraviolet/ Visible/ VUV)	Photo phosphors	Photoluminescence	Lamps and display
Heat	Thermo phosphor	Thermoluminescence	Radiation dosimetry; Environment protection
Cathode rays (Electrons)	Cathodophosphors	Cathodoluminescence	Cathode ray tube (CRT); Field emission display (FED); Vacuum fluorescence display
X-rays	X-ray phosphors	X-ray luminescence	Storage panels; Interfacing screens; Computation tomography
Ions (particles)	Ionophosphors	Ionoluminescence	Device fabrication
Mechanical forces	Tribophosphors	Triboluminescence	Crash prevention (car crash)
Electric field strength	Electro phosphors	Electroluminescence	Electronic discharge; EL panels
Biochemical	Biophosphors	Bioluminescence	Chemical assay; Oxygen detection

Table reproduced with permission from Rohit Saraf [6].

molecule in the excited state, at some random moment it will spontaneously return to the ground state. This return process is called decay, deactivation or relaxation. Under some special conditions, the energy absorbed during the excitation process is released during the relaxation in the form of a photon. This type of relaxation is called emission. Emission of a photon

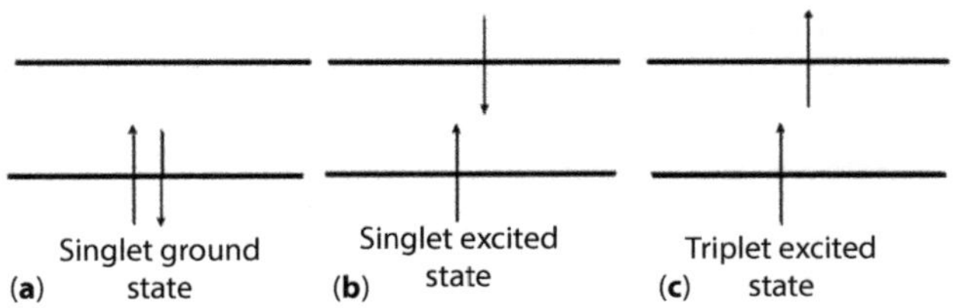

Figure 9.1 Different excited states.

from a singlet excited state to a singlet ground state, or between any two energy levels with the same spin, is called fluorescence. The probability of a fluorescent transition is very high, and the average lifetime of the electron in the excited state is only 10^{-5}–10^{-8} s. Fluorescence, therefore, decays rapidly after the excitation source is removed [8, 9].

9.1.2.2 *Delayed Fluorescence or Phosphorescence*

Delayed fluorescence or phosphorescence is a type of photoluminescence where, even after the cut off of source, luminescence (glow) occurs for some time. It is a photoluminescent process that originates from the triplet electronic state. Emissions from the triplet state are from 10 to 10,000 times longer than fluorescence; therefore, to the eye these radiators appear to emit after the excitation radiation is removed. In some cases an electron in a singlet excited state is transformed to a triplet excited state (Figure 9.1c) in which its spin is no longer paired with that of the ground state. Emission between a triplet's excited state and a singlet ground state, or between any two energy levels that differ in their respective spin states, is called phosphorescence. Because the average lifetime for phosphorescence ranges from 10^{-4} to 10^{4} s, phosphorescence may continue for some time after removing the excitation source [8, 9].

9.1.2.3 *Jablonski Diagram*

The processes that occur between the absorption and emission of light are usually illustrated by the Jablonski diagram. Jablonski diagrams are often used as the starting point for discussing light absorption and emission. They are used in a variety of forms to illustrate various molecular processes that can occur in excited states. These diagrams are named after Professor Alexander Jablonski, who is regarded as the father of fluorescence spectroscopy because of his many accomplishments, including descriptions of concentration depolarization and defining the term "anisotropy" to describe the polarized emission from solutions [10].

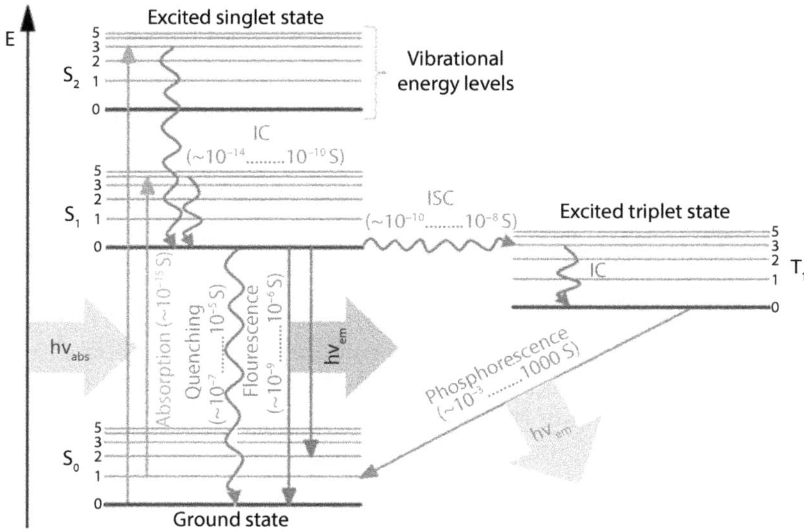

Figure 9.2 One form of a Jablonski diagram.

A typical Jablonski diagram is shown in Figure 9.2. The singlet ground, first, and second electronic states are depicted by S_0, S_1, and S_2, respectively. At each of these electronic energy levels the fluorophores can exist in a number of vibrational energy levels, depicted by 0, 1, 2, etc. In this Jablonski diagram we excluded a number of interactions such as quenching, energy transfer, and solvent interactions. The transitions between states are depicted as vertical lines to illustrate the instantaneous nature of light absorption. Transitions occur in about 10^{-15} s, a time too short for significant displacement of nuclei. This is the Franck-Condon principle [11, 12].

Following light absorption, several processes usually occur. A fluorophore is usually excited to some higher vibrational level of either S_1 or S_2. With a few rare exceptions, molecules in condensed phases rapidly relax to the lowest vibrational level of S_1. This process is called internal conversion and generally occurs within 10^{-12} s or less. Since fluorescence lifetimes are typically near 10^{-8}s, internal conversion is generally complete prior to emission. Hence, fluorescence emission generally results from a thermally equilibrated excited state, that is, the lowest energy vibrational state of S_1. Return to the ground state typically occurs to a higher excited vibrational ground state level, which then quickly (10^{-12}s) reaches thermal equilibrium (Figure 9.2). An interesting consequence of emission to higher vibrational ground states is that the emission spectrum is typically a mirror image of the absorption spectrum of the $S_0 \rightarrow S_1$ transition. This similarity occurs because of an electronic excitation does not greatly alter the nuclear geometry. Hence, the spacing of the vibrational energy levels of the excited states is similar to that of the ground state. As a result, the vibrational structures seen in the absorption and the emission spectra are similar [13].

Molecules in the S_1 state can also undergo a spin conversion to the first triplet state T_1. Emission from T_1 is termed phosphorescence, and is generally shifted to longer wavelengths (lower energy) relative to the fluorescence. Conversion of S_1 to T_1 is called intersystem crossing. Transition from T_1 to the singlet ground state is forbidden, and as a result the rate constants for triplet emission are several orders of magnitude smaller than those of fluorescence. Molecules containing heavy atoms such as bromine and iodine are frequently phosphorescent. The heavy atoms facilitate intersystem crossing and thus enhance phosphorescence quantum yields [13].

9.1.2.4 Phosphors

A phosphor is a substance that exhibits the phenomenon of luminescence. This includes both phosphorescent materials, which show a slow decay in brightness (> 1 ms), and fluorescent materials, where the emission decay takes place over tens of nanoseconds. A phosphor is a chemical material that, when stimulated by absorption of energy—often in the form of photons—will emit photons, usually at lower energy (longer wavelength) than the stimulating source. The electromagnetic radiation emitted by a luminescent material is usually in the visible range. Depending upon the nature of the excitation energy, the resulting phosphors and luminescence are given in Table 9.1.

A material can emit light either through incandescence, where all atoms radiate, or by luminescence, where only a small fraction of atoms, called emission centers or luminescence centers, emit light. In inorganic phosphors, these inhomogeneities in the crystal structure are usually created by addition of a trace amount of dopants, impurities called activators. In rare cases dislocations or other crystal defects can play the role of the impurity. The wavelength emitted by the emission center is dependent on the atom itself, and on the surrounding crystal structure.

A phosphor consists of a host lattice in which activator ions are incorporated. The activator creates a center, which absorbs excitation energy and converts it into visible radiation (Figure 9.3a). When an activator with the desired emission does not have a significant absorption for the available excitation energy, a sensitizer has to be used. The sensitizer absorbs the excitation energy and then transfers this energy to the activator, which can then emit its characteristic luminescence (Figure 9.3b). Table 9.2 lists

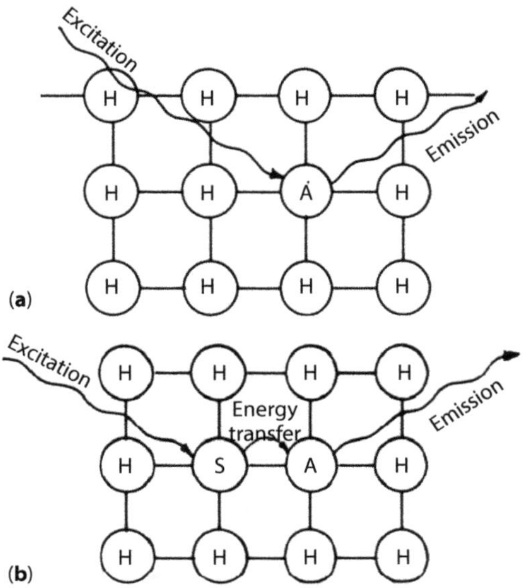

Figure 9.3 Representation of luminescence process: (a) activator (A) in a host (H) and (b) sensitizer (S) and activator (A) in a host (H). (Reproduced with permission from Rohit Saraf [6])

Table 9.2 Transition and emission of activators.

Type	Activators	Color range
$s^2 \rightarrow sp$ band	Sb^{3+}, Tl^{+}, Ga^{+} Sn^{2+}, Bi^{3+}, In^{+}	blue-green visible
$d \rightarrow f$ broad (50nm)	Eu^{2+} Ce^{3+}	blue-green UV-green
$O \rightarrow M$ very broad (100 nm)	WO_4^{2-}, MoO_4^{2-} VO_4^{3-}, NbO_4^{3-}	460–520 nm 480–580 nm
$d_t \rightarrow d_e$ broad and narrow	Mn^{2+}, Mn^{4+}, Fe^{3+} Cr^{3+}, Ni^{2+}	510–580 nm green-orange
$f \rightarrow f$ narrow	Eu^{3+}, Pr^{3+}, Nd^{3+} Tb^{3+} Tm^{3+}, Dy^{3+}, Er^{3+}, Ho^{3+}	red green

Table reproduced with permission from Rohit Saraf [6].

some commonly used activator ions, the transitions involved and emission observed.

Phosphor (lamp phosphor) in fluorescence lamps (tube lights) converts UV to visible (white) light. Several other phosphors like TV (cathode ray) phosphors, X-ray and γ-ray phosphors (scintillators) are also known. These phosphor materials find wide applications in lighting, white LEDs, scintillators, communication and as a diagnostic tool in medicine and biology. Some of the technologically important applications of phosphors are summarized in Table 9.3. General perspectives of the luminescence of solids can be found in various textbooks [14–17].

Table 9.3 Applications of phosphors.

Phosphors		Emission color	Applications
Lamp phosphors			
$Ca_5(PO_4)_3(F,Cl):Sb^{3+},Mn^{2+}$		white	
$Y_2O_3:Eu^{3+}$		red	
$Gd_2O_3:Eu^{3+}$	UV into	red	Fluorescent lamp
$Gd_2O_2S:Eu^{3+},Ti^{4+},Mg^{2+}$	Visible	red	
$(Ce, Gd)MgB_5O_{10}:Mn^{2+}$		red	
$CeMgAL_{11}O_{19}:Tb^{3+}$		green	
$BaMgAl_{10}O_{17}:Eu^{2+}$		blue	
Cathode ray phosphors			
$Y_2SiO_5:Tb^{3+}$		white	
$Y_2O_2S:Eu^{3+}$		red	Oscilloscope and
$YVO_4:Eu^{3+}$		red	Radar tubes
$KMgF_3:Mn^{2+}$	Electrons	orange	Monitor tubes
$Zn_2SiO_4:Mn^{2+}$	into	green	Color television
$ZnS:Cu^{+}$	Visible	green	
$(Ca,Mg)SiO_3:Ti^{4+}$		blue	
$ZnS:Ag^{+}$		blue	
X-ray phosphors	X-rays into	green	Fluoroscopic screens
$Gd_2O_2S:Tb^{3+}$	Visible	blue	Storage panels
$LaOBr:Tb^{3+}$		blue	Intensifying screens
$CaWO_4$			Computed tomography
γ-ray phosphors	γ-rays into	orange	Electromagnetic
$CsI:Tl^{+}$	Visible	blue	Calorimeters
$NaI:Tl^{+}$	UV into	blue	Medical
$Bi_4Ge_3O_{12}$	Visible	blue	Diagnostics
$ZnWO_4$		green	
$Y_3A_{15}O_{12}$			
Eu^{3+}- Cryptates		red	Luminescence Immunoassay

Table reproduced with permission from Rohit Saraf [6].

9.1.2.4.1 Requirements for Phosphors

Luminescent properties of lamp phosphors are affected by the structure of hosts, activator, sensitizer, flux, etc. In order to get good luminescence, phosphor hosts should have the following characteristics:

- They should be compatible with the luminescent centers, i.e., accommodate the luminescent ions and allow them to become involved in the luminescence process.
- They should be stable during fabrication of lamps.
- They should be stable to highly energetic electrons and ultraviolet radiation.

Recent research on luminescent nanomaterials provides challenges to both fundamental and advanced development of technologies in various areas such as electronics, photonics, displays, lasing, detection, optical amplification, and fluorescent sensing in biomedical engineering and environmental control [18]. Nanophosphors may have a number of potential advantages over traditional micron-sized phosphors. Such nanosized phosphor particles are reported to be somewhat different in their electrical, optical and structural characteristics. It is reported that these differences in electrical and optical characteristics of very small particles are caused by quantum effects due to their high surface to volume ratio, which increases the band gap by reduction of the number of allowable quantum states in the small particles and improves surface and interfacial effects [19]. In addition, quantum confinement in nanocrystalline materials may result in an enhancement of their luminescence.

Rare earths (REs) are well known for their extensive use in luminescent materials. The RE ions doped inorganic nanophosphor is one of the most promising materials for a variety of applications in solid-state lighting, solid-state lasers, lighting and displays and optical communication fields such as fluorescent lamps, cathode ray tubes and field emission displays. Many luminescent particles of different chemical compositions, shapes and size distributions have been prepared by different kinds of methods. In comparison, luminescent particles prepared by combustion synthesis are more attractive than many other methods for producing fine particle size, synthesizing homogeneous phosphor at relatively low temperature and reduced processing time, as well as being a low-cost method for production of various industrially useful materials. In the energy level diagram of the rare earths, luminescence processes often correspond to electronic transitions within the incompletely filled 4f shell. Consequently, these phosphors have narrow band spectra which are to a great extent independent of the nature of the host lattice. Because of the low interaction with the crystal lattice, the luminescence quantum yield of phosphors activated with rare earths is often high compared to other phosphors. Quenching occurs only at higher temperatures or higher activator concentrations.

9.1.2.5 *Photoluminescence of Samarium Ion (Sm^{3+})*

Among the rare earth ions, Sm^{3+} is an important activator, due to both the narrow-band emission spectra and the long lifetime of the optically active states. Luminescence of lanthanide ions has found application in medical diagnostics, lasers, optical fiber, night vision goggles, sunglass lenses, cathode ray tube (CRT), etc. Early CRT of color television had poor quality red color. Samarium as a phosphor dopant made the quality of the red color good [20]. When introduced into a host material, lanthanide ions effectively deactivated non-radiatively in organic host. On the other hand, inorganic hosts, such as glasses and crystals, have less effective non-radiative deactivation channels for excited lanthanide ions. However, absorption band of lanthanide ions is weak in inorganic hosts. A means to obtain efficient light absorption is the incorporation of lanthanide ions into semiconducting nanocrystals [21–24]. Therefore, selections of inorganic hosts which are semiconducting and nanocrystalline in nature may best suit the photoluminescence study of lanthanide ions doped materials.

9.1.3 Scope and Objectives of the Present Study

Rare earth ion doped phosphor materials have tremendous application in displays, due its short decay time, high quantum efficiency and good color coordination. Most of the earlier materials are toxic as well as carcinogenic and show limited thermodynamic stability against oxidation. These significant disadvantages hinder the practical application. Therefore, the search for nontoxic, stable luminescent oxide nanoparticles [25–27] or other rare earth doped oxides is currently on track. Looking at potential applications, there is a strong need for nontoxic oxide luminescent nanocomposites that are stable in ambient conditions. Luminescence of rare earth oxide nanoparticles is subject to rapid aging caused by formation of hydroxides at the surface of oxide nanoparticles, quenching luminescence. Coating of the surface of the particles with a polymer protects the surface of the oxide nanoparticles against ambient air, even against water, which avoids or at least reduces this problem [28,29]. Additionally, polymer-coated oxide nanoparticles can be suspended in water, which is most important in biological applications.

Polymer-nanomaterial composites (PNCs) have been an area of great interest in the research field in both academia and industry [30] due to their unique optical and mechanical properties and potential commercial application. PNC materials are reported to have improved mechanical, corrosion protection, flame retardant, thermal, electrical and optical characteristics [31-34]. An increase in thermal decomposition temperature, glass transition temperature and mechanical strength of PVA polymer is achieved by incorporation of nanoclay into PVA polymeric matrix [35]. The first ever polymer nanocomposite-PVA/MMT PNC material fabrication using solvent method was reported by Greenland in 1963 [36, 37].

In recent years there has been considerable interest in pyrochlore-type rare earth zirconium oxides ($RE_2Zr_2O_7$, RE = rare earth). These oxides are isomorphous with each other and isomorphous with the naturally occurring mineral pyrochlore of composition $(CaNa)Nb_2O_6F$ [38]. These rare earth zirconium oxides have complex chemistry, low thermal conductivity, high melting point, high thermal expansion coefficient, high stability and the ability to accommodate defects [39,40]. The compound $La_2Zr_2O_7$ doped with samarium can be produced by various methods such as ceramic [41], co-precipitation [42], sol-gel [43], inorganic sol-gel [44], hydrothermal [45], hydrazine methods [46] and nitrate-citrate gel combustion [47].

In order to achieve this goal, an appropriate synthesis strategy is required for synthesis of luminescent material powders with good luminescence characteristics at a low cost. All the synthetic methods, such as conventional solid-state synthesis, room temperature co-precipitation, sol-gel synthesis, hydrothermal synthesis, microwave synthesis, sonochemical synthesis, etc., either have or require high temperature, long processing time and sophisticated equipment with high maintenance costs. They also lead to the formation of deleterious phases. However, nitrate-citrate gel combustion method overcomes the above problems and offers many advantages such as low heating temperature, short reaction time, high purity of products, less expensive, low energy requirements, better compositional control and relative simplicity of the process. Gel-combustion routes are based on the gelling and subsequent combustion of an aqueous solution containing salts of the desired metals, La and Zr (usually nitrates), and citric acid, which also act as fuel. Exothermic reaction takes place between nitrate ions and the fuel. A large volume of gases are produced during the process, which promotes inflation of the gel produced and helps in formation of fine lanthanum zirconate crystals after calcination.

Though $La_2Zr_2O_7$ alone is not photoluminescent, when doped with samarium it can show photoluminescent property. Synthesized high quality

Sm^{3+}-doped $La_2Zr_2O_7$ nanophosphors have excellent redispersibility, high yield of luminescence, low cost and abundance of precursors and, above all, make the process cost-effective. It is well known that luminescent efficiency of phosphors can be enhanced or decreased by modifying the compositions. It has been reported that sometimes even very small quantities of co-dopants can lead to a great improvement of luminescent efficiency of phosphors [48]. Photoluminescent property generally increases with an increase in dopant concentration. Thus, this compound can be used to impart photoluminescent property to the PVA polymeric matrix. Hence, by doing so, the doped pyrochlore will also affect the other properties of polymer, which can be observed and elucidated by characterization and by comparing the PVA-$Sm^{3+}/La_2Zr_2O_7$ composite with a pure PVA polymer, which is the main objective of this study.

9.2 Experimental

9.2.1 Synthesis of Sm^{3+}-Doped $La_2Zr_2O_7$

$Sm^{3+}/La_2Zr_2O_7$ Nanophosphors were prepared by nitrate-citrate gel combustion method. All chemicals used were of analytical grade. Stoichiometric quantities of the reagents were taken in a beaker. La_2O_3 and Sm_2O_3 powders were used as starting materials and dissolved in a minimum amount

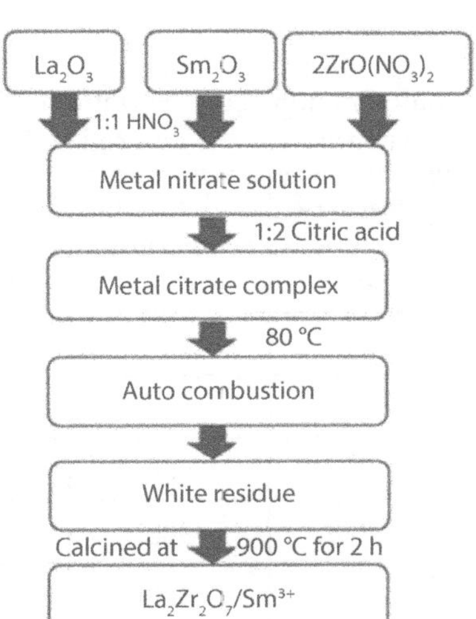

Figure 9.4 Flow chart for synthesis of Sm^{3+}-doped $La_2Zr_2O_7$ and $La_2Zr_2O_7$.

of nitric acid in separate beakers. Zirconium nitrate powder was dissolved in deionized water and kept on a hot plate with continuous stirring. Lanthanum and samarium oxides were added gradually to the beaker with continuous stirring. Citric acid was dissolved in deionized water and added in the end. Heating was continued for an hour until transparent viscous gel-like fluid was formed and the latter reaction was self-propogatory, resulting in the formation of porous brown mass. Some of it was taken as a sample for TGA and the rest was crushed in a crucible and calcined at 900 °C/2h in a muffle furnace. Whitish grey powder of $La_2Zr_2O_7$:Sm^{3+} was obtained and characterized. Accordingly, $La_2Zr_2O_7$ with different doped concentrations of samarium were prepared (Table 9.4). All these prepared compounds were characterized by XRD, UV-visible absorption spectroscopy and photoluminescence spectroscopy.

9.2.2 Preparation of PVA Polymer Thin Films

Dry PVA pellets of weight 1 g were taken and added to a glass beaker containing distilled water of 15 ml with continuous stirring maintained at room temperature. With continuous vigorous stirring and sonication, a clear viscous solution of PVA was obtained. Sonication of solution helped in the evolution of air bubbles. The solution was then transferred into a plastic Petri dish and kept in a hot air oven for overnight drying at 50 °C. After drying, the polymer was slowly peeled out from the Petri dish and cut into required shape.

Table 9.4 $La_2Zr_2O_7$ with different doped concentrations of samarium; lanthanum atoms are replaced by samarium in the unit cell at the sites [0.5 0.5 0.5].

Label	Composition	Ratio of citric acid used	Calcination temperature (°C)	Calcination time (hour)	Amount of samarium doped
1	$La_{1.98}Sm_{0.02}Zr_2O_7$	1:2	900	2	1%
2	$La_{1.92}Sm_{0.08}Zr_2O_7$	1:2	900	2	4%
3	$La_{1.84}Sm_{0.16}Zr_2O_7$	1:2	900	2	8%
4	$La_{1.76}Sm_{0.24}Zr_2O_7$	1:2	900	2	12%
5	$La_{1.68}Sm_{0.32}Zr_2O_7$	1:2	900	2	16%
6	$La_{1.60}Sm_{0.40}Zr_2O_7$	1:2	900	2	20%

9.2.3 Preparation of Sm^{3+}-Doped $La_2Zr_2O_7$ with PVA-Polymer Composite Films

The same procedure as above can be followed for producing polymer nanocomposite thin films. Compound with 1% doped samarium was added to the PVA solution matrix at the time of sonication, which helped in uniform distribution. The solution was dried overnight in a plastic Petri dish, peeled out and cut into required shape. Likewise, PVA-Sm^{3+}-$La_2Zr_2O_7$ polymer composites with different concentrations of the doped pyrochlore were produced (Table 9.5).

9.2.4 Characterization

Powder X-ray diffraction (XRD) patterns of the samples were recorded on a PANalytical X'Pert Pro Powder diffractometer operated at 40 kV and 30 mA using Ni-filtered Cu Ka radiation (λ = 1.5418 Å). For Rietveld refinement, the data were collected at a scan rate of 1°/min with a 0.02° step size for 2θ from 10° to 80°. Rietveld refinement method was employed to refine the structural parameters using FullProf Suite-2000 software program. Field emission scanning electron microscopy (FESEM) measurements were performed with FEI Quanta 200. UV–Vis diffuse reflectance spectra were recorded on a PerkinElmer Lambda 750 spectrophotometer using $BaSO_4$ as the reference. Thermogravimetric analysis (TGA) was performed using a Mettler-Toledo system in the presence of N_2 as a carrier gas up to 900 °C. The photoluminescence (PL) spectra were measured using a Fluorolog-3 spectrofluorometer (Jobin Yvon USA) at room temperature [49, 50].

Table 9.5 PVA with different concentrations of compound are combined to produce PNCs.

Label	PVA (mg)	Sm^{3+}-$La_2Zr_2O_7$ (mg)	Percentage of PVA-Sm^{3+}-$La_2Zr_2O_7$ in PVA
0	1000	0	0% (Blank)
1	1000	50	5%
2	1000	100	10%
3	1000	200	20%

9.3 Results and Discussion

9.3.1 Structural Analysis by X-ray Diffraction

Phase purity of Sm^{3+}-doped $La_2Zr_2O_7$ compounds were identified by powder X-ray diffraction. Figure 9.5a,b shows the indexed XRD patterns of a) $La_{2-x}Sm_xZr_2O_7$ (x = 0, 0.2, 0.8, 0.16, 0.24, 0.32 and 0.40) and b) y wt% $La_{2-x}Sm_xZr_2O_7$-PVA composite (X = 0.2) (y = 0, 5, 10 and 20). The strong diffraction peaks revealed the crystalline nature of the samples. The observed diffraction peaks of Sm^{3+}-doped $La_2Zr_2O_7$ compounds match well with the reported JCPDS Card No. 71-2363. No traces of additional peaks were observed in the XRD patterns, which confirmed that the Sm^{3+} ions have been uniformly incorporated into the host lattice of $La_2Zr_2O_7$. This implies that the powders obtained by nitrate-citrate gel combustion method were single phase materials and in composites this material was uniformly mixed.

The average crystallite size was estimated using Scherrer's equation:

$$D = \frac{k\lambda}{\beta\cos\theta} \tag{9.1}$$

where λ is the wavelength (1.5418 Å) of X-rays, β is the full width at half maximum (FWHM), θ is the diffraction angle, k is the shape factor (0.9) and D is the average crystallite size.

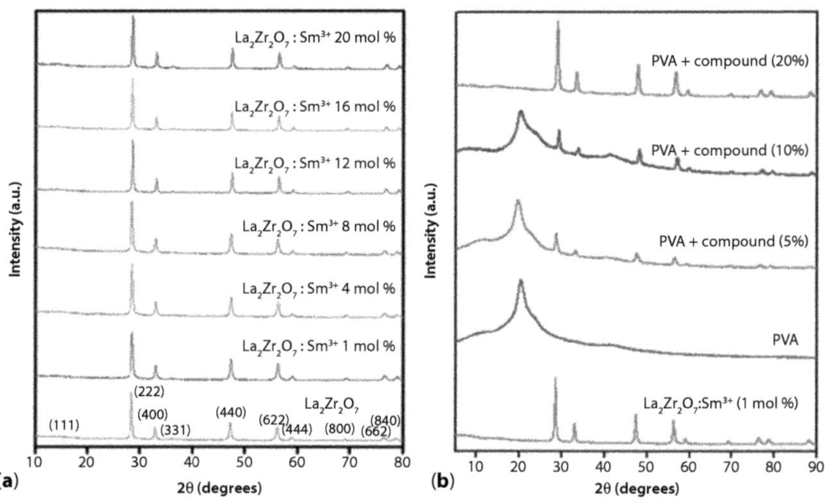

Figure 9.5 (a) Powder XRD of Sm^{3+}-doped $La_2Zr_2O_7$ and $La_2Zr_2O_7$ calcined at 950 °C for two hour; (b) XRD patterns for Sm^{3+}-doped $La_2Zr_2O_7$, PVA and Sm^{3+}-doped $La_2Zr_2O_7$ with PVA polymer composites.

Based on this equation, the average crystallite sizes for these compounds were found to be 80–107 nm. The structural parameters were refined by the Rietveld method using powder XRD data. The patterns were typically refined for lattice parameters, scale factor, backgrounds, pseudo-Voigt profile function (u, v and w), atomic coordinates and isothermal temperature factors (B_{iso}). The refinement results confirmed that all the compounds were crystallized in the cubic phase with space group *Fd-3m* (No. 227). The observed, calculated XRD patterns of $La_2Zr_2O_7$ and Sm^{3+} (0, 8, 12 and 20 mol%) doped $La_2Zr_2O_7$ compounds and the difference between them are shown in Figure 9.6a–d. The difference between XRD pattern profiles experimentally observed and calculated data display near zero in the intensity scale as illustrated by a line ($Y_{obs}-Y_{calc}$). The refined structural parameters for host $La_2Zr_2O_7$, Sm^{3+} 8 and 20 mol% doped $La_2Zr_2O_7$ compounds are summarized in Table 9.6. The results in Table 9.6 reveal that there are

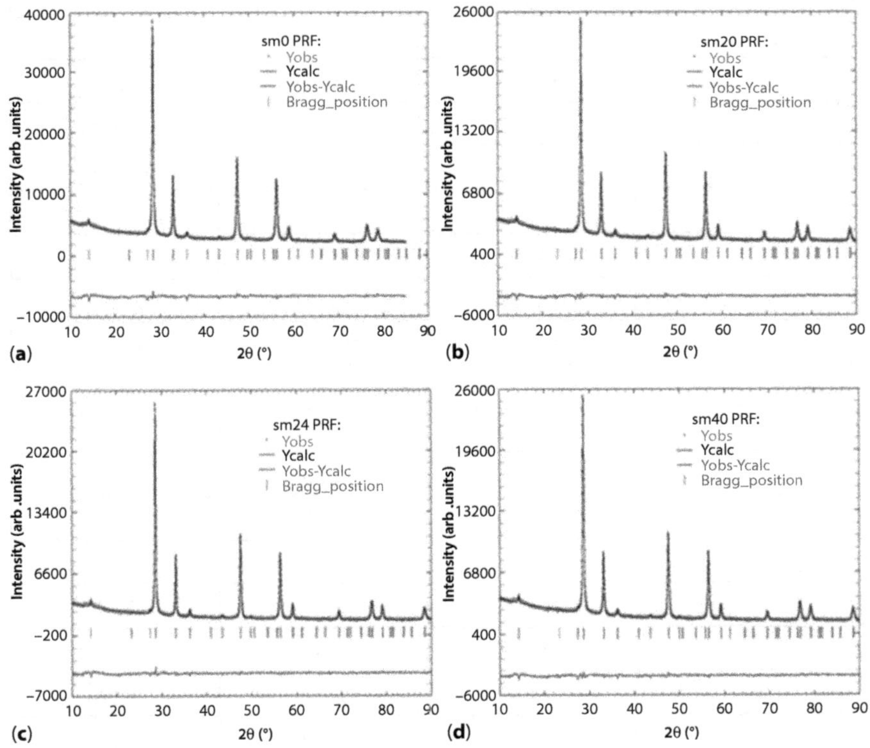

Figure 9.6 Observed, calculated XRD patterns and the difference between them for (a) $La_2Zr_2O_7$, (b) $La_2Zr_2O_7:Sm^{3+}$ (10 mol%), (c) $La_2Zr_2O_7:Sm^{3+}$ (12 mol%) and (d) $La_2Zr_2O_7:Sm^{3+}$ (20 mol%).

Table 9.6 Rietveld refined structural parameters for $La_2Zr_2O_7$:Sm^{3+}.

Compounds	**$La_2Zr_2O_7$**	**$La_2Zr_2O_7$:Sm^{3+} (8 mol%)**	**$La_2Zr_2O_7$:Sm^{3+} (20 mol%)**
Crystal system	Cubic	Cubic	Cubic
Space group	Fd-3m (No. 227)	Fd-3m (No. 227)	Fd-3m (No. 227)
Lattice parameters			
a (Å)	10.868(7)	10.845(7)	10.814(4)
Cell volume (Å^3)	1283.75(2)	1275.45(9)	1264.56(8)
Atomic positions			
La (16d)			
x	0.5000	0.5000	0.5000
y	0.5000	0.5000	0.5000
z	0.5000	0.5000	0.5000
Zr (16c)			
x	0.0000	0.0000	0.0000
y	0.0000	0.0000	0.0000
z	0.0000	0.0000	0.0000
O1 (48f)			
x	0.3468(2)	0.3451(8)	0.3448(16)
y	0.1250	0.1250	0.1250
z	0.1250	0.1250	0.1250
O2 (8a)			
x	0.3750	0.3750	0.3750
y	0.3750	0.3750	0.3750
z	0.3750	0.3750	0.3750
$R_{Factors}$			
R_p	2.16	2.09	2.01
R_{wp}	2.89	2.67	2.62
R_{exp}	1.66	2.12	1.96
c^2	3.01	1.59	1.79
R_{Bragg}	3.16	3.38	4.27
R_F	4.16	4.48	6.77

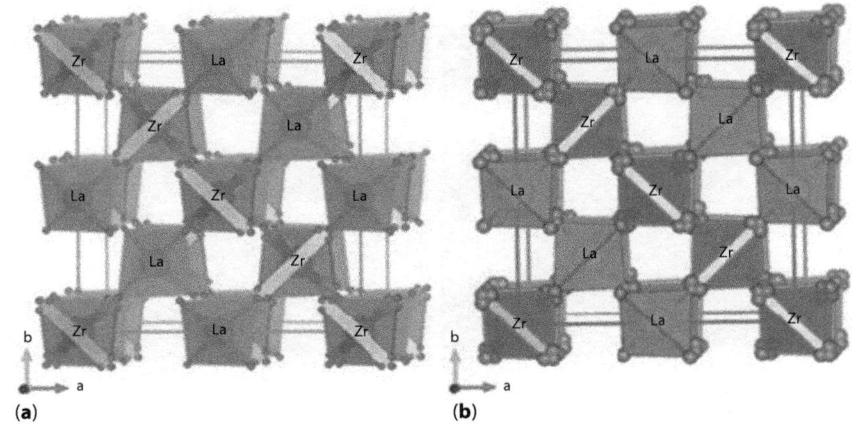

Figure 9.7 Crystal structure of (a) $La_2Zr_2O_7$ and (b) $La_2Zr_2O_7$:Sm^{3+} (20 mol%).

changes in the lattice parameters and cell volumes upon doping with Sm^{3+} and La^{3+} ions due to differences in ionic radii. Figure 9.7a,b shows the crystal structure of pyrochlore $La_2Zr_2O_7$ compound.

9.3.2 SEM Analysis

Figure 9.8a shows the surface morphology of $La_{1.98}Sm_{0.02}Zr_2O_7$ compound obtained by nitrate-citrate gel combustion method. For $La_2Zr_2O_7$, 9.8b and 9.8c are enlarged views of 9.8a at 50, 100 kx magnification. Figure 9.8d shows the $La_{1.98}Sm_{0.02}Zr_2O_7$-PVA composite (20 wt%) compounds; the micrograph Figure 9.8c revealed agglomerated morphology. Figure 9.8d confirms the homogeneity of compound in the polymer.

9.3.3 UV-Vis Spectroscopy

Figure 9.9a,b shows the diffused reflectance spectra of (a) $La_{2-x}Sm_xZr_2O_7$ (x = 0, 0.2, 0.8, 0.16, 0.24, 0.32 and 0.40) and (b) $La_{2-x}Sm_xZr_2O_7$-PVA composite (x = 0.2). The above spectra indicates that these compounds have band gap values of approximately 4–5 eV.

9.3.4 Thermogravimetric Analysis (TGA)

The stability/decomposition of $La_{2-x}Sm_xZr_2O_7$ precursor compound was measured using thermogravimetric analysis up to 1000 °C in the presence of N_2 gas. We observed an initial weight loss up to 100 °C due to

Figure 9.8 SEM images of (a) $La_{1.98}Sm_{0.02}Zr_2O_7$ [(b) and (c) are enlarged views of (a)] at different magnifications and (d) $La_{1.98}Sm_{0.02}Zr_2O_7$-PVA composite 20 wt% compounds.

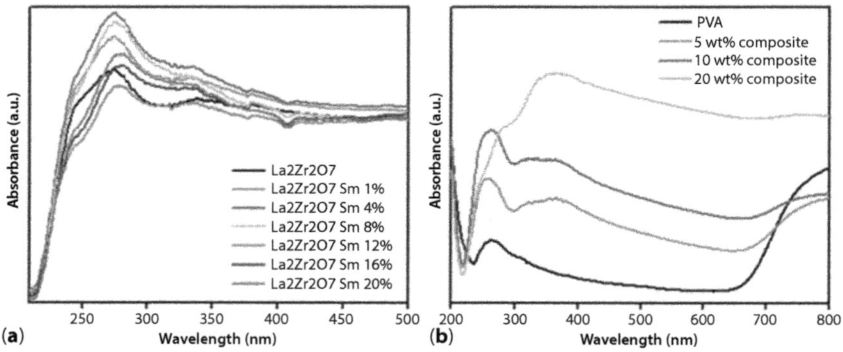

Figure 9.9 Diffused reflectance spectra of (a) $La_{2-x}Sm_xZr_2O_7$ and (b) $La_{2-x}Sm_xZr_2O_7$-PVA composite.

the evaporation of moisture/dehydration, as shown in Figure 9.10. Further, weight loss was also due to decomposition of nitrate, carbonate and residual organic moieties. Above 900 °C, we did not see any weight loss in the TGA plot, revealing that the stable pyrochlore phase is formed around 900 °C.

9.3.5 Photoluminescence Properties

Figure 9.11a shows the PL excitation and emission spectra of Sm^{3+}-doped $La_2Zr_2O_7$ and Figure 9.11b,c shows the emission spectra of Sm^{3+}-doped

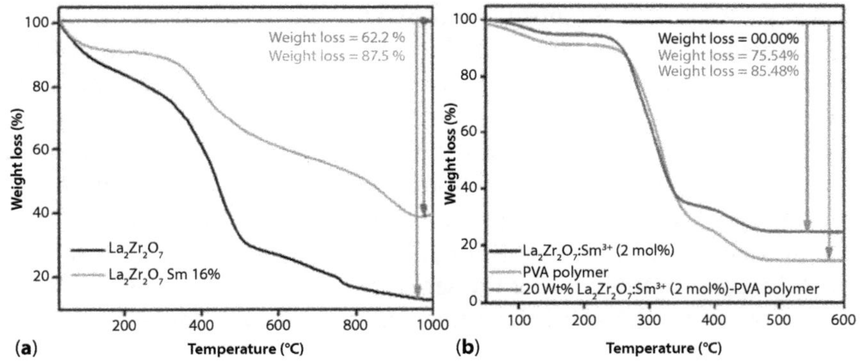

Figure 9.10 TGA curve for (a) $La_2Zr_2O_7$ and (b) $La_{2-x}Sm_xZr_2O_7$ (x = 0.32) compounds.

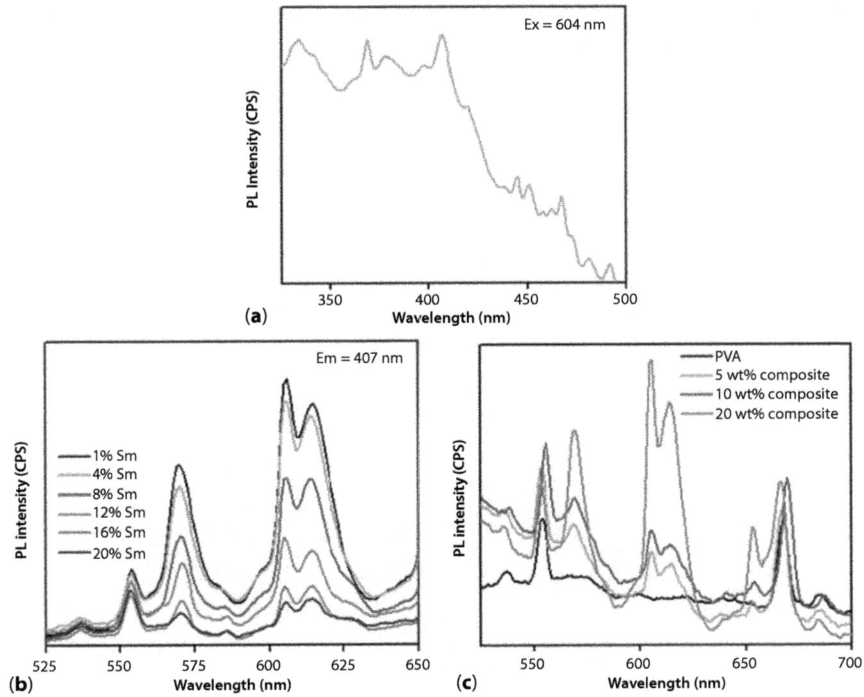

Figure 9.11 (a) PL excitation spectra of $La_{1.98}Sm_{0.02}Zr_2O_7$ and (b) PL emission spectra of $La_{2-x}Sm_xZr_2O_7$ (x = 0, 0.2, 0.8, 0.16, 0.24, 0.32 and 0.40); (c) y wt% $La_{2-x}Sm_xZr_2O_7$-PVA composite (x = 0.2) (y = 0, 5, 10 and 20).

$La_2Zr_2O_7/Sm^{3+}$-PVA composite phosphors. The emission spectrum of Sm^{3+}-doped $La_2Zr_2O_7$ phosphor-PVA composites under excitation of 407 nm wavelength are recorded in Figure 9.11b,c. The sharp peaks detected at 555, 570, 605 and 620 nm are due to the fact that there is more red luminescence compared to other transitions. It was observed that the emission intensity decreases with the increase in Sm^{3+} doping concentration (up to 20 mol%), due to internal quenching process. Further photoluminescence properties of x wt% $La_{1.98}Sm_{0.02}Zr_2O_7$-PVA composite (x = 0, 5, 10 and 20) were examined under the same excitation of 407 nm wavelength and we observed the same transitions as those in Sm^{3+}-doped $La_2Zr_2O_7$ phosphors.

9.3.6 Chromaticity Color Coordinates

The Commission International de I'Eclairage (CIE) chromaticity coordinates of $La_{1.98}Sm_{0.02}Zr_2O_7$ and x wt% $La_{1.98}Sm_{0.02}Zr_2O_7$-PVA composite (x = 0, 5, 10 and 20) were found from the PL spectra at 407 nm wavelength. Figure 9.12 shows the CIE 1931 chromaticity diagram

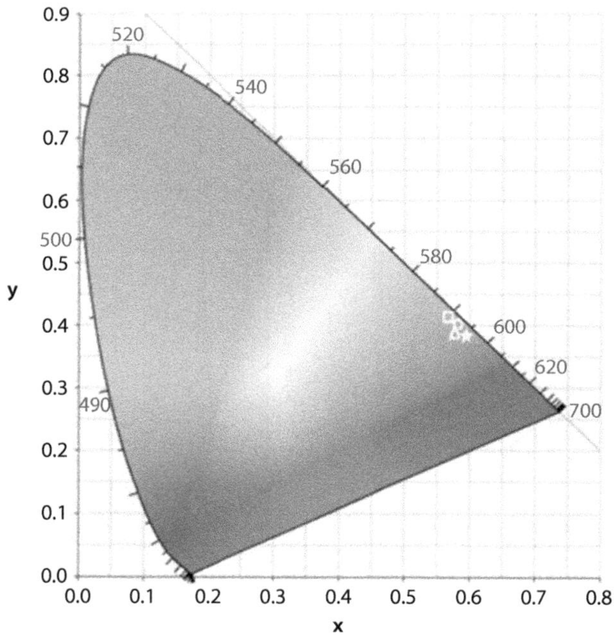

Figure 9.12 CIE 1931 chromaticity diagram of $La_{1.98}Sm_{0.02}Zr_2O_7$ and x wt% $La_{1.98}Sm_{0.02}Zr_2O_7$-PVA composite (x = 0, 5, 10 and 20) excited at 407 nm.

Table 9.7 The CIE coordinates and CCT values of $La_{1.98}Sm_{0.02}Zr_2O_7$ and x wt% $La_{1.98}Sm_{0.02}Zr_2O_7$-PVA composites (x = 0, 5, 10 and 20).

Symbol	Compounds	CIE coordinates		CCT Values (K)
		x	y	
Triangle	$La_{1.92}Sm_{0.08}Zr_2O_7$	0.5926	0.4066	1639.23
Circle	$La_{1.92}Sm_{0.08}Zr_2O_7$-5%PVA	0.5956	0.4037	1629.73
Star	$La_{1.92}Sm_{0.08}Zr_2O_7$-10% PVA	0.5930	0.4069	1635.61
Square	$La_{1.92}Sm_{0.08}Zr_2O_7$-20% PVA	0.5938	0.4055	1638.32

excited at 407 nm wavelength. The color coordinates of $La_{1.98}Sm_{0.02}Zr_2O_7$ and x wt% $La_{1.98}Sm_{0.02}Zr_2O_7$-PVA composite (x = 0, 5, 10 and 20) are listed in Table 9.7. From Figure 9.12, we can clearly observe that the color coordinates lie in the orange-red region of the CIE chromaticity diagram.

Correlated color temperature (CCT) is another important parameter to assess the phosphor performance, which specifies the color appearance of light emitted by the light source. CCT is defined as the color temperature corresponding to the point on the Planckian locus which is nearest to the point representing the chromaticity of the illuminant considered on the (u', v'). CCT is calculated by transforming the (x, y) coordinates of the light source to (u', v') using the equations:

$$v' = \frac{9y}{-2x+12y+3}, u' = \frac{4x}{-2x+12y+3} \tag{9.2}$$

According to McCamy's approximation [50], the CCT value can be derived from CIE color coordinates using the third power polynomial and is given by the expression:

$$T = -449n^3 + 3525n^2 - 6823.3n + 5520.33 \tag{9.3}$$

where $n = (x - 0.3320)/(y - 0.1858)$. The calculated CCT values for $La_{1.98}Sm_{0.02}Zr_2O_7$ and x wt% $La_{1.98}Sm_{0.02}Zr_2O_7$-PVA composite (x = 0, 5, 10 and 20) are listed in Table 10.7, which is regarded as a cool red light. The observed high PL brightness along with excellent CCT values indicate that these composites can be useful for display and other optical device applications.

9.4 Conclusion

We have synthesized a series of Sm^{3+}-doped $La_2Zr_2O_7$ phosphors by nitrate-citrate gel combustion method. These compounds were characterized using a variety of experimental techniques. Further, among these compounds, 1 mol% samarium-doped $La_2Zr_2O_7$ was mixed with PVA matrix in different weight percentage and optical and mechanical properties were examined. The results indicate that these composites can be useful for display and other optical device applications.

Aknowledgment

The author, Pramod Halappa, sincerely thanks CSIR, Government of India, for their financial support through CSIR-JRF.

References

1. Goddard III, W.A., Brenner, D.W., Lyshevski, S.E., and Iafrate, G.J. (Eds.), *Handbook of Nanoscience, Engineering, and Technology*, CRC Press: Boca Raton, 2003.
2. Wang, J., Lin, C., Lai, C., Hsu, J., and Ai, C., Improved characteristics of Gd_2O_3 nanocrystal memory with substrate high–low junction. *Solid State Electron.* 54, 1493, 2010.
3. Engström, M., Klasson, A., Pedersen, H., Vahlberg, C., Käll, P., and Uvdal, K., High proton relaxivity for gadolinium oxide nanoparticles. *MAGMA* 19, 180, 2006.
4. Viswanathan, V., Laha, T., Balani, K., Agarwal, A., and Seal, S., Challenges and advances in nanocomposite processing techniques. *Mater. Sci. Eng. R* 54, 121, 2006.
5. Dosev, D., Kennedy, I.M., Godlewski, M., Gryczynski, I., Tomsia, K., and Goldys, E.M., Fluorescence upconversion in Sm-doped Gd_2O_3. *Appl. Phys. Lett.* 88, 011906, 2006.
6. Saraf, R., Facile synthesis of metal tungstates, Eu^{3+}-activated metal molybdates and bismuth oxyhalides nanophosphors: Applications in white LEDs and photocatalysis. MTech thesis, University of Rajasthan and Indian Institute of Science, 2014.
7. The Fluorescent Mineral Society, Luminescence and Fluorescence (accessed 11/30/2015).
8. Valeur, B., *Molecular Fluorescence: Principles and Applications*, Wiley-VCH Verlag GmbH, 2001.
9. Harvey, D., *Modern Analytical Chemistry*, McGraw Hill, 2000.

10. Jablonski, A., Über den Mechanisms des Photolumineszenz von Farbstoffphosphoren, *Z. Phys.* 94, 38-46, 1935.
11. Szudy, J. (Ed.), *Born 100 Years Ago: Aleksander Jabłoński (1898–1980)*, Uniwersytet Mikołaja Kopernika – UMK: Toruń, Poland, 1998.
12. Polska Akademia Nauk Instytut Fizyki, Acta Physica Polonica. *Europhys. J.* A65(6), 1978.
13. Lakowicz, J.R., *Principles of Fluorescence Spectroscopy*, 3rd ed., Springer, 2006.
14. Blasse, G., and Grabmaier, B.C., *Luminescent Materials*, Springer Verlag: Berlin, 1994.
15. Ropp, R.C., *Luminescence and the Solid State*, Elsevier Science: Amsterdam, 1991.
16. Kroger, F.A., *Some Aspects of the Luminescence of Solids*, Elsevier: Amsterdam, 1948.
17. Marfanin, A.S., *Spectroscopy, Luminescence and Radiation Centers in Minerals*, Springer-Verlag: Berlin Heidelberg, New York, 1979.
18. Wang, F., Tan, W.B., Zhang, Y., Fan, X., and Wang, M., Luminescent nanomaterials for biological labelling. *Nanotechnology* 17, R1, 2006.
19. Trave, A., Buda, F., and Fasolino, A., Band-gap engineering by III-V infill in sodalite. *Phys. Rev. Lett.* 77, 5405, 1996.
20. Halappa, P., Shivakumara, C., Saraf, R., and Nagabhushana, H., Synthesis, structure and photoluminescence properties of Sm^{3+}-doped BiOBr phosphor. *DAE Solid State Physics Symposium* 1731, 140064, 2016.
21. Stouwdam, J.W., Lanthanide-doped nanoparticles as the active optical medium in polymer-based devices. PhD Thesis, University of Twente, The Netherlands, 2003.
22. Blasse, G., and Grabmaier, B.C., *Luminescent Materials*, Springer: Berlin, London, 1994.
23. Ronda, C., *Luminescence: From Theory to Applications*, Wiley-VCH: Weinheim, Germany, 2008.
24. Werts, M.H.V., Making sense of lanthanide luminescence. *Sci. Prog.* 88, 101, 2005.
25. Chen, Y., Cao, Y., Bai, Y., Yang, W., Yang, J., Jin, H., and Li, T., *J. Vac. Sci. Technol. B* Study on photoelectric properties of a TiO_2 nanoparticle, 15, 1442, 1997.
26. Monticone, S., Tufeu, R., and Kanaev, A.V., Complex nature of the UV and visible fluorescence of colloidal ZnO nanoparticles. *J. Phys. Chem. B* 102, 2854-2862, 1998.
27. Guo, L., Yang, S., Yang, C., Yu, P., Wang, J., Ge, W., and Wong, G.K.L., Synthesis and characterization of poly(vinylpyrrolidone)-modified zinc oxide nanoparticles. *Chem. Mater.* 12, 2268, 2000.
28. Vollath, D., German patent application, DE 101 54988.1, 2001.
29. Vollath, D., Lamparth I., and Szabó, D.V., Fluorescence from coated oxide nanoparticles. *Mat. Res. Soc. Symp. Proc.* V7, 703, 2002.
30. Awad, W.H., Beyer, G., Benderly, D., Ijdo, W.L., Songtipya, P., Jimenez-Gasco, M.M., Manias, E., and Wilkie, C.A., Material properties of nanoclay PVC composites. *Polymer* 50, 1857, 2009.

31. Tyan, H.-L., Liu, Y.-C., and Wei, K.-H., Thermally and mechanically enhanced clay/polyimide nanocomposite via reactive organoclay. *Chem. Mater.* 11, 1942, 1999.
32. Yeh, J.-M., Liou, S.-J., Lai, C.-Y., Wu, P.-C., and Tsai, T.-Y., Enhancement of corrosion protection effect in polyaniline via the formation of polyaniline–clay nanocomposite materials. *Chem. Mater.* 13, 1131, 2001.
33. Alexandre, M., and Dubois, P., Polymer-layered silicate nanocomposites: Preparation, properties and uses of a new class of materials. *Mater. Sci. Eng. R* 28, 1, 2000.
34. Khanna, P.K., Gokhale, R.R., Subbarao, V.V.V.S., Singh, N., Jun, K.-W., and Das, B.K., Synthesis and optical properties of CdS/PVA nanocomposites. *Mater. Chem. Phys.* 94, 454, 2005.
35. Yu, Y.-H., Lin, C.-Y., Yeh, J.-M., and Lin, W.-H., Preparation and properties of poly(vinyl alcohol)–clay nanocomposite materials. *Polymer* 44, 3553, 2003.
36. Chen, Y., Cao, Y., Bai, Y., Yang, W., Yang, J., Jin, H., and Li, T., Study on photoelectric properties of a TiO_2 nanoparticle. *J. Vac. Sci. Technol. B* 15, 1442, 1997.
37. Guo, L., Yang, S., Yang, C., Yu, P., Wang, J., Ge, W., and Wong, G.K.L., Synthesis and characterization of poly(vinylpyrrolidone)-modified zinc oxide nanoparticles. *Chem. Mater.* 12, 2268, 2000.
38. Wells, A.F., *Structural Inorganic Chemistry*, 4th ed., p. 499, Oxford University Press: Oxford, 1975.
39. Vassen, R., Cao, X., Tietz, F., Basu, D., and Stover, D., Zirconates as new materials for thermal barrier coatings. *J. Am. Ceram. Soc.* 83, 2023, 2000.
40. Roth, R.S., Pyrochlore-type compounds containing double oxides of trivalent and tetravalent ıons. *J. Res. Natl. Bur. Stand. (U.S.)* 56, 17, 1956.
41. Yashima, M., Kakihana, M., Ishii, K., Ikuma, Y., and Yoshimura, M., Synthesis of metas[10 tetragonal (t') zirconia-calcia solid solution by pyrolysis of organic precursors and coprecipitation route. *J. Mater. Res.* 11, 1410, 1996.
42. Kido, H., Komarneni, S., and Roy, R., Preparation of $La_2Zr_2O_7$ by sol–gel route. *J. Am. Ceram. Soc.* 74, 422, 1991.
43. Bhattacharya, A.K., Hartridge, A., Mallick, K.K., and Woodhead, J.L., Preparation and characterization of $Ln_2Zr_2O_7$ microspheres by an inorganic sol-gel route. *J. Mater. Sci.* 29, 6076, 1994.
44. Chen, D., and Xu, R., Hydrothermal synthesis and characterization of $La_2M_2O_7$ (M = Ti, Zr) powders. *Mater. Res. Bull.* 33, 409, 1998.
45. Matsumura, Y., Yoshinaka, M., Hirota, K., and Yamaguchi, O., Formation and sintering of $La_2Zr_2O_7$ by the hydrazine method. *Solid State Commun.* 104, 341, 1997.
46. Juárez, R.E., Lamas, D.G., Lascalea, G.E., and Walsöe de Reca, N.E., Synthesis of nanocrystalline zirconia powders for TZP ceramics by a nitrate–citrate combustion route. *J. Eur. Ceram. Soc.* 20, 133, 2000.

47. Liu, X., Han, K., Gu, M., Xiao, L., Ni, C., Huang, S., and Liu, B., Effect of codopants on enhanced luminescence of $GdTaO_4:Eu^{3+}$ phosphors. *Solid State Commun.* 142, 680, 2007.
48. Shivakumara, C., Saraf, R., and Halappa, P., White luminescence in Dy^{3+} doped BiOCl phosphors and their Judd–Ofelt analysis. *Dyes Pigm.* 126, 154, 2016.
49. Devakumar, B., Halappa, P., and Shivakumara, C., Dy^{3+}/Eu^{3+} co-doped $CsGd(MoO_4)_2$ phosphor with tunable photoluminescence properties for near-UV WLEDs applications. *Dyes Pigm.* 137, 244, 2017.
50. McCamy, C.S., Correlated color temperature as an explicit function of chromaticity coordinates. *Color Res. Appl.* 17, 142, 1992.

10

Study of Gas Transport Phenomenon in Layered Polymer Nanocomposite Membranes

A.K. Patel and N.K. Acharya*

Applied Physics Department, Faculty of Technology and Engineering, The Maharaja Sayajirao University of Baroda, Vadodara, India

Abstract

The transport of mass through a variety of polymer membranes has been reported over the last few years. Researchers have attempted a range of samples from pure polymer to modify its structure by blending, filling micro/nanoparticles in the polymer matrix, synthesizing mixed matrix membranes, etc. The change in free volume due to modification in polymer chains directly alters the transport mechanism. However, very little attention has been paid to layered polymer composites. In this work, the transport behavior of different gases across the metal coated membrane has been studied and compared with that of standard polycarbonate membrane. Deposition of nanolayer affects the transport properties due to the modification in sorption-diffusion process. Permeability coefficient is the product of diffusivity coefficient (D) and solubility coefficient (S). A platinum-palladium (Pt-Pd) alloy thin film of around 8 nm was coated on thick polycarbonate substrate using sputtering technique to make layered polymer nanocomposite membrane. The tests were carried out at upstream pressure of 30 psia. Gases used for the study herein were He, H_2, CO_2, O_2, N_2 and CH_4 and the selectivity was calculated for a particular gas pair. Pure gas permeability coefficients were calculated using constant volume/variable pressure method at 35 °C temperature.

***Keywords*:** Gas transport, thin film coating, selectivity, composite membrane

**Corresponding author*: sarnavee@gmail.com

Sanjay K. Nayak *et al.* (eds.) Trends and Applications in Advanced Polymeric Materials, (191–206) 2018 © Scrivener Publishing LLC

10.1 Introduction

Membrane science and technology has led to the development of industrial applications by technologies such as microfiltration, ultrafiltration, reverse osmosis, electrodialysis, gas transport and pervaporation. In addition to industry, membrane technology has also focused attention on applications in the medical field such as artificial lungs, artificial kidneys and controlled drug delivery systems [1, 2]. The separation process concerns material having some specific properties (chemical or physical) that can exist under the operating conditions [3]. Polymer membranes are widely used to perform mass transport process in the chemical industry due to a wide range of applications in fluid separation. Additionally, the polymer membrane can be produced cost-effectively and the separation process based on membrane is environmentally friendly [4]. The flow of gas molecules across the membrane is influenced by several factors: gas properties, material used to design the membrane, morphology of membrane structure and operating conditions [5]. The permeation as well as separation properties can be improved by chemically modifying the structure or by physical modifications. Different modifications are practiced to improve the performance of membranes like mixed matrix, cross-linking, grafting, polymer blending, making composite or hybrid membrane [6–9].

To fulfill the high gas barrier property, polymer surface can be modified by depositing thin film coating. Cellulose coating on silica-carbide improves gas flux permeation for H_2, CO_2, O_2 and N_2, and even gives better selectivity compared to virgin material [10]. A silica layer coated on polymer membrane reduces permeation flux, which improves the selectivity [11]. Metal coating can reduce permeability, which enhances the separation factor, or in some cases it may rise due to a defect in the coating layer [12]. Ti coating on polycarbonate membrane reduces the permeability of H_2 and CO_2 and simultaneously increases the selectivity of H_2 [13]. In the last decades, H_2 has become a major energy source, which needs to be extracted from water, hydrocarbons, coal or natural gases [14]. In addition, H_2 fuel is considered to be a clean energy source because the by-product of its combustion produces water. Its production from hydrocarbon forms CO_2, which is one of the gases responsible for the greenhouse effect. Pd is considered highly selective material to attract H_2 molecule. Even its composite form performs better for separation of H_2 from mixed state with other gases [15]. For this purpose Pd membrane have received growing consideration due to its separation ability with respect to all other gases [16]. Moreover, Pd membrane is unambiguously selective to H_2, which is a favorable response to separate it from other gases [17]. But several factors limit the gas permeation application for pure Pd membrane such as its high cost and irreversible damage by sulfide formation. Moreover, above 450 °C temperature it is subjected to deactivation by carbon compounds and below 300 °C temperature, when exposed to pure H_2, hydride formation takes place. To overcome this difficulty, Pd can be alloyed with other metals. For this purpose Pt gives better performance with Pd [18]. Herein, Pt-Pd alloy has been coated over polymeric support.

The supportive polymer has been chosen in such a way that it can provide good thermal resistance and mechanical strength. For this purpose polycarbonate (PC) is coated by Pt-Pd alloy and suspended for the gas permeation process, as explained in [19, 20]. Polycarbonate provides better backbone rigidity so that it permeates faster rate flow of gas flux, therefore it was chosen as polymer matrix [21]. H_2 and CO_2 permeability increase with an increase in temperature for polycarbonate as the diffusivity increases [22, 23]. Moreover, CO_2 treated polycarbonate enhances the solubility, which is credited to the plasticization effect [24]. In this study, gas permeability of pure PC as well as Pt-Pd coated PC has been compared with respect to their kinetic diameter for various gases. The key parameter for permeability measurements is the gas flux through the given membranes, which has been measured by a constant volume/variable pressure system [25]. Our goal was to develop a membrane which would have improved permeability along with selectivity. In this study, a new type of layered hybrid membrane was synthesized, consisting of thin inorganic bridged platinum-palladium layer deposited onto the surface of polycarbonate.

10.1.1 Transport Phenomenon

The permeability of gas through dense polymer membrane depends on the solubility and diffusivity of the reentrant. Figure 10.1 explains the steady-state permeation of gas molecules through polymer membrane. This process is classified into three steps:

- Absorption of gas molecules at upstream side;
- Diffusion through membrane with respect to pressure gradient;
- Desorption at downstream side.

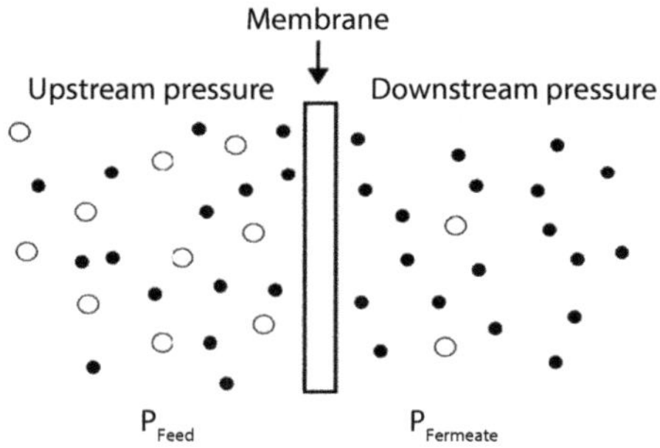

Figure 10.1 Transport of gas through membrane.

The upstream side at high pressure provides feed gas and the downstream side at low pressure, permeate gas. The gas molecules first dissolve into upstream side, diffuse across the membrane and desorb from downstream side.

Gas permeation is the flux of gas molecules through an object normalized by the pressure gradient. The driving force behind the transport of gas molecules which involves sorption, diffusion and permeation, is mainly due to the pressure difference between two phases. The principle of transport phenomenon is explained by Fick's first law of diffusion according to which the flux J in the direction of flow is directly proportional to the concentration gradient $\left(\frac{dC}{dX}\right)$ given by the equation:

$$J = -D\left(\frac{dC}{dX}\right) \tag{10.1}$$

where D is the diffusion coefficient. The equation is only applicable for steady-state diffusion process. Fick's law gives the linear relationship between the flux of substance diffusing through the membrane and the concentration gradient between two phases [4, 12].

Gas permeability coefficient P of a pure gas for polymeric membrane is given by the following formula:

$$P = \frac{\mathbf{JL}}{\mathbf{P_2 - P_1}} \tag{10.2}$$

where J is the steady state flux across the membrane, L is the membrane thickness, P_2 is the upstream pressure and P_1 is the downstream pressure [4]. Permeability is expressed in barrers, where 1 barrer = 10^{-10} [cm^3 (STP) cm/(cm^2 s cm-Hg)]. For dense polymeric membrane, the downstream pressure is much less than the upstream pressure. Gas permeation is explained by a simplest solution-diffusion model as follows:

$$P = D\,S \tag{11.3}$$

where D is the diffusion coefficient and S is the solubility coefficient. Diffusion coefficient can be estimated by the intercept of steady state line and the time axis of downstream pressure versus time curve:

$$D = L^2/6\theta \tag{11.4}$$

where θ is the time lag, L is the membrane thickness and D is diffusion coefficient. Solubility coefficient can be determined by the ratio of permeability over diffusion coefficient [29].

The selectivity of a polymer membrane for gas pair A and B is defined as the ratio of their permeability:

$$\alpha = P_A/P_B \tag{11.5}$$

Permeation across membrane takes place via solution diffusion process, i.e., feed gas absorbed at upstream pressure decided by chemical affinity of penetrated gas with polymer material of membrane. Further, the reentrant gas diffuses inside the polymer matrix within the available free volume and finally desorption occurs at downstream side. For dense membranes permeability is the product of diffusion coefficient and solubility coefficient; hence modification in either diffusion coefficient or in solubility coefficient alters directly to the overall gas permeability. Therefore, selectivity of gas A relative to gas B can be expressed in terms of ideal selectivity defined by the relation:

$$a_{AB} = \frac{P_A}{P_B} = \frac{D_A}{D_B} \cdot \frac{S_A}{S_B} \tag{11.6}$$

where the ratio D_A/D_B is described as selectivity of diffusion and ratio S_A/S_B is viewed as solubility selectivity. Generally, glassy polymers are hard and brittle and have restricted chain mobility. The diffusion coefficient decreases with an increase in the size of molecule. This decrease of diffusivity depends on the flexibility of polymer backbone. For more rigid polymer structure, higher selectivity is achieved for a given gas pair. The diffusion selectivity is a controllable factor for most of the glassy polymers. Therefore, smaller molecular transport is favored. Whereas, the solubility of gases increases with molecular size due to increase in intermolecular force between gas and polymer [1].

10.1.2 Metal Coating

Metal coating over polymer membrane uplifts the permeation as well as performance of the membrane. This hybridization improves surface interaction of gas molecules, which leads to better thermal properties and mechanical strength [26]. The gas transport parameter is also influenced by polymer chain flexibility, segmental mobility and interaction with penetrant. These three factors can be controlled by surface coating as the feed gas does not interact directly with the host material. Before its transport through the polymer film, it diffuses through the metal coating which serves barrier properties. Coating can improve the selectivity of CO_2/CH_4 separation performance under different operating conditions when exposed to polysulfon mixed matrix membrane with silicate coating [30]. Palladium is extremely selective for H_2 gas but at low temperature, if Pd is exposed to H_2, it can become locked inside the Pd lattice, which causes stress inside the membrane and failure of membrane. The remedy to this problem is to dope Pd with other metals such as silver or platinum. The doping will improve the interaction with gases and protect the lattice structure. The transport of H_2 through Pd membrane occurs in six steps [31]:

- Dissociation of H_2 molecule at gas/metal interface;
- Sorption of atomic hydrogen on membrane surface;
- Dissolution into palladium matrix;
- Diffusion through membrane;
- Recombination of atomic hydrogen into molecule at gas/metal interface;
- Desorption of H_2.

Moreover, charge transfer occurs as the gases are interacted with metal surface, which may play a vital role in improving sorption of gas molecules [8, 27].

10.2 Experimental

10.2.1 Fabrication of Nanocomposite Membrane

Solution cast method, which is easy and reliable, was used to develop polycarbonate membrane [28]. Initially, the desired quantity of bispheneol A (polycarbonate), which is commercially available, was dissolved in dichloromethane and stirred overnight at room temperature using a magnetic stirrer. The mixture was poured onto a flat-bottomed glass surface covered with glass ring. The system was allowed to evaporate the solvent overnight for complete removal of the solvent. The next day, the membrane was taken for the measurements and commercially available film was also used for study. Uniformity in membrane thickness was confirmed by several thickness measurements using a digital thickness meter. Average thickness 180 (±2) μm was purchased for permeability calculation. Standard vapor deposition sputtering technique has been used to make layered composite membrane at ICMB facility of University of Texas at Austin, TX, USA. In this technique the target and substrate material were placed in a vacuum chamber. Pt-Pd thin film of about 8 nm thicknesses was deposited over PC thin film. In this technique, the polycarbonate substrate and target material were placed in high vacuum of 10^{-6} torr [32].

10.2.2 Gas Permeability Test

Figure 10.2 shows a schematic diagram of cell permeation in which membrane is placed normal to the upstream pressure. The membrane is placed under metal support inside the permeation cell and its diameter is the same size as the inner diameter of the cell. As the flat membrane is placed inside the cell, the system is divided into upstream and downstream compartments. The upstream side is at higher pressure from which feed gas enters into the cell and the downstream side is at lower pressure from which gas is permeated. O-rings are placed inside the cell at the upstream

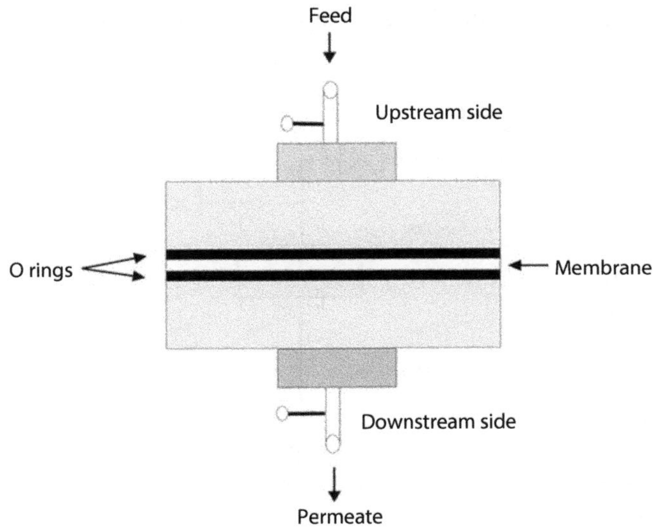

Figure 10.2 Schematic diagram of gas permeation cell.

and downstream sides of the membrane to prevent gas leakage as well as exchange of air particles from outer atmosphere. Thus, only feed gas from the higher pressure side can permeate within the membrane which is collected towards the downstream side.

The key parameter for permeability measurements is the gas flux through the given membranes. As illustrated in Figure 10.3, a constant-volume variable pressure system measures permeate flux by monitoring the pressure increase of collected permeate gas in a closed volume using a pressure transducer.

First of all, the whole system is exposed to vacuum prior to perform the experiment. This action can remove any volatile impurities inside the system. To degas the film, the upstream as well as the downstream volume are evacuated. Then the valves connecting the permeation cell to the vacuum pump are closed. Now a slow pressure rising towards the downstream volume is observed, which is called the leak rate of the system. The leak rate should be around 10 times less than the steadily permeating gas flow. Then the feed gas is introduced within the cell towards the upstream side of the membrane and at the downstream side the pressure rise is recorded as function of time [25]. The permeability of He, H_2, CO_2, O_2, N_2 and CH_4 was measured using a constant volume/variable pressure method. The tests were carried out at constant 30 psia upstream pressure and constant temperature of 35 °C. For pure polycarbonate, the surface area and thickness of membrane were 1.97 cm^2 and 180 μm respectively. For Pt-Pd coated PC, the surface area was 0.6529 cm^2. A constant-volume/variable pressure system measures permeate flux by monitoring the pressure increase of collected permeate gas in a closed

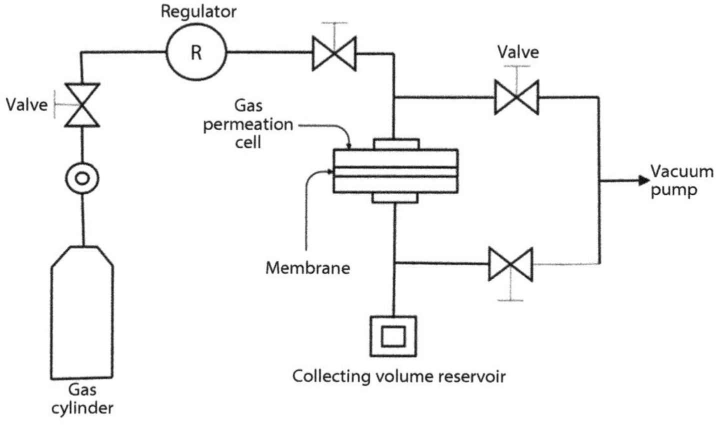

Figure 10.3 Constant volume/variable pressure system setup.

volume using a pressure transducer. Membrane was placed in such a way that the coated side was at the upstream, which was purged with penetrate gas, and the downstream side of the permeation cell was evacuated with the help of a rotary pump before taking the individual measurements. The following formula was used to calculate the permeability:

$$P = \frac{Vd}{pART}(m1) - (m2) \tag{10.7}$$

where P is the pure gas permeability, V the total volume of the downstream side, i.e., 39.43 cm^3, d the thickness of the membrane, p the upstream absolute pressure in psig, m1 the steady state slope of the downstream pressure versus time calculated in cm-Hg/s, and m^2 the leak rate calculated in cm-Hg/s which was maintained at 10^{-7} cm-Hg/s during the experiment [25]. Permeability was measured in barrers, where 1 barrer = 10^{-10} {cm^3 (STP) cm/(cm^2 s cm-Hg)} [25].

10.3 Results and Discussion

In Table 10.1, the comparison of permeability of all experimental gases within their kinetic diameters is shown. It is obvious that the decrease of gas permeability is proportional to the degree of kinetic parameter of gas molecule. The gases with larger kinetic diameter experience lower permeability. Permeation behavior was found in the following order: He > H_2 > CO_2 > O_2 > N_2 > CH_4, for both the pure and Pt-Pd layered PC membrane. All the data were statistically analyzed after steady state towards the downstream side. Hydrogen and helium have excessively fast permeation even at low upstream pressure, hence it is difficult to find the steady state for such gases. Compared with reference data for pure polycarbonate, the permeability of He has decreased in this experiment, which can promote its selectivity. It can be seen that He gives the highest permeability with respect to other gases that have a lot smaller kinetic diameter, while for N_2 and CH_4 it reduces permeability. Hence, according to the literature survey and the performance of this experiment, the transport of molecules from matter totally depends on their size. Compared with H_2 reference data, its permeability increases, which enhances its application as an energy tool. Additionally, as shown in Table 10.1, the percentage increase of kinetic diameter and the permeability also increase in the same manner. There is no further change in the permeability for CO_2, O_2, and N_2 gases with respect to the references given in Table 10.1. Moreover, for H_2 and He the percentage due to leakage

Table 10.1 Permeability of pure PC compared with leak test.

Gases	Kinetic diameter (Å)	Permeability of PC (barrer)			% change due to leakage
		With leak test	Without leak test	Reference data	
He	2.6	13.56	13.64	17.6 [21]	0.6%
H_2	2.89	13.23	13.31	11.5 [21]	0.6%
CO_2	3.3	8.64	8.72	8.4 [19]	0.9%
O_2	3.46	3.26	3.35	3.8 [20]	2.6%
N_2	3.64	0.36	0.44	0.44 [20]	18%
CH_4	3.8	–	–	0.4 [19]	–

is the same, as they are smaller in magnitude. But again, a rise in molecular size affects the leakage. For CO_2, 0.9% permeability arises without leakage and for O_2, its 2.6%, which provides evidence for the effect of molecular size. Finally, N_2 gives the highest percent change in permeability due to leakage, which proves that larger molecular size restricts its leakage. The permeability coefficient decreases as the kinetic diameter increases.

The same trade-off is followed by Pt-Pd coated PC membrane. There is marginal change which can be acceptable due to leakage in the permeability of pure PC. Moreover, under the same operating conditions Pt-Pd coated PC shows enhanced permeability for all gases with regards to their kinetic dimension given in Table 10.2. For all gases, Pt-Pd coating tends to increase the permeability compared to pure PC. This is due to an increase in solubility by surface coating and it is assumed that a rise in solubility may be due to the catalytic effect of Pt-Pd coating. For H_2 and He, permeability rises with nearly the same percentage. As previously stated, for H_2 interaction with Pt-Pd surface, Pd is considered an attractive tool for H_2 molecule.

The polycarbonate used as host material in this study was glassy polymer. In glassy state, the molecular motion is due to chain segments and substitute group. As the segmental motions are more confined, the selectivity is performed by enhanced mobility selectivity. It is shown in Table 10.3 that the permeability decreases and in Table 10.4, the diffusion selectivity rises due to coating. The selectivity for almost all gases decreases due to the coating effect, which follows the trade-off relationship for glassy polymers. Selectivity of different gas pairs is shown in Figure 10.3, which shows a decrease for layered composite membrane as compared to pure

Table 10.2 Permeability of pure and Pt-Pd coated polymers.

Gases	Kinetic diameter (Å)	Permeability (barrer)		% change by coating
		PC	PC/Pt-Pd	
He	2.6	13.64	17.37	27.38%
H_2	2.89	13.31	16.91	27.06%
CO_2	3.3	8.72	12.95	48.46%
O_2	3.46	3.35	10.66	218.64%
N_2	3.64	0.44	2.3	427%
CH_4	3.8	0.4	1.59	297%

polycarbonate; however, permeability of all the measured gases increases for nanocomposite membrane, which follows the trade-off relation of glassy polymer. For gas pairs H_2/N_2, CO_2/N_2 and O_2/N_2 the selectivity is reduced by 76%, 72% and 39% respectively.

In all three of these gas pairs, N_2 is the common gas even though the reduction in selectivity for the first two gas pairs has a change of barely 4%, whereas compared to these two pairs the O_2/N_2 selectivity reduction is 39%. This is due to the solubility selectivity, which is reduced by 77% for CO_2/N_2 and 49% for O_2/N_2. Here the reduction in solubility selectivity plays a vital role in changing the overall selectivity compared to pure polycarbonate membrane. As in gas pairs H_2/O_2 and CO_2/O_2, O_2 is a mutual gas which shows reduction in selectivity by 60% and 53% respectively. This can reduce the selectivity of H_2 and CO_2 by nearly the same amount. Selectivity by solubility for CO_2/O_2 is reduced by 54%. In the gas pairs H_2/CH_4 and CO_2/CH_4 the reduction in selectivity is 68% and 69% respectively, whereas for O_2/CH_4 and N_2/CH_4 it is 20% and 32% respectively, although CH_4 is a mutual gas. This may be due to the small kinetic diameter of H_2 and CO_2 compared to O_2 and N_2.

The selectivity for H_2/CO_2 was not more affected due to metal coating, as shown in Table 10.3. Pt-Pd coating plays an important role in enhancing the permeability of hydrogen and carbon dioxide by maintaining their selectivity. This is especially true for Pd, which absorbs H_2 in large amounts, but there are no more specific changes in the H_2 permeation. It can be seen from Figure 10.4 that selectivity was highly affected by H_2/N_2 gas pairs, which shows that the selectivity for PC/Pt-Pd is reduced 76% compared to pure PC due to coating. Selectivity for H_2/O_2 gas pair does not decrease

Table 10.3 Selectivity due to permeation.

Gas pair	Selectivity		% Change*
	PC	PC/Pt-Pd	
H_2/CO_2	1.52	1.3	14%
H_2/O_2	3.97	1.58	60%
H_2/N_2	30.25	7.35	76%
H_2/CH_4	33.27	10.63	68%
CO_2/O_2	2.6	1.21	53%
CO_2/N_2	19.81	5.63	72%
CO_2/CH_4	21.8	6.7	69%
O_2/N_2	7.61	4.63	39%
O_2/CH_4	8.38	6.7	20%
N_2/CH_4	1.1	1.45	32%

*For all gas pairs the selectivity decreases and only for N_2/CH_4 does it increase by 32%.

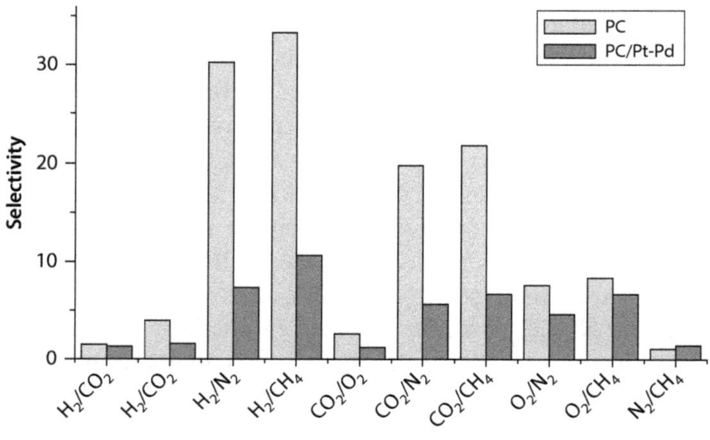

Figure 10.4 Selectivity comparison of pure PC and Pt-Pd coated PC.

in large amounts, i.e., 14%, which shows that it even is nearly stable due to surface modification. There is a prominent coating effect which promotes both permeability as well as selectivity for N_2 and N_2/CH_4 respectively. This results in the application of separation of CH_4 from natural gas.

Table 10.4 Selectivity due to diffusion and solubility.

	Selectivity by diffusion			Selectivity by solubility		
Gas pair	**PC**	**PC/Pt-Pd**	**% Change***	**PC**	**PC/Pt-Pd**	**% Change***
CO_2/O_2	0.88	0.9	2%	3.01	1.36	54%
CO_2/N_2	1.4	1.56	16%	17.33	3.98	77%
O_2/N_2	1.6	1.73	8%	5.76	2.93	49%

*Selectivity by diffusion increases whereas solubility selectivity is reduced.

10.4 Conclusion

The hybrid membrane fabricated by Pt-Pd metal coating permeates comparatively more gas flux than pure polycarbonate membrane. The constant volume/variable pressure system used for this study provides accurate data for calculation.

The conclusion is explained as follows:

- The above study shows that the permeability from hybrid membrane is directly proportional to the kinetic diameter of gas molecules.
- The leakage does not affect the gases having smaller kinetic diameter; as the kinetic diameter rises, the permeability difference due to leakage also increases.
- Permeability increases for all the gases but for N_2 there is a large difference in the permeability achieved by coating.
- The highest change of permeability due to coating has been recorded for N_2 gas compared to other gases and it is selective for N_2, as its selectivity increases for N_2/CH_4 gas pair as compared to other gases. This result is applicable for separation of N_2 with CH_4 from natural gas.

Acknowledgment

The author NKA is thankful to Dr. Freeman's research group at the University of Texas at Austin, Texas, USA for providing experimental facilities and DST, New Delhi for financial assistance under BOYSCAST scheme (BY/P-01/2010-11) and Fast Track Young Scientist Scheme (FTP/PS-32/2006).

The author AKP is thankful to the University Grants Commission, New Delhi, India, for providing fellowship (GCU-4/UGCBSR Fel/App.Phy./13-14/#1263/943) under the BSR RF SMS scheme during the course of the work.

References

1. Nunes, S.P., and Peinemann, K.-V., *Membrane Technology in the Chemical Industry*, pp. 3–8, Wiley-VCH, 2015.
2. Strathmann, H., Giorno, L., and Drioli, E., *An Introduction to Membrane Science and Technology*, pp. 339–357, Consiglio Nazionale delle Ricerche: Rome, 2006.
3. Clarizia, G., Polymer-based membranes applied to gas separation: Material and engineering aspects. *Desalination* 245, 763–768, 2009.
4. George, S.C., and Thomas, S., Transport phenomenon through polymeric system. *Prog. Polym. Sci.* 26, 985–1017, 2001.
5. Nwogu, N.C., Gobina, E., and Nasir Kajama, M.N., Improved carbon dioxide capture using nanostructured ceramic membranes. *Low Carbon Economy* 4, 125–128, 2013.
6. Lee, J.-H., Interface engineering in zeolite-polymer and metal-polymer hybrid materials, PhD dissertation, Georgia Institute of Technology, 2010.
7. Kulshrestha, V., Acharya, N.K., Awasthi, K., Nathawat, R., Singh, M., and Vijay, Y.K., Characterization of asymmetric polymeric membranes by gas permeation. *Micron* 38, 326–329, 2007.
8. Jain, A.K., Acharya, N.K., Kulshreshtha, V., Awasthi, K., Singh, M., and Vijay, Y.K., Study of hydrogen transport through porous aluminium and composite membranes. *Int. J. Hydrogen Energy* 33, 346–349, 2008.
9. Vijay, Y.K., Wate, S., Acharya, N.K., and Garg, J.C., The titanium-coated polymeric membranes for hydrogen recovery. *Int. J. Hydrogen Energy* 27, 905–908, 2002.
10. Srikanth, R., Siva Kumar, K., Satyanarayana, S.V., and Madhusoodana, C.D., Study of permeation of gases through ceramic supported polymeric and zeolite membranes. *IJREAT* 3, 478–482, 2014.
11. Zulhairun, A.K., and Ismail, A.F., The role of layered silicate loadings and their dispersion states on the gas separation performance of mixed matrix membrane. *J. Membr. Sci.* 468, 20–30, 2014.
12. Hamad, F., Chowdhury, G., and Matsuura, T., Effect of metal cations on the gas separation performance of sulfonated poly (phenylene oxide) membranes. *Desalination* 145, 365–370, 2002.
13. Acharya, N.K., Kulshrestha, V., Awasthia, K., Kumar, R., Jain, A.K., Singh, M., Avasthic, D.K., and Vijay, Y.K., Gas permeation study of Ti-coated, track-etched polymeric membranes. *Vacuum* 81, 389–393, 2006.

14. Muradov, N., Hydrogen via methane decomposition: An application for decarbonization of fossil fuels. *Int. J. Hydrogen Energy* 26, 1165–1175, 2001.
15. Lu, G.Q., Diniz da Costa, J.C., Duke, M., Giessler, S., Socolow, R., Williams, R.H., and Kreutz, T., Inorganic membranes for hydrogen production and purification: A critical review and perspective. *J. Colloid Interface Sci.* 314, 589–603, 2007.
16. Liguori, S., Iulianelli, A., Dalena, F., Pinacci, P., Drago, F., Broglia, M., Huang, Y., and Basile, A., Performance and long-term stability of Pd/PSS and Pd/Al2O3 membranes for hydrogen separation. *Membranes* 4, 143–162, 2014.
17. Jordal, K., Bredesen, R., Kvamsdal, H.M., and Bolland, O., Integration of H_2-separating membrane technology in gas turbine processes for CO_2 capture. *Energy* 29, 1269–1278, 2004.
18. Kajama, M.N., Nwogu, N.C., Okon, E.O., Shehu, H., Orakwe, I., and Gobina, E., Hydrogen separation using Pt-alumina impregnated membrane. *Energy Power Eng.* 7, 412–417, 2015.
19. Sridhar, S., Aminabhavi, T.M., and Ramakrishna, M., Separation of binary mixtures of carbon dioxide and methane through sulfonated polycarbonate membranes. *J. Appl. Polym. Sci.* 105, 1749–1756, 2007.
20. Hu, C.-C., Fu, Y.-J., Lee, K.-R., Ruaan, R.-C., and Lai, J.-Y., Effect of sorption behavior on transport properties of gases in polymeric membranes. *Polymer* 50, 5308–5313, 2009.
21. Şen, D., Kalıpçılar, H., and Yılmaz, L., Gas separation performance of polycarbonate membranes modified with multifunctional low molecular weight additives. *Separ. Sci. Technol.* 41, 1813–1828, 2006.
22. Acharya, N.K., Yadav, P.K., and Vijay, Y.K., Study of temperature dependent gas permeability for polycarbonate membrane. *Indian J. Pure Ap. Phy.* 42, 179–181, 2004.
23. Vijay, Y.K., Acharya, N.K., Wate, S., and Avasthi, D.K., Characterization of track etched membranes by gas permeation. *Int. J. Hydrogen Energy* 29, 515–519, 2004.
24. Chen, S.-H., Huang, S.-L., Yu, K.-C., Lai, J-Y., and Liang, M.-T., Effect of CO_2 treated polycarbonate membranes on gas transport and sorption properties. *J. Membr. Sci.* 172, 105–112, 2000.
25. Czichos, H., Saito, T., and Smith, L. (Eds.), *Springer Handbook of Materials Measurement Methods*, Springer, 2006.
26. Lee, J.-H., Interface engineering in zeolite-polymer and metal-polymer hybrid materials, PhD thesis, Georgia Institute of Technology, 2010.
27. Kanan, S.M., El-Kadri, O.M., Abu-Yousef, I.A., and Kanan, M.C., Semiconducting metal oxide based sensors for selective gas pollutant detection. *Sensors* 9, 8158–8196, 2009.
28. Nathawat, R., Kumar, A., Acharya, N.K., and Vijay, Y.K., XPS and AFM surface study of PMMA irradiated by electron beam. *Surf. Coat. Tech.* 203, 2600–2604, 2009.

29. Flaconneche, B., Martin, J., and Klopffer, M.H., Transport properties of gases in polymers: Experimental methods, oil and gas science and technology. *Rev. IFP* 56, 245–259, 2001.
30. Zulhairun, A.K., Ng, B.C., Ismail, A.F., Surya Murali, R., and Abdullah, M.S., Production of mixed matrix hollow fiber membrane for CO_2/CH_4 separation. *Sep. Purif. Technol.* 137, 1–12, 2014.
31. Basile, A., Iulianelli, A., Longo, T., Liguori, S., and De Falco, M., Pd-based selective membrane state-of-the-art, in: *Membrane Reactors for Hydrogen Processes*, De Falco, M., Marrelli, M., and Iaquaniello, G. (Eds.), pp. 21–57, Springer, 2011.
32. Andreeta, M.R.B. (Ed.), *Crystallization - Science and Technology*, InTech, 2012.

11

Synthesis and Ion Transport Studies of K^+ Ion Conducting Nanocomposite Polymer Electrolytes

Angesh Chandra[1]*, Alok Bhatt[2] and Archana Chandra[3]

[1]Solid State Physics Research Laboratory, Department of Applied Physics, Shri Shankaracharya Institute of Professional Management and Technology, Raipur, Chhattisgarh, India
[2]Department of Applied Physics, Bharti College of Engineering and Technology, Durg, Chhattisgarh, India
[3]Dr. C. V. Raman University, Kargi Road Kota, Bilaspur, Chhattisgarh, India

Abstract

Reported in this chapter are synthesis and ion transport studies of K^+ ion conducting nanocomposite polymer electrolyte (NCPE) membranes:(1–x) [70PEO:30KBr] + x SiO_2, where 0 < x < 20 wt%. In place of the traditional solution cast method, NCPE membranes were casted using a novel hot-press technique. The conventional solid polymer electrolyte (SPE) composition: (70PEO:30KBr), identified as the highest conducting film at room temperature, was used as Ist-phase host matrix and nanosize (~ 8 nm) particles of SiO_2 as IInd-phase dispersoid. As a consequence of dispersal of SiO_2 in SPE host, a conductivity enhancement of an order of magnitude resulted in NCPE film: [95(70PEO: 30KBr) + 5SiO_2]. This is referred to as optimum conducting composition (OCC). The ion transport behavior in NCPE OCC is discussed on the basis of experimental measurements of some basic ionic parameters, viz., conductivity (σ), ionic mobility (μ), mobile ion concentration (n), ionic transference number (t_{ion}), etc. The conductivity as a function of temperature was also studied to compute the activation energy (E_a) by least square linear fitting of 'log σ – 1/T' Arrhenius plot. A new K+ ion conducting solid-state

Corresponding author: chandrassi@gmail.com

Sanjay K. Nayak *et al.* (eds.) Trends and Applications in Advanced Polymeric Materials, (207–218) 2018 © Scrivener Publishing LLC

polymer battery was also fabricated using the present NCPE OCC as an electrolyte and some important cell parameters were calculated at room temperature.

***Keywords*:** Nanocomposite polymer electrolytes, ionic conductivity, ionic mobility, activation energy, polymer battery

11.1 Introduction

The ion conducting polymer electrolytes show great technological potential to develop all-solid-state mini/micro-electrochemical devices, viz., batteries, fuel cells, supercapacitors, etc. [1–6]. Pure polymeric materials are known for their poor electrical conductivity and hence they are often referred to as insulators. However, the polymers can be made conducting by complexing/dissolving electronic and/or ionic salts. The first ion conducting solid polymer electrolyte (SPE) was reported in 1973 [7]. Since then, a wide variety of SPE film materials, involving different mobile ionic species, viz., H^+, Ag^+, Li^+, Na^+, Mg^{2+}, etc., have been investigated [8–13]. The poly ethylene oxide (PEO)-based SPEs exhibited lower ionic conductivity at room temperature, hence, are not very useful for practical device applications. Nevertheless, the room temperature conductivity of SPEs can be significantly increased by fractional dispersal of low dimension (μm or nm) particles of insulating/inert filler materials such as Al_2O_3, SiO_2, TiO_2, etc. [8–12]. Such systems are known as composite polymer electrolytes in which the enhancement in the conductivity is predominantly due to the increase of amorphous region in PEO as a consequence of the dispersal of the filler particles. When nano-size filler particles are dispersed, they are referred to as nanocomposite polymer electrolytes (NCPEs). The dispersal of nanoparticles also brings substantial improvements in several other material properties of NCPE films, viz., mechanical stability, intimate electrode/electrolyte contacts as well as enhanced interfacial reactivity during battery applications. SPEs/NCPEs films are usually synthesized by the traditional solution cast/sol-gel methods. However, an alternate procedure, popularly referred to as hot-press (extrusion) technique, has been developed for casting SPE/NCPE films [9–13]. Recently, this technique has been gaining wider acceptance due to the fact that it has several merits. It is not only relatively quicker and inexpensive but a completely dry/solution-free film casting procedure as compared to the traditional method.

The present chapter reports synthesis of NCPEs: (1–x) (70PEO: 30KBr) + x SiO_2, where $0 < x < 20$ wt (%), by hot-press technique. The ion transport property has been studied in terms of basic ionic parameters, viz., conductivity (σ), mobility (μ), mobile ion concentration (n) and activation energy (E_a). We are also introducing a new ionic mobility-based τ model for conductivity enhancement in NCPEs.

11.2 Experimental

The as-supplied AR grade precursor chemicals were: poly(ethylene oxide) PEO (10^5 Mw, Sigma-Aldrich, USA), KBr (purity > 98%, Merck, India) and nanofiller SiO_2 (> 99.8%, size ~ 8 nm, Sigma-Aldrich, USA) have been used for synthesis of K^+-ion conducting NCPEs: (1–x) (70PEO: 30KBr) + x SiO_2, where 0 < x < 20 wt%. The present NCPE has been synthesized by hot-press technique. The details about the hot-press casting method have been reported in our earlier communications [10,12,13]. The conductivity (σ) measurements were carried out by the following equation:

$$\sigma = \frac{1}{R_b \cdot A} \tag{11.1}$$

where R_b is the bulk resistance, l is the thickness and A is the cross-sectional area of the polymeric sample material. The bulk resistance R_b was determined on different polymeric samples by using an LCR bridge (model: HIOKI 3520-01, Japan). The ionic mobility (μ), mobile ion concentration (n) and ionic transference number (t_{ion}) were determined with the help of the following well-known equations:

$$\mu = \frac{d^2}{V \cdot \tau} \tag{11.2}$$

$$\sigma = n \cdot q \cdot \mu \tag{11.3}$$

and

$$t_{ion} = 1 - \frac{I_e}{I_T} \tag{11.4}$$

where I_e is the electronic current and I_T is the total current of the cell: [SS // NCPE OCC // SS], d is the thickness of NCPE films, V is the applied external dc potential and τ is the time of flight. The time of flight (τ) was determined by directly employing dc polarization transient ionic current (TIC) [14]. To determine the activation energy (E_a) value of the optimum conducting composition of NCPE, a temperature-dependent conductivity study was done. A new solid-state polymeric battery was fabricated using hot-press synthesized NCPE OCC as an electrolyte in the configuration of the following cells:

K-metal	NCPE OCC	(C+I_2+Electrolyte)
(Anode)	(Electrolyte)	(Cathode film)

The cathode film was prepared by hot-pressing the physical mixture of elemental iodine (I_2), conducting graphite (C) and SPE in 1:1:1 weight ratios at ~50 °C. The cell performances were studied under different load conditions at room temperature and some important cell parameters were also calculated from the discharge profiles.

11.3 Results and Discussion

The salt concentration-dependent conductivity (σ) plot at room temperature for different hot-pressed solid polymer electrolyte (SPEs): (PEO:KBr), is shown in Figure 11.1 [15]. Ionic conductivity (σ) increased abruptly just after the addition/complexation of salt in the host polymer PEO. σ-Value remained practically stable on further addition of salt in PEO. However, a moderate sized σ-maxima appeared at 30 wt% of salt KBr. SPE films beyond 50 wt% salt concentration were brittle and appeared physically less stable/flexible. SPE film: (70PEO:30KBr), with room temperature conductivity $\sigma \sim 5.01 \times 10^{-7}$ S.cm^{-1}, have been identified as better host and this has been used as a first phase host matrix for synthesis of the present NCPEs. Figure 11.2 shows the room temperature conductivity variation of nanocomposite polymer electrolytes (NCPEs): (1−x) (70PEO:30KBr) + x SiO_2, where $0 < x < 20$ wt (%). The ionic conductivity (σ) of NCPE increases with increasing the nanofiller SiO_2 concentration and two maximum values of conductivity 2.5×10^{-5} S.cm^{-1} and 1.5×10^{-5} Scm^{-1} were observed at 5 wt% and 12 wt% of SiO_2. The approximate two and half orders of conductivity enhancement were observed from that of pure SPE host: (70PEO:30KBr). However, the maximum conductivity of 1.0×10^{-4} Scm^{-1} has been observed at 5 wt% of SiO_2 and this composition is known as optimum conducting composition (OCC). The existence of two conductivity maxima in the present NCPE can be attributed to two separate percolation thresholds involving two different kinds of mobile species: cation (K^+) and anion (Br^-). The ionic conductivity enhancement in NCPE is also due to the Lewis acid-base reaction [8, 10, 16, 17].

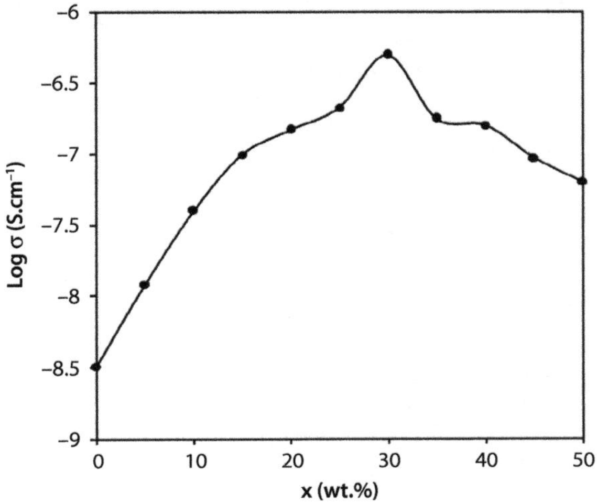

Figure 11.1 Room temperature conductivity plot of SPEs: (PEO:KBr) [15].

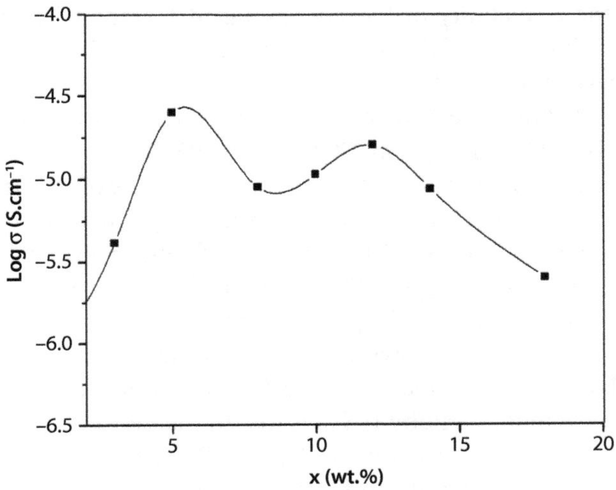

Figure 11.2 'Log σ – x' plot for the hot-pressed NCPEs: (1–x) (70PEO:30KBr) + x SiO_2.

Figure 11.3 shows 'log μ – x' and 'log n – x' plots for NCPEs: (1–x) [70PEO:30KBr] + x SiO_2. It is obvious from the figure that both variations are almost analogous to 'log σ – x' and the highest values of 'μ' and 'n' at room temperature are obtained at x = 5 wt%. These results clearly indicate the fact that the increase in room temperature conductivity of OCC NCPE

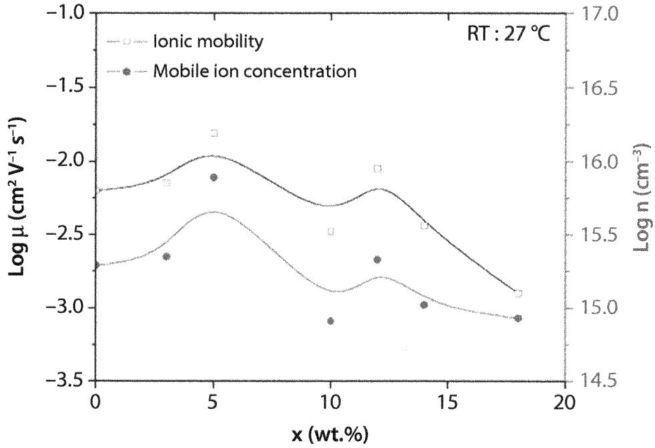

Figure 11.3 'Log μ – x' and 'log n – x' plots for the hot-pressed NCPEs: (1–x)(70PEO:30KBr) + x SiO_2.

film is also due to the increase in both 'μ' and 'n'. The increase in 'n' would be due to the availability of loosely entangled K^+ ions with the polymer chains, while increase in 'μ' would be a consequence of increased degree of amorphousity in SPE host.

Figure 11.4 shows the XRD analysis of pure KBr, pure PEO, pure SiO_2, SPE host: (70PEO:30 KBr) and NCPE OCC: 95[70PEO: 30KBr] + 5SiO_2. On closer inspection, it can be clearly noted that some of the peaks of pure PEO became relatively broader as well as less prominent/feeble after salt complexation/SiO_2 dispersal. The intensity of PEO main peaks has been decreased substantially, especially in NCPE film. This is indicative of a decrease in degree of crystallinity and/or increase of amorphousity in PEO and it also confirmed the polymer salt complexation as well as some formation of NCPE, as reported in our earlier communications [12, 13].

The increase in ionic mobility (μ) is due to the decrease in time of light (τ) in SPEs and NCPEs. The time of light in PEs decreases after the addition of nano inert materials and hence the number of free conducting paths increases. Therefore, the ionic conductivity as well as mechanical and thermal stability increases in filler doped PEs or nanocomposite polymer electrolytes. This kind of τ-based conductivity enhancement in PEs is also known as τ-model for ionic systems. The schematic representation of generation of free conducting paths (indicated by the arrow) in τ-model is shown in Figure 11.5.

To explain the ionic nature of the present NCPE, the ionic transference number (t_{ion}) of NCPE OCC has been evaluated using dc polarization

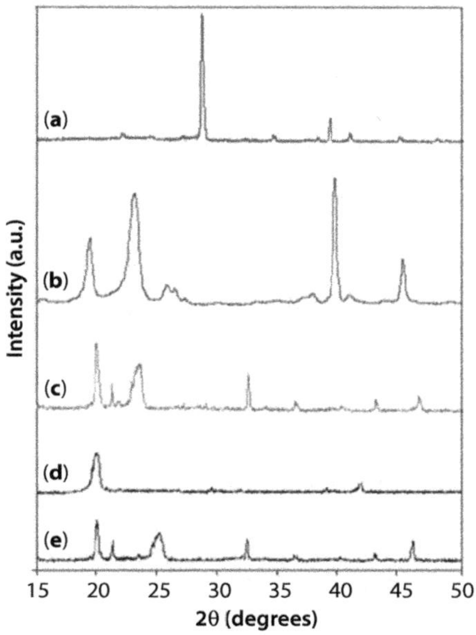

Figure 11.4 XRD patterns: (a) KBr, (b) pure PEO, (c) hot-pressed SPE host: (70PEO:30KBr), (d) pure nano-sized SiO_2 and (e) hot-pressed NCPE OCC: [95 (70PEO:30KBr) + 5 SiO_2].

TIC technique. The variation of current with time is shown in Figure 11.6, where t_{ion} ~ 0.95 is obtained for both the electrolytes and is indicative that the majority of charge carriers (~ 95%) are cations K^+ and only a very small contribution (~ 5%) of anion Br^-.

The temperature-dependant conductivity variations for hot-pressed NCPE OCC: [95(70PEO: 30KBr)+ 5SiO_2] are shown in Figure 11.7. 'Log σ – 1/T' plot exhibited straight line behavior with an upward change in slope at ~ 60 °C, which is due to the well-known semicrystalline-amorphous phase change/melting point temperature of PEO-based polymers. The slight decrease in the activation energy for NCPE OCC from that of SPE host is indicative of increased mobility as a consequence of increase in degree of amorphicity of PEO after SiO_2 dispersal. The straight line portion of the plot below this transition temperature can be linearly fitted-in very well in the following Arrhenius equation:

$$\sigma (T) = 1.14 \times 10^{1} \exp (-0.34/ kT) \qquad [S.cm^{-1}] \qquad (11.5)$$

where E_a = 0.34 eV is the activation energy value for hot-pressed NCPE OCC.

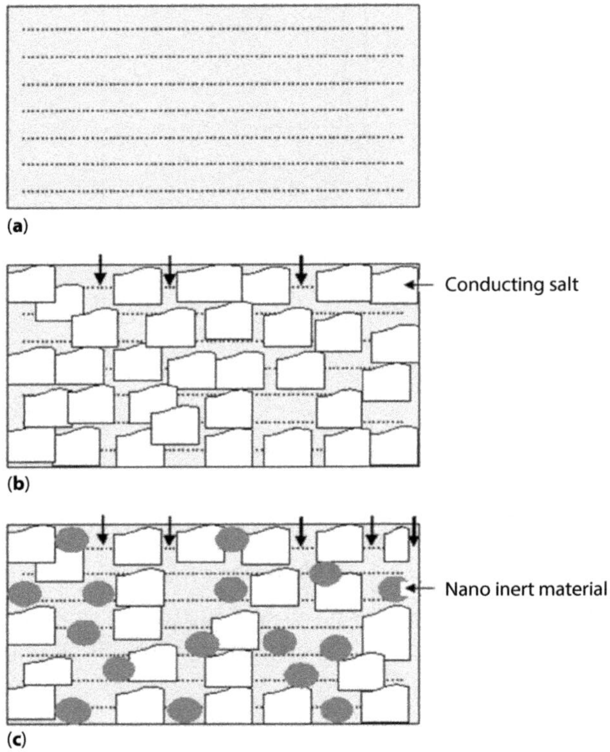

Figure 11.5 Schematic representation of conducting paths (↓) in PEs: (a) pure polymer, (b) polymer with ion conducting salt and (c) NCPE: (polymer + ion conducting salt + nano inert material).

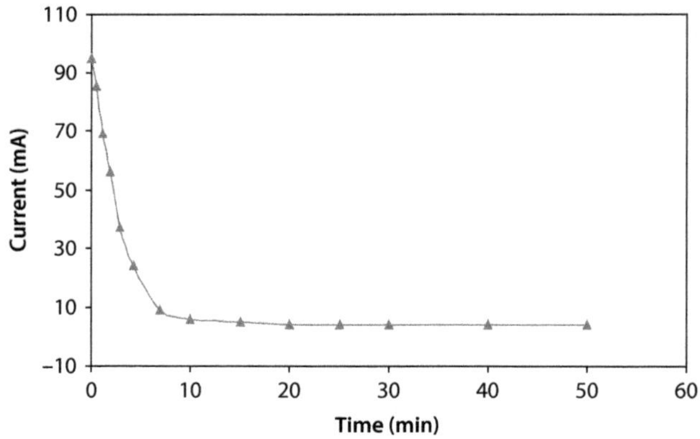

Figure 11.6 'Current – time' plot for the hot-pressed NCPE OCC: [95(70PEO:30KBr) + 5 SiO_2].

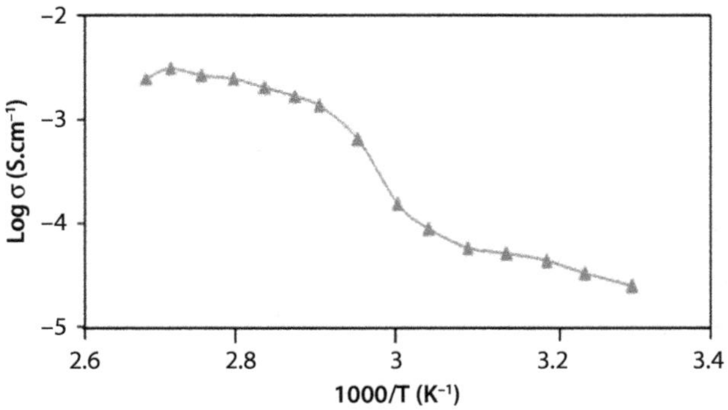

Figure 11.7 'Log σ – 1/T' plot for the hot-pressed NCPE OCC: [95(70PEO: 30KBr) + 5 SiO_2].

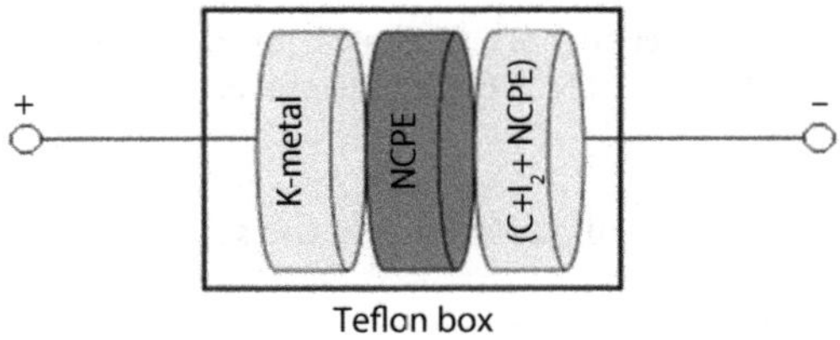

Figure 11.8 Schematic structure of solid-state polymeric cell.

A new K^+ ion conducting solid-state polymer battery was fabricated, as mentioned in the Experimental section. Figure 11.8 shows the schematic representation of solid-state polymeric cell. The cell potential discharge characteristics for the solid-state polymeric cell under two different loads, viz., 100 kΩ and 50 kΩ, at room temperature are shown in Figure 11.9. The obtained open circuit voltage (OCV) is ~2.9 V. It can be clearly noticed that except for the initial potential drop, OCV value remained practically stable for ~ 90 h (i.e., during a low current drain state). However, the cell potential decreased relatively faster when discharged through 50 kΩ load (i.e., during higher current drain states). The initial sharp decrease in voltage may be due to polarization and/or the formation of thin layer of potassium salt at electrode-electrolyte interface. Some important cell parameters were also calculated from the discharge profiles for direct comparison and are listed in Table 12.1. On the basis of these studies, it can be observed that the present polymeric cell, performed fairly satisfactorily, especially during low current drains at room temperature.

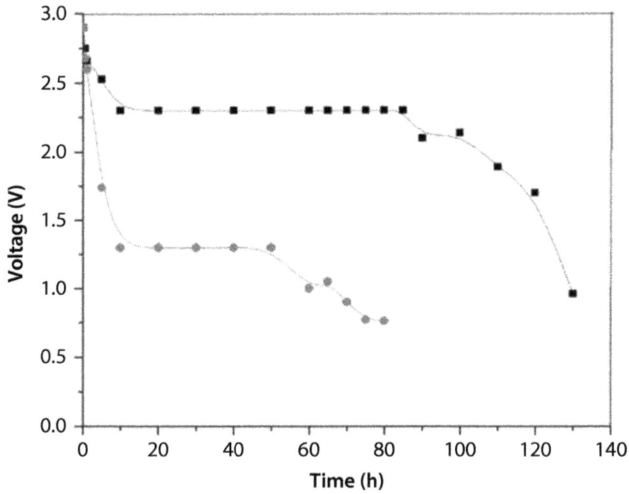

Figure 11.9 Discharge characteristics of a solid-state polymeric cell: [K || 95(70PEO:30KBr)+ 5 SiO_2 || (C+I_2+NCPE)] at different loads: 100 kΩ (▪) and 50 kΩ (•)

Table 11.1 Some important cell parameters of the solid-state polymeric cell at room temperature.

Load (kΩ)	Working voltage (V)	Current density (μA.cm^{-2})	Discharge capacity (μA.h)	Power density (mW.kg^{-1})	Energy density (mWh.kg^{-1})
100	2.3	1.34	204	31.06	2785
50	1.3	0.70	45	8.10	340.9

11.4 Conclusion

A new K^+ ion conducting nanocomposite polymer electrolyte (NCPE): [95(70PEO:30KBr) + 5SiO_2] has been synthesized by hot-press method. The two orders of conductivity enhancement ($\sigma \sim 2.5 \times 10^{--}$ S.cm^{-1}) were achieved at room temperature. Polymer-salt/filler complexations were confirmed by XRD analysis. The ionic mobility and mobile ion concentration are both responsible for the conductivity and stability enhancement in the present NCPEs. A new ionic mobility-based τ-model has been proposed for conductivity enhancement in PEs. The low activation energy in the present study indicates that it can be potentially used for solid-state device applications. The cell potential discharge characteristics were studied under varying load conditions at room temperature. The cell performed fairly satisfactorily under low current drain states.

Acknowledgment

We gratefully acknowledge SERB DST, New Delhi, for providing financial assistance through the Research Project (No. SR/ FTP/ PS-23/ 2009).

References

1. Chandra, S., *Superionic Solids: Principles and Applications*, Elsevier Science Ltd, 1981.
2. Owen, J.R., Fast ion conducting polymers, in: *Superionic Solids and Solid Electrolytes Recent Trends*, Laskar, A. (Ed.), Academic Press: New York, 1989.
3. Armand, M.B., Polymers with ionic conductivity. *Adv. Mater.* 2, 278, 1990.
4. Gray, F.M., *Solid Polymer Electrolytes: Fundamentals and Technological Applications*, Wiley-VCH: New York, 1991.
5. Dissanayak, M.K.L., Nanocomposite polymer electrolytes for device applications. *Ionics* 10, 221, 2004.
6. Chandra, A., and Chandra, A., Superionic solids in energy device applications, in: *Advanced Energy Materials*, Tripathi, A., and Valyukh, S. (Eds.), pp. 167–206, Wiley-Scrivener Publishing: USA, 2014.
7. Fenton, D.E., Parker, J.M., and Wright, P.V., Complexes of alkali metal ions with poly(ethylene oxide). *Polymer* 14, 589, 1973.
8. Argawal, R.C., and Pandey, G.P., Solid polymer electrolytes: Materials designing and all-solid-state battery applications: An overview. *J. Phys. D: Appl. Phys.* 41, 223001, 2008.
9. Agrawal, R.C., and Mahipal, Y.K., Study of electrical and electrochemical behaviour on hot-press synthesized nano-composite polymer electrolyte (NCPE) membranes: [(70PEO: 30 KNO_3) + x SiO_2]. *Int. J. Electrochem. Sci.* 6, 867, 2011.
10. Chandra A., and Chandra A., *Hot-Pressed Polymer Electrolytes: Synthesis and Characterization*, Lambert Academic Publishing: Germany, 2010.
11. Appetecchi, G.B., Croce, F., Hassoun, J., Scrosati, B., Salomon, M., and Cassel, F., Hot-pressed, dry, composite, PEO-based electrolyte membranes: I. Ionic conductivity characterization. *J. Power Sources* 114, 105, 2003.
12. Chandra, A., Chandra, A., and Thakur, S.S., Hot-pressed nano-composite polymer electrolytes: (1–x) [70PEO:30$KBrO_3$]: x SiO_2. *Indian J. Pure Appl. Phys.* 49, 698, 2011.
13. Chandra, A., Chandra, A., and Thakur, K., Synthesis, characterization and polymer battery fabrication of hot-pressed nano composite polymer electrolytes. *Compos. Part B* 60, 292, 2014.

14. Chandra, S., Tolpadi, S.K., and Hashmi, S.A., Transient ionic current measurement of ionic mobilities in a few proton conductors. *Solid State Ionics* 28–30, 651, 1988.
15. Chandra, A., Hot pressed K^+ ion conducting solid polymer electrolytes: Synthesis, ion conduction and polymeric battery fabrication. *Indian J. Phys.* 90, 759, 2016.
16. Lakshmi, N., and Chandra, S., Proton conducting composites of heteropolyacid hydrates (phosphomolybdic and phosphotungstic acids) dispersed with insulating Al_2O_3. *Phys. Status Solidi A* 186, 383, 2001.
17. Dey, A., Karan, S., and De, S.K., Effect of nanofillers on thermal and transport properties of potassium iodide–polyethylene oxide solid polymer electrolyte. *Solid State Commun.* 149, 1282, 2009.

12

Recent Studies in Polyurethane-Based Drug Delivery Systems

Archana Solanki* and Sonal Thakore*

Department of Chemistry, Faculty of Science, Maharaja Sayajirao University of Baroda, Vadodara, India

Abstract

Polyurethanes (PUs) are polymers tailored to specific material applications by varying the structure of their components, viz., polyol, diisocyanate and cross-linker, to yield a wide range that varies from thermoplastic elastomers to rigid thermoset foams. An approach to develop biocompatible PUs utilizes natural ingredients such as vegetable oils (as polyols) and carbohydrates (as cross-linkers). The variety in structure of carbohydrates, degree of cross-linking and NCO:OH ratio can lead to materials with tunable mechanical properties and biodegradability ideal for the controlled release systems. PUs have also been developed for pH-responsive drug delivery by using α-hydroxy acids. This chapter discusses an approach for development of biocompatible polyurethanes for drug delivery applications.

Keywords: Carbohydrate, pH sensitive, drug delivery, polyurethanes

12.1 Introduction

12.1.1 Polyurethane Chemistry: A Brief Overview

Polyurethanes are important versatile polymers, varying from flexible thermoplastics to rigid thermosets. Polyurethanes (PUs) originated in 1937 at I.G. Farbenindutrie when Otto Bayer (Figure 12.1), also called the "father" of polyurethanes, and his coworkers performed addition polymerization reaction between diisocyanates and diols. Since their discovery,

**Corresponding authors*: archu.hrt@gmail.com; drsonalit@gmail.com

Sanjay K. Nayak *et al.* (eds.) Trends and Applications in Advanced Polymeric Materials, (219–244) 2018 © Scrivener Publishing LLC

Figure 12.1 Otto Bayer, the "father" of polyurethanes (reprinted from [5]).

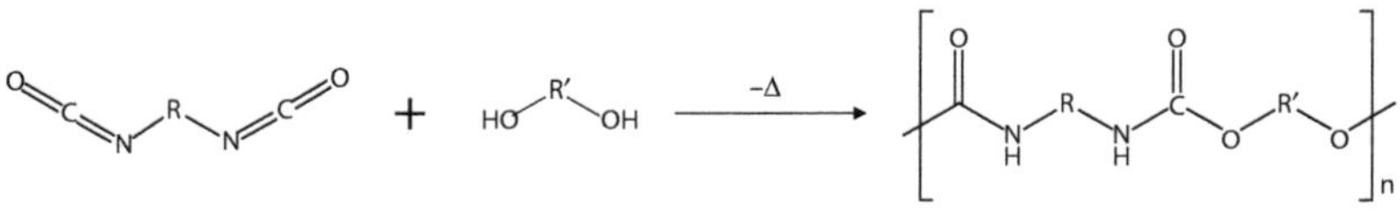

Figure 12.2 Chemical reaction of polyurethane formation [7, 8].

the demand for PUs has continued to increase and is expected to attain a production of 18 million tons in 2016 [1–4].

The term "polyurethane" is confusing since the name may mean that PUs are either polymers obtained by polymerization of ethane or are polymers containing urethane compounds as monomers; neither meaning is true! Regardless of the chemical composition of the rest of the chain, PU polymers contain a plurality of urethane groups in the molecular backbone. PUs are obtained by step-growth polymerization between diisocyanate/polyisocyanate with isocyanate (-NCO) functional group and hydroxyl-terminated oligomer (polyol) having at least two reactive hydrogen atoms [6]. Thus, in addition to urethane, a typical PU may contain aliphatic and aromatic hydrocarbons, esters, ethers, amides, urea and isocyanurate functional groups.

The chemistry of PUs involves reactions between organic isocyanates and polyols, i.e., compounds containing active hydrogen at proper ratio in the presence of suitable catalyst (Figure 12.2). The reaction is exothermic and leads to the formation of prepolymer. If the initial concentration of isocyanate is higher than polyol, the prepolymer obtained by the reaction between them is said to be NCO-terminated prepolymer, which can further react with other cross-linker/chain extender or other chemicals that are prone to react with NCO functional groups.

Depending upon their structure, PUs can be either thermoplastic or irreversibly thermoset. For example, PUs synthesized from diisocyanate and a bifunctional polyol result in thermoplastic PU, whereas introduction of cross-linkers with more than two functional groups will yield thermoset PUs with three-dimensional cross-linking [9]. The common components used in PU synthesis are listed in Table 12.1.

The synthesis of PUs requires two essential ingredients: isocyanate (typically diisocyanate) and a bi- or multifunctional polyol with two hydroxyl (OH) terminal groups [10]. The step-growth polymerization reaction between these two reactants yields the polyurethane polymer. Polyols are generally di-hydroxyl-terminated polymers, viz., polyethers, polyesters of polycarbonates. The type, chain length, functionality and molecular weight of polyols have a significant effect on the structural and physical properties of PUs. For example, relative cohesive energy of oligo-polyols increases as we move from polyester polyol to polyether polyols. Since a stronger cohesive energy leads to higher physicochemical properties, the PUs based on polyester polyols impart better properties than those based on polyether polyols [8]. Similarly, short-chain polyol-based PUs have a higher concentration of urethane and urea bonds, leading to high cohesive interactions between these bonds, which lead to rigid structure. On the contrary, long-chain polyols lead to flexible PUs for the same reasons. If we compare functionalities of polyols, the high functionality polyol leads to high cross-linking in PU, which ultimately results in enhanced rigidity, softening points and modulus of elasticity, while reduced elongation and molecular mobility.

On the other hand, low functionality polyols give PUs with linear structure showing an opposite trend of properties. Moreover, low molecular weight polyols yield hard PU, whereas high molecular weight polyols yield elastic, flexible PUs.

The PUs obtained through reaction between only bifunctional polyols and isocyanates is linear in nature and may show poor mechanical properties. The introduction of chain extenders/cross-linkers (Table 12.1) into the PU network can yield PUs with improved physical properties. This is due to enhanced distinction of micro-phase separation between hard segments (composed of –NH-COO- and the chain extender/cross-linker) and soft segments (the polyol part), as shown in Figure 12.3.

Table 12.1 Detailed list of common components in PUs.

Component	Type	Examples	Structure
Diisocyanate	Aromatic	Toluene-2,4-diisocyanate (TDI)	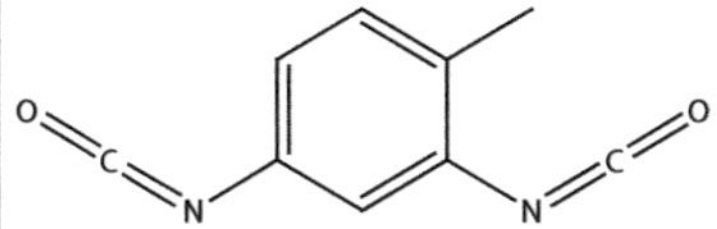
		4,4'-Methylene-bis(phenylisocyanate) (MDI)	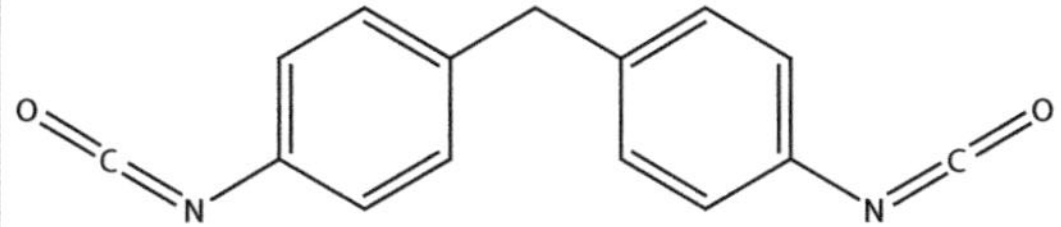
	Aliphatic	Hexamethylene diisocyanate (HMDI)/1,6-diisocyanatohexane	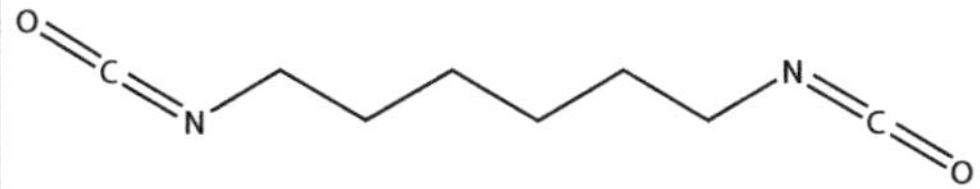
	Alicyclic	Isophorone diisocyanate (IPDI)	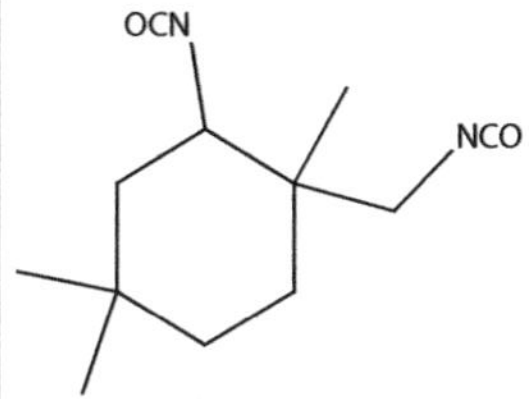
Polyols	Polyether	Polypropylene glycols	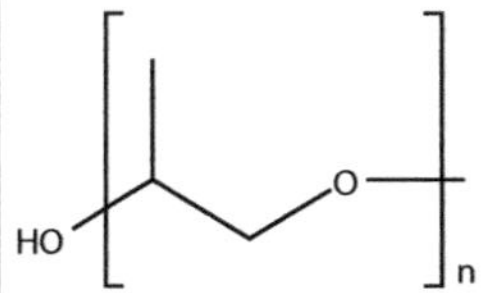

	Polyester	Polyester polyol of diacid and diol	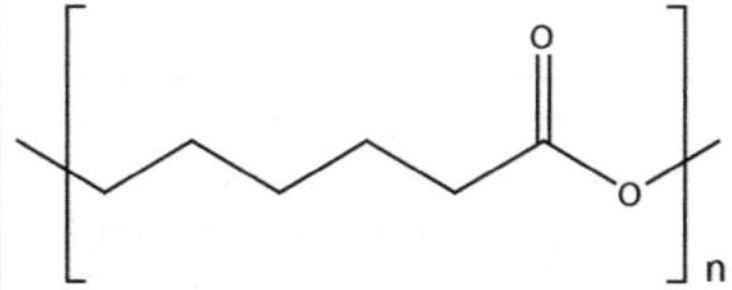
	Natural oils	Castor oil	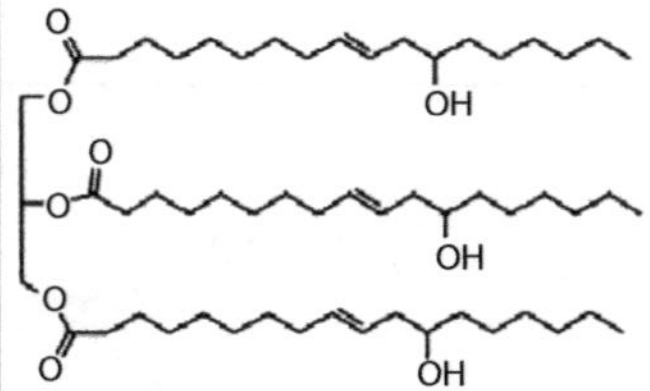
Chain extenders	Hydroxylated	Diethylene glycol	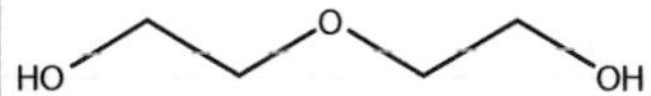
		1,4-Butane diol	
	Diamines	Ethylene diamine	
Cross-linkers	Hydroxylated	Trimethylolpropane	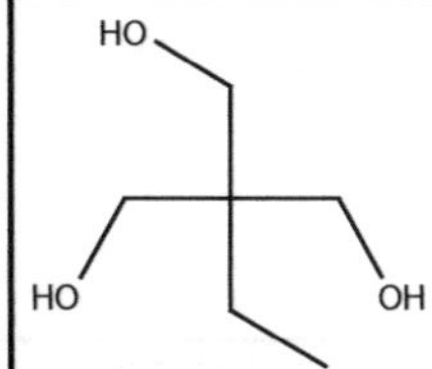

(Continued)

Table 12.1 Cont.

Component	Type	Examples	Structure
		Carbohydrates Monosaccharide (Glucose)	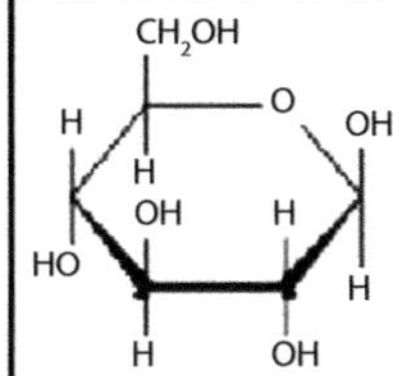
		Disaccharide (Sucrose)	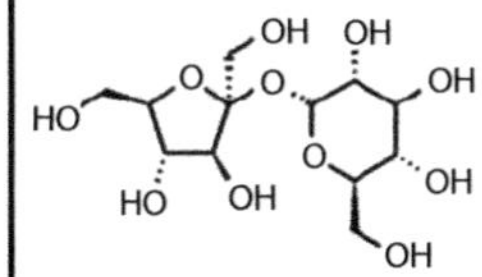
		Polysaccharide (Starch)	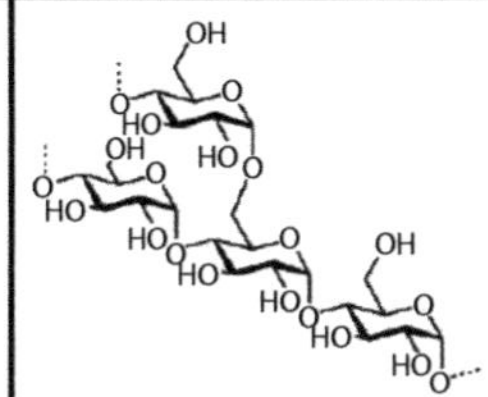
		Polysaccharide (Cellulose)	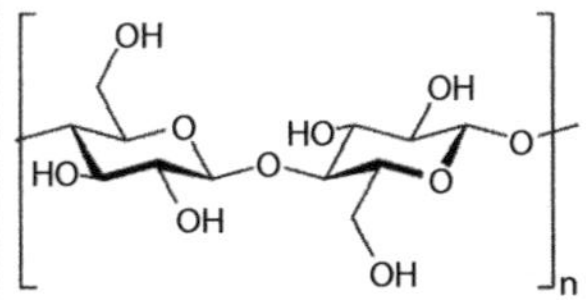

	Diamine	4,4'-Methylene-bis-(2-chloroaniline)	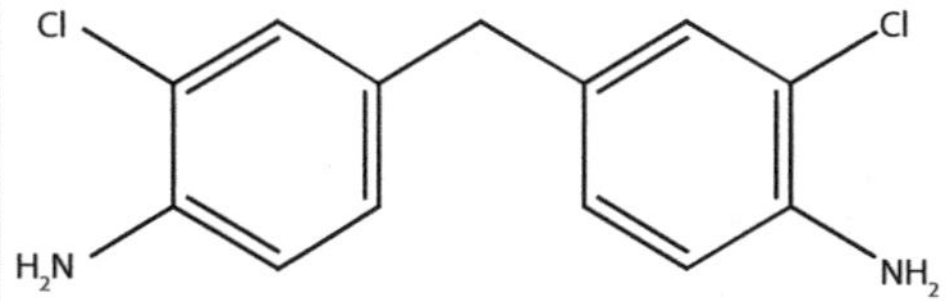
Catalysts	Organometalic	Stannous octate	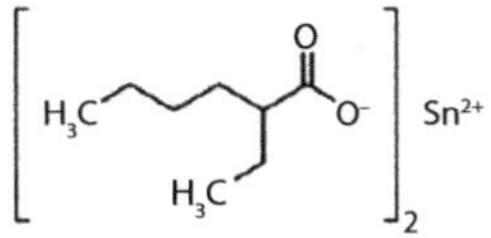
		Dibutyl-tin-dilaurate	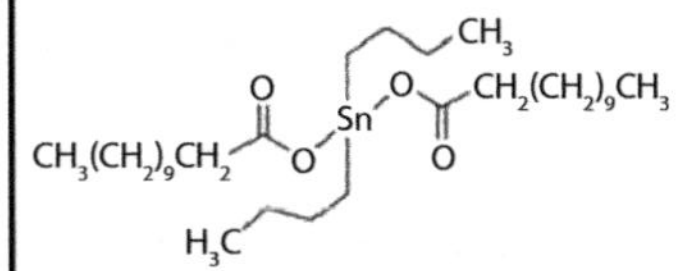
	Amine	N,N-Dimethylethanolamine	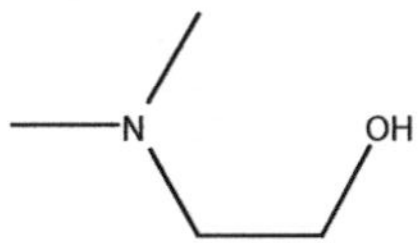

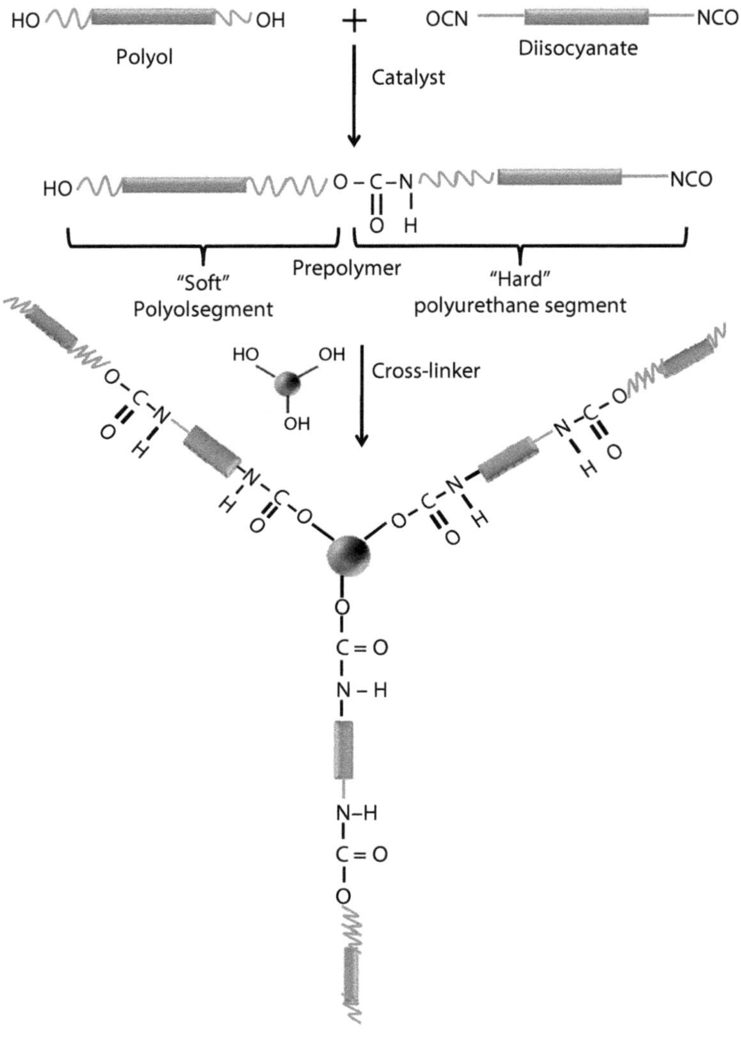

Figure 12.3 Synthesis of cross-linked PU elastomer.

The hard domains are responsible for mechanical strength while the soft domain provides flexibility [11, 12]. The extent of hard and soft segment content directly influences the final properties of PUs. For example, PUs with hard segment content of 40–65% are reported to be tough elastomeric materials while those with hard segment contents of 15–40 wt% are rather soft and rubbery elastomers [13].

12.1.2 Carbohydrate Cross-Linked Polyurethanes

Carbohydrates are an abundant and inexpensive renewable resource. Being multi-hydroxyl compounds, they can easily serve the purpose of a cross-linker due to the reactivity of hydroxyl groups with isocyanate groups. Carbohydrates, such as starch, cellulose, chitosan, carrageenan and alginate, are finding increasing application as biomaterials due to their properties like availability, biodegradability, sustainability, lower toxicity and biocompatibility. However, lack of thermal stability, poor solubility and difficult processability are serious limitations for these materials. The incorporation of carbohydrates into the structure of PU can result in synergistic polyurethane with enhanced mechanical properties, thermal stability and biodegradability. The multiple hydroxyl groups in the carbohydrate structure can be utilized in the synthesis of PUs to confer biomaterial qualities. Hence, we decided to use multi-hydroxyl carbohydrates as cross-linkers in PU system, which is likely to offer the following advantages:

1. Enhance both degradability and mechanical properties of polymer;
2. Provide multifunctional sites to form randomly cross-linked networks;
3. Nontoxic, inexpensive, readily available from renewable resources.

Among carbohydrates, the use of starch and cellulose has been reported in the synthesis of PUs [14–17].

Starch is a major carbohydrate reserve in plants. It is found in the natural resources of plants like leaves, fruits, flowers, seeds, stems and roots [18]. Chemically, starch consists of mainly two units, viz., several million amylopectin molecules and a much larger number of smaller amylase molecules. The biochemical chain responsible for starch synthesis involves glucose molecules produced in plant cells by photosynthesis. Amylopectin consists of linear chains of glucose units linked by α-1,4-glycosidic bonds. It is highly branched at the α-1,6 positions by small glucose chains at intervals of 10 nm along the molecule's axis. Amylopectin constitutes between 70 to 85% of common starch [19]. On the other hand, amylose is essentially a linear chain of α-1,4 glucans with limited branching points at the α-1,6 positions. Amylose constitutes between 15 to 30% of common starch. Starch's structural units, amylose and amylopectin, are shown in Figure 12.4.

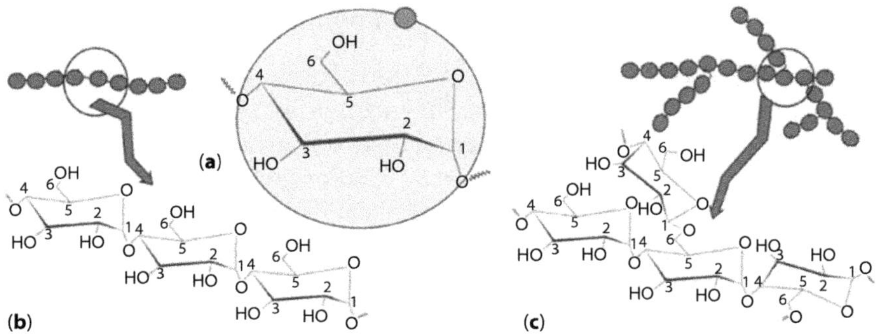

Figure 12.4 Chemical structures of (a) Starch, (b) Amylose and (c) Amylopectin.

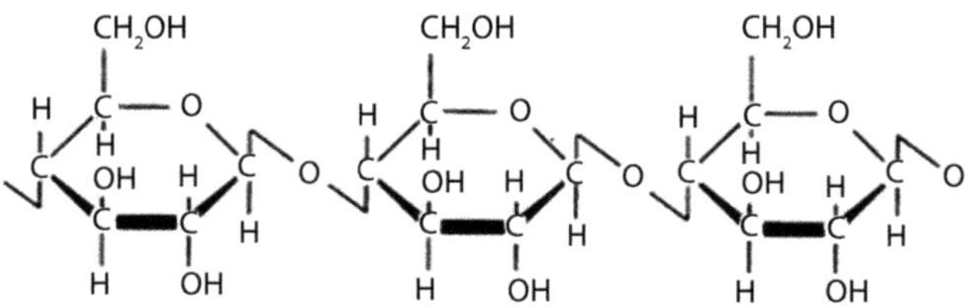

Figure 12.5 Chemical structure of cellulose.

Plasticized starch has been reported for enhancement of mechanical properties, biodegradation and water resistance of poly(ε-caprolactone)-based PUs [20]. For enhancement in mechanical properties, starch nanocrystals have been used as fillers to prepare nanocomposite materials by preparing its composites with PU [14].

Cellulose is found in plants as microfibrils [21, 22] with dimension of around 2–20 nm diameter and length of 100–40,000 nm. The cellulose is responsible for the formation of a strong framework in the cell walls. Structurally, cellulose is a linear polymer of β-(1,4)-D-glucopyranose units, as shown in Figure 12.5. The fully equatorial conformation of β-linked glucopyranose residues stabilizes the chair structure, minimizing its flexibility (for example, relative to the slightly more flexible α-linked glucopyranose residues in amylose).

Many studies have been reported using cellulose in different forms. Small quantities of cellulose nanocrystals were loaded in as fillers in waterborne PU matrix in order to obtain a reinforcing effect on resulting PU. The strong hydrogen bonding between PU matrix and cellulose nanocrystals, as well as the rigid nature, hydrophilicity and high aspect ratio, were considered as main factors contributing excellent reinforcement of properties

[23, 24]. In another study, cellulose nanocrystals were dispersed in thermoplastic polyurethane as matrix. It was observed that the matrix formation can have a significant effect on PU microstructure. The presence of a high number of hydroxyl groups on the surface of cellulose nanocrystals were deemed responsible for interactions between soft segments and hard segment [25]. Moreover, nanocomposites prepared from cellulose nanocrystals and PU have been reported for wool antifelting agents intended for application in the textile field [26]. Cellulose acetate and PU were blended to prepare nanofibrous scaffolds. The drug loading on PU scaffolds was carried out by electrospinning process for intended applications in burn, chronic and diabetic wound infections [27].

Incorporation of modified forms of starch and cellulose nanocrystals into PUs resulted in enhancement in mechanical strength, biodegradability as well as biocompatibility [23, 28, 29]. However, a systematic investigation and comparison of the performance of various carbohydrates as cross-linkers of PUs is not available. Moreover, a major concern is the thermal stability of the resulting PUs.

A systematic comparative study was attempted by synthesizing polyurethanes using different polyols (castor oil, polypropylene glycols (PPGs) and polycaprolactone diol [PCL-diol]), aromatic and aliphatic isocyanates, namely toluene diisocyanate (TDI) and Hexamethylene diisocyanate (HMDI) respectively, and various carbohydrate cross-linkers [30–32]. Scheme 12.1 describes the synthesis of a series of PUs with carbohydrate cross-linkers with or without chain extender [30]. The PUs were characterized for various properties, as listed in Table 12.2.

The study reveals that carbohydrates can be promising cross-linkers for synthesis of polyurethanes. The IR spectrum of PU elastomer (Figure 12.6) confirms the cross-linking between the carbohydrate and polyurethane phase. A comparison of IR spectrum of PU, drug, and drug-loaded PU serves as a tool to ascertain the loading of drug in PU. The mechanical property of PUs was also dependent on the type of polyol. With a combination of polyols and carbohydrates a wide spectrum of significant mechanical properties can be achieved, which can be further tuned by varying the NCO/OH ratio and polyol/cross-linker ratio (Figure 12.7). The resulting PUs also possessed reasonable thermal stability and interesting sorption properties.

12.1.3 Biomedical Applications of PUs

Polyurethanes can be synthesized from a wide variety of materials exhibiting extremely different properties that can lead to a broad spectrum of biomedical applications, as illustrated in Figure 12.8.

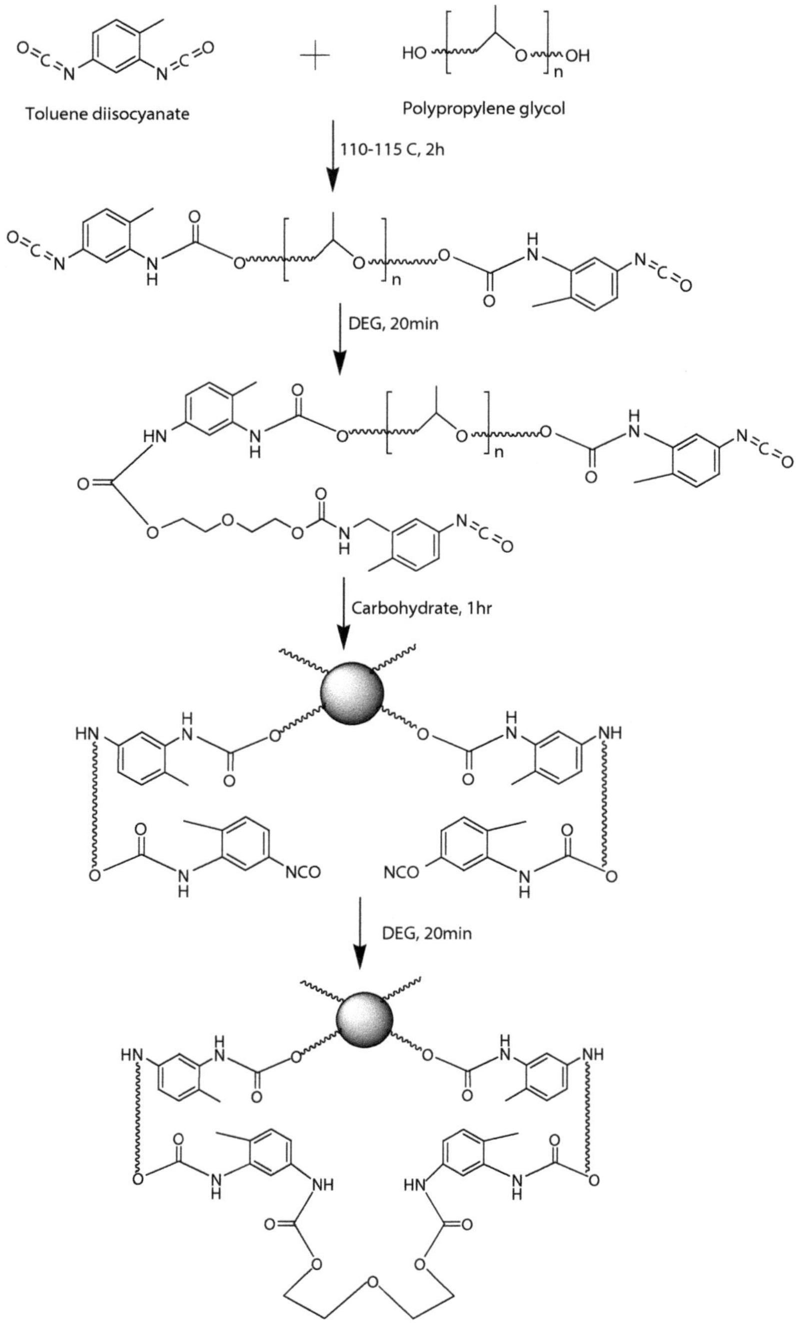

Scheme 12.1 Synthesis of carbohydrate cross-linked PU (reprinted from [30]).

Table 12.2 List of techniques used for determination of properties of PUs.

Sl. no.	Techniques/instrument used	Properties
1	Universal testing machine	Mechanical properties
2	Thermo gravimetric analysis (TGA) and Differential scanning calorimetry (DSC)	Thermal properties
3	Scanning electron microscopy (SEM)	Morphology
4	Fourier transform infrared spectroscopy (FTIR)	Structural elucidation
5	UV-Vis-NIR spectroscopy	Drug release profile

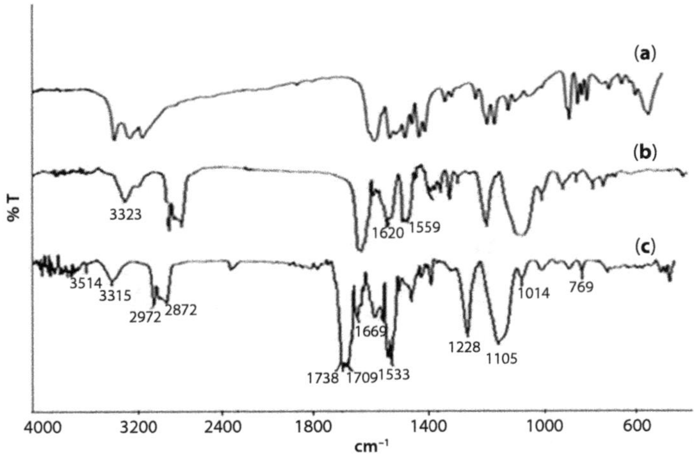

Figure 12.6 FTIR spectra of (a) Lamotrigine drug sample, (b) Lamotrigine loaded PU elastomer and (c) PU elastomer. (Reproduced from [30]).

A broad spectrum of properties essential for biomedical applications can be duly achieved by maintaining the type and quantity of ingredients used for PU synthesis. The PUs are largely explored for various biomedical applications, including drug delivery devices. Sivak *et al.* reported catalyst-dependent drug loading and controlled release of PU foams for implantable devices. The study revealed the role of catalyst selection in maintaining biocompatibility of medical-grade PU [33]. Biocompatible PU based on lysine diisocyanate was prepared and covalently bonded with 7-tert-butyldimethylsilyl-10-hydroxy-camptothecin (DB-67) for potential

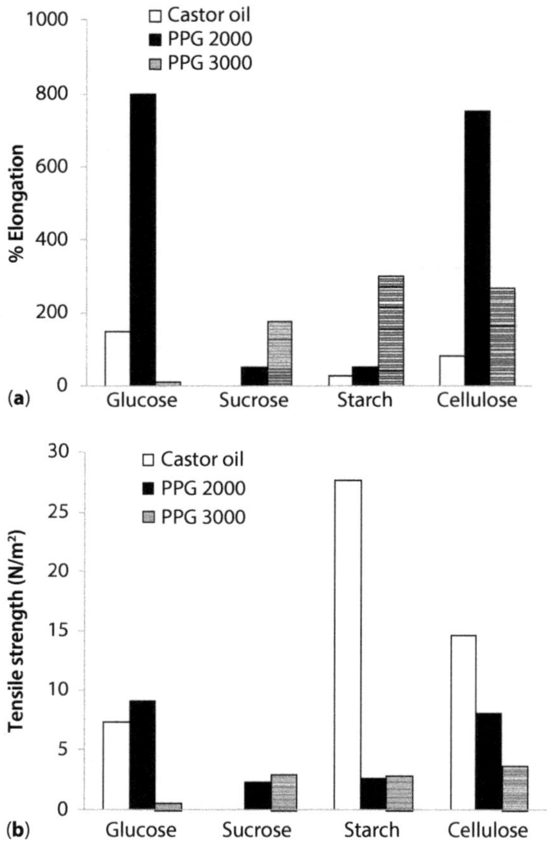

Figure 12.7 Effect of various polyols and carbohydrates on (a) % elongation and (b) tensile strength of polyurethanes. (Reproduced from [31]).

application in postoperative intracranial chemotherapy of cancers. The drug was released through slow, passive hydrolysis of urethane bonds [34].

12.2 Experimental

12.2.1 Impact of PU Chemistry on Drug Delivery Profiles

As discussed in the previous section, PUs are characterized by phase segregation due to the presence of hard and soft segments. This specific property is a boon for loading and release of drugs, as will be discussed in this

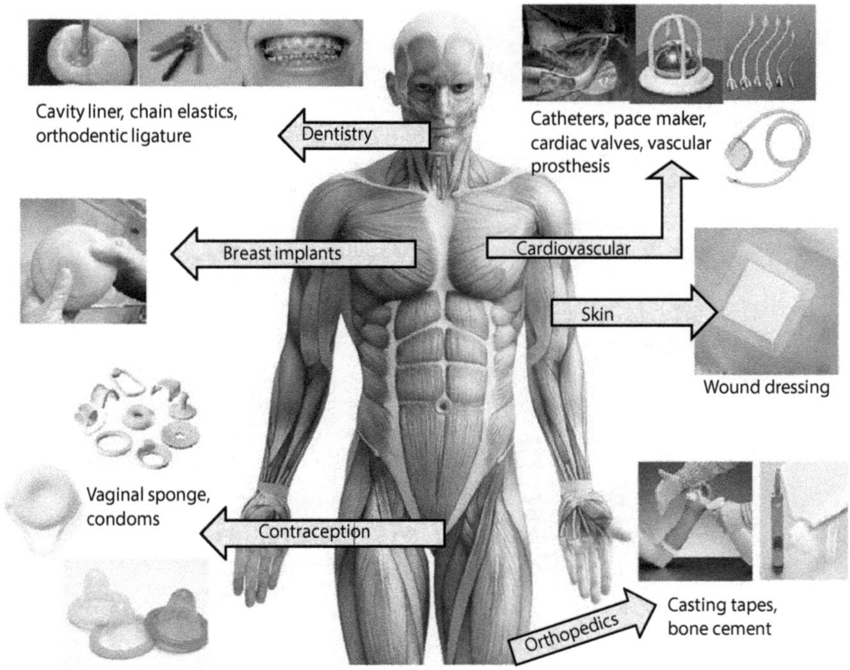

Figure 12.8 Applications of PUs in the biomedical field.

section. Drug loading in the PU matrix can be obtained due to the following three phenomena (Figure 12.9) [13]:

1. Drug entrapment inside the cross-linked structure of PU (Figure 12.9a);
2. Incorporation of drug into soft segments (Figure 12.9b);
3. Covalent attachment of drug with the hard segment of PU (Figure 12.9c).

The first phenomenon is as true for all polymers with cross-linked structure as it is for PUs. What is special about PUs are the last two phenomena. It is therefore said that phase segregation can be a boon for drug delivery applications in the case of PUs. The migration process of drugs from their initial position inside the polymer matrix to the release medium is responsible for the release of drugs from the polymeric matrix. The drug molecules might also be adsorbed onto the surface of material that might get released immediately after they are brought in contact with the release medium. Increasingly complex and multifunctional PUs have been explored in order

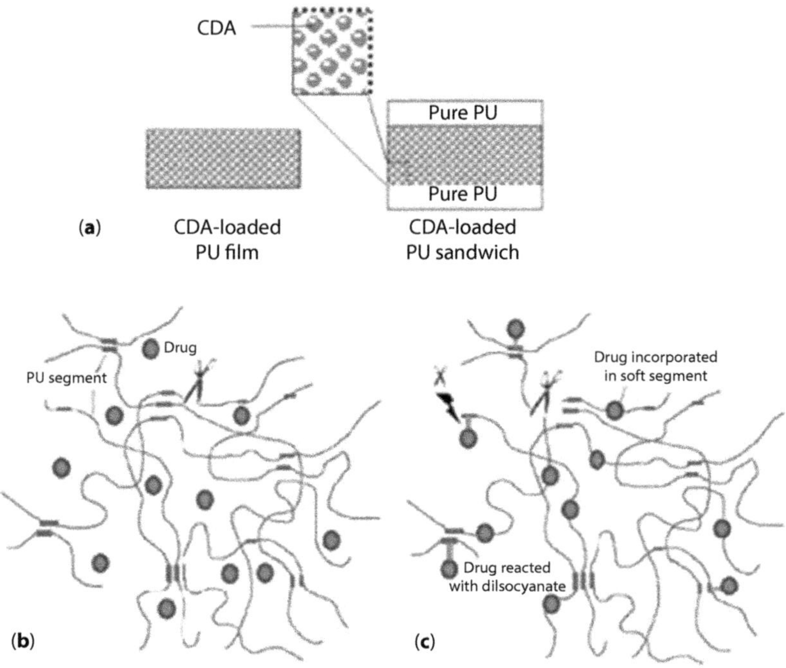

Figure 12.9 Mechanism of drug loading in PU matrix. (Reproduced from [35]).

to control the release of drugs with respect to external factors like pH [36], temperature [24, 37, 38] and magnetic field [39]. Moreover, the surface of PU nanoparticles can be conjugated with targeting units to promote cellular recognition, binding and internalization [40].

Another interesting fact about the structural dependency of drug release in the case of PUs is that the drug release from the PU network is expected to follow different mechanisms depending upon the degradability of PUs. For example, drug release in the case of nondegradable PUs is mainly governed by diffusion and the rate of drug release is controlled by thickness and permeability of polymer network as well as the solubility of the drug in the polymer matrix [41]. The nondegradable PUs have found biomedical applications, such as orthopedic implants [42], due to their good blood compatibility, durability and mechanical strength over a long peiod of time. On the other hand, the drug release from degradable PU matrix is often discussed in relation to their composition, drug loading, swelling and rate of degradation. The pH of external media, presence of enzymes and temperature of release media also affect the rate of drug release. The major areas of application for degradable PUs include nanoparticulate systems [43], membrane systems [37] and matrix/scaffolds [44, 45].

Moreover, the stoichiometry of PUs does impact the drug release profiles. This is because the mechanical, thermal and morphological properties can be tuned by varying PU stoichiometry [15, 31, 46, 47]. Such tuning can lead to control over swelling properties of PUs, thereby possible command over the drug loading and release profiles of PUs. The variation in ratio of hard/soft segment content can have a considerable effect on the properties of PU. This characteristic of PU can be implemented for essential modification of drug diffusion and release profiles. Hence, by altering the stoichiometry of PUs, the desired drug payload and release rate can be established.

12.2.2 Drug Loading and Release Kinetics

The drug loading and release kinetics were studied by using lamotrigine as a model drug. The polymers were designed in such a way that the drug release was tailored by differences in the stoichiometry of polymers. The kinetics and release mechanisms were observed to be a function of stoichiometric parameters such as type of cross-linker, polyol/cross-linker ratio and polyol/chain extender ratio (Figure 12.10).

Scanning electron microscopy (SEM) is an appropriate tool to study the effect of drug loading and drug release on the morphology of a polymer matrix. The SEM of PU film (Figure 12.11) without drug shows uniform

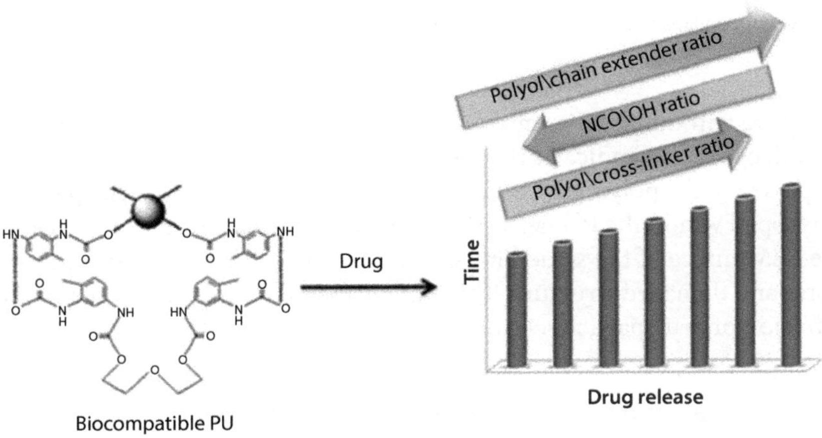

Figure 12.10 Schematic representation of drug release rate as a function of stoichiometry of PUs. (Reproduced from [30])

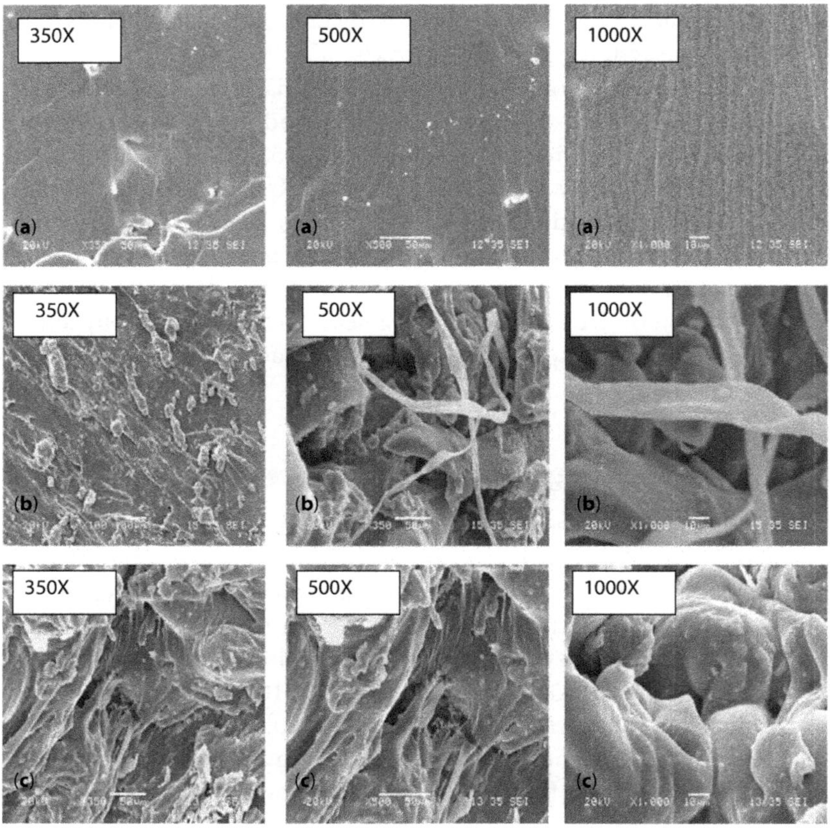

Figure 12.11 SEM micrographs of PU III-Cel at various magnifications: (a) only PU, (b) PU with encapsulated drug, (c) PU after release of drug. (Reproduced from [30])

structure without the presence of voids or cavities. The same film analyzed after drug loading indicates the presence of drug particles with varying size, dispersed in the polyurethane. This ascertains the presence of drug particles entrapped within the PU network. When the drug was completely released, the SEM image of the same film showed the layered bruises, non-uniform pores and damaged structure. This was attributed to the vacated space after diffusion of drug particles, leading to severe damage to the PU surface.

12.2.3 Waterborne pH-Responsive Polyurethanes

Polyurethanes are one of the most adaptable polymeric materials with regard to both processing methods and mechanical properties [48]. However, the

PU products synthesized by conventional process usually contain a significant amount of hazardous air pollutants (HOPs) and sometimes even free isocyanate monomers [49]. Besides this, conventional PU resin systems usually contain a high proportion of volatile organic components (VOC), leading to serious environmental problems. Therefore, in response to the increasing environmental legislation demand for raw materials with low VOCs, solvent-based PUs are being gradually replaced by waterborne polyurethanes (WPUs). The increased use of WPUs nowadays is attributed to their exceptional set of properties such as [50–52]:

- Resistance to chemicals, solvents and water;
- Flexibility and toughness;
- Abrasion resistance;
- Adhesion to various substrates.

In recent years, considerable attention has been focused on controlled drug delivery systems capable of releasing their drug payload in response to external stimuli. Such systems have found increasing applications in tissue targeting, achieving specific intracellular locations, or promoting controlled drug release. Currently there is an increasing demand for stimuli-sensitive drug devices having a modulated drug release in response to external factors such as pH [53, 54], temperature [37, 55], magnetic field [56, 57], etc. The release of drug from the polymer can be controlled by external stimuli such as temperature, pH, ionic strength, an electric field, magnetic field or metal, etc. (Figure 12.12). Polymers that respond to these external stimuli can be used as controlled-release devices. This type of polymer is called a stimuli-responsive polymer or "smart polymer." For a brief overview of such systems, examples of temperature and pH-responsive polymers are considered below.

The pH-responsive polymers are synthesized by incorporation of acidic or basic pendant groups that either accept or donate protons in response to the environmental pH. The swelling of polymer is directly proportional to the pH of external media in the case of polymers with weakly acidic (anionic) groups, but inversely proportional in the case of polymers with weakly basic (cationic) groups [58] (Figure 12.13).

To prepare stimuli-sensitive polyurethanes or the so-called "smart polymers," α-hydroxy acids were introduced into the PU structure, which acted as drug release modifiers cum chain extenders. Cellulose cross-linked waterborne polyurethanes (PUs) based on poly ε-caprolactone with lactic acid (LAPU)/glycolic acid (GAPU)/dimethylol-propionic acid (DAPU) were prepared [32]. This resulted in the pH-responsive

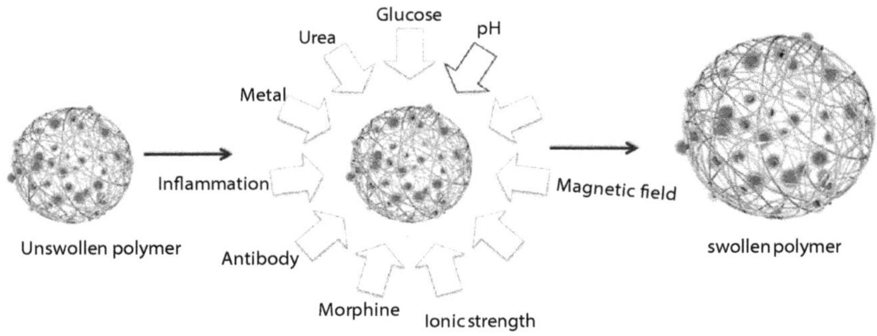

Figure 12.12 A schematic diagram showing swelling of stimuli-responsive polymers.

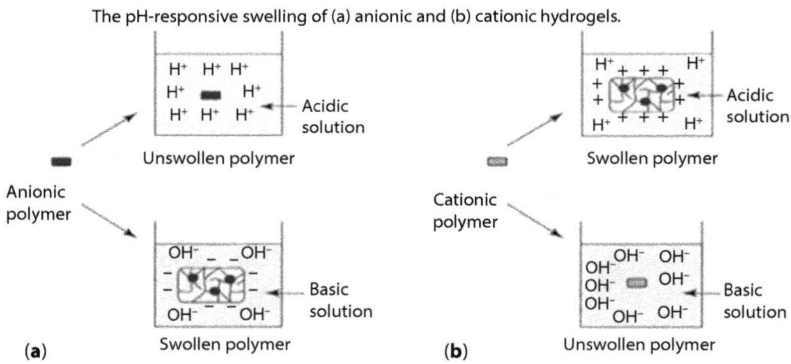

Figure 12.13 Effect of pH of external media on the swelling properties of polymer.

characteristic of synthesized PUs. The thermal degradation, structural and morphological properties of all PUs were identical, irrespective of type of chain extender.

The rate of release of the drug was increased when the pH of buffer increased from 1.2 to 4.5 and again increased when pH was 7.4 (Figure 12.14). The pH response of the PUs was attributed to acid chain extenders that form anionic polymer when subjected to phosphate buffer solution (PBS). At pH 1.2, the release of the drug was observed due to swelling in the PBS. The response with respect to pH is triggered due to the presence of insoluble -COOH groups that get ionized at pH 4.5. The presence of a large number of anionic groups on the polymer makes the polymer chains repel one-another, leading to expanded dimension or higher swelling of the polymer. The phenomenon becomes more effective when the pH increases to 7.4 (Scheme 12.2). Thus, all PUs showed a significantly

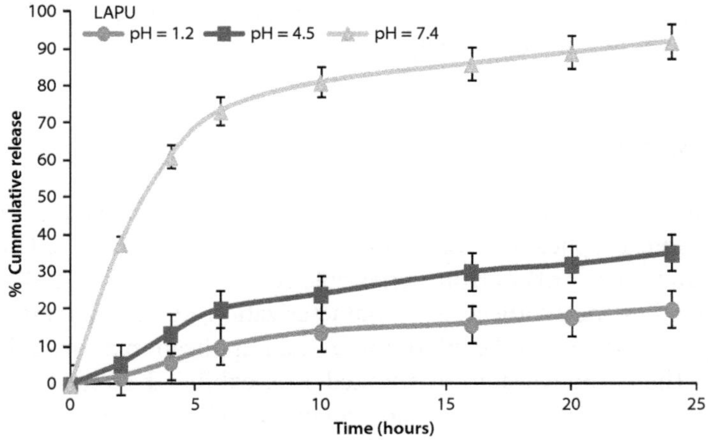

Figure 12.14 Drug release profiles for pH-responsive polyurethanes. (Reproduced from [32])

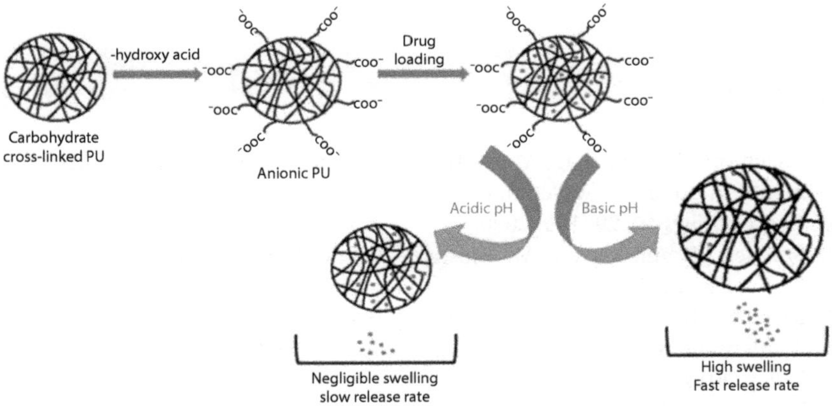

Scheme 12.2 Mechanism of pH-responsive PUs. (Reproduced from [32])

fast rate of release of the drug in basic pH 7.4, considered to be the pH of human organs such as the colon as well as feminine genitals, compared to acidic gastric pH.

As reported by Bajpai *et al.*, the swelling of anionic polymer increases with an increase in external pH due to the increased ionic strength of external media [59]. The polymer, in such stimuli, can exchange ions with the solution; hence maintaining the charge neutrality. The concentration of free counterions inside polymer increases. This results in increased osmotic pressure difference between the polymer and the solution. This finally causes swelling of the polymer.

12.3 Conclusion

Polyurethanes are versatile polymers with numerous applications, including biomedical applications. Biocompatible PUs can be synthesized by using polyols and cross-linkers from renewable resources. Several studies have been aimed at the synthesis and characterization of carbohydrate cross-linked polyurethanes and exploration of their application as drug delivery devices. The carbohydrate cross-linked PUs exhibited good mechanical strength, tunable release rates and biocompatibility that can be utilized in biomedical applications like wound dressing, biomedical implants and drug delivery carriers. The pH-responsive PUs, being "smart" polymers, can find attractive application in the field of drug delivery, for example, as microbicides and vaginal pessaries for prevention of HIV-AIDS and other sexually transmitted diseases.

References

1. Melchiors, M., Sonntag, M., Kobusch, C., and Jürgens, E., Recent developments in aqueous two-component polyurethane (2K-PUR) coatings. *Prog. Org. Coatings* 40, 99–109, 2000.
2. Szycher, M., and Reed, A.M., Biostable polyurethane elastomers. *Med. Device Technol.* 3, 42–51, 1992.
3. Szycher, M., *Szycher's Handbook of Polyurethanes*, 2nd ed., CRC Press, 2013.
4. Sonnenschein, M.F., *Polyurethanes: Science, Technology, Markets, and Trends*, Wiley, 2014.
5. Versteegen, R.M., Sijbesma, R.P., and Meijer, E.W., [n]-polyurethanes: Synthesis and characterization. *Angew. Chem. Int. Ed.* 38, 2917–2919, 1999.
6. Chattopadhyay D.K., and Raju, K.V.S.N., Structural engineering of polyurethane coatings for high performance applications. *Prog. Polym. Sci.* 32, 352–418, 2007.
7. Sonnenschein, M.F., Introduction to polyurethane chemistry, in: *Polyurethanes: Science, Technology, Markets, and Trends*, pp. 105–126, Wiley, 2014.
8. Ionescu, M., *Chemistry and Technology of Polyols for Polyurethanes*, Rapra Technology, 2007.
9. Ilavsk, M., Ulmer, H.W., te Nijenhuis, K., and Mijs, W.J., Network formation in polyurethanes based on triisocyanate and diethanolamine derivatives. *Eur. Polym. J.* 37, 887–896, 2001.

10. Król, P., Synthesis methods, chemical structures and phase structures of linear polyurethanes. Properties and applications of linear polyurethanes in polyurethane elastomers, copolymers and ionomers. *Prog. Mater. Sci.* 52, 915–1015, 2007.
11. Ping, P., Wang, W., Chen, X., and Jing, X., The influence of hard-segments on two-phase structure and shape memory properties of PCL-based segmented polyurethanes. *J. Polym. Sci. Part B* 45, 557–570, 2007.
12. Caracciolo, P.C., Buffa, F., and Abraham, G.A., Effect of the hard segment chemistry and structure on the thermal and mechanical properties of novel biomedical segmented poly(esterurethanes). *J. Mater. Sci. Mater. Med.* 20, 145–155, 2009.
13. Cherng, J.Y., Hou, T.Y., Shih, M.F., Talsma, H., and Hennink, W.E., Polyurethane-based drug delivery systems. *Int. J. Pharm.* 450, 145–62, 2013.
14. Zou, J., Zhang, F., Huang, J., Chang, P.R., Su, Z., and Yu, J., Effects of starch nanocrystals on structure and properties of waterborne polyurethane-based composites. *Carbohydr. Polym.* 85, 824–831, 2011.
15. Lalwani, R., and Desai, S., Sorption behavior of biodegradable polyurethanes with carbohydrate cross-linkers. *J. Appl. Polym. Sci.* 115, 1296–1305, 2010.
16. Liu, H., Cui, S., Shang, S., Wang, D., and Song, J., Properties of rosin-based waterborne polyurethanes/cellulose nanocrystals composites. *Carbohydr. Polym.* 96, 510–515, 2013.
17. Han, W., Tu, M., Zeng, R., Zhao, J., and Zhou, C., Preparation, characterization and cytocompatibility of polyurethane/cellulose based liquid crystal composite membranes. *Carbohydr. Polym.* 90, 1353–1361, 2012.
18. Buléon, A., Colonna, P., Planchot, V., and Ball, S., Starch granules: Structure and biosynthesis. *Int. J. Biol. Macromol.* 23, 85–112, 1998.
19. Durrani, C.M., and Donald, A.M., Physical characterisation of amylopectin gels. *Polym. Gels Networks* 3, 1, 1–27, 1995.
20. Cao, X., Chang, P.R., and Huneault, M.A., Preparation and properties of plasticized starch modified with poly(ε-caprolactone) based waterborne polyurethane. *Carbohydr. Polym.* 71, 119–125, 2008.
21. Brown, R.M., Cellulose structure and biosynthesis. *Pure Appl. Chem.* 71, 767–775, 1999.
22. O'Sullivan, A., Cellulose: The structure slowly unravels. *Cellulose* 4, 173–207, 1997.
23. Gao, Z., Peng, J., Zhong, T., Sun, J., Wang, X., and Yue, C., Biocompatible elastomer of waterborne polyurethane based on castor oil and polyethylene glycol with cellulose nanocrystals. *Carbohydr. Polym.* 87, 2068–2075, 2012.
24. Wu, G., Chen, J., Huo, S., Liu, G., and Kong, Z., Thermoset nanocomposites from two-component waterborne polyurethanes and cellulose whiskers. *Carbohydr. Polym.* 105, 207–13, 2014.
25. Rueda-Larraz, L., d'Arlas, B.F., Tercjak, A., Ribes, A., Mondragon, I., and Eceiza, A., Synthesis and microstructure–mechanical property relationships

of segmented polyurethanes based on a PCL–PTHF–PCL block copolymer as soft segment. *Eur. Polym. J.* 45, 2096–2109, 2009.
26. Zhao, Q., Sun, G., Yan, K., Zhou, A., and Chen, Y., Novel bio-antifelting agent based on waterborne polyurethane and cellulose nanocrystals. *Carbohydr. Polym.* 91, 169–174, 2013.
27. Unnithan, A.R., Gnanasekaran, G., Sathishkumar, Y., Lee, Y.S., and Kim, C.S., Electrospun antibacterial polyurethane-cellulose acetate-zein composite mats for wound dressing. *Carbohydr. Polym.* 102, 884–92, 2014.
28. Zia, F., Zia, K.M., Zuber, M., Kamal, S., and Aslam, N., Starch based polyurethanes: A critical review updating recent literature. *Carbohydr. Polym.* 134, 784–798, 2015.
29. Liu, H., Cui, S., Shang, S., Wang, D., and Song, J., Properties of rosin-based waterborne polyurethanes/cellulose nanocrystals composites. *Carbohydr. Polym.* 96, 510–515, 2013.
30. Solanki, A.R., Kamath, B.V., and Thakore, S., Carbohydrate cross-linked biocompatible polyurethanes: Synthesis, characterization, and drug delivery studies. *J. Appl. Polym. Sci.* 132, 42223, 2015.
31. Solanki, A., Mehta, J., and Thakore, S., Structure-property relationships and biocompatibility of carbohydrate cross-linked polyurethanes. *Carbohydr. Polym.* 110, 338–344, 2014.
32. Solanki, A., and Thakore, S., Cellulose cross-linked pH-responsive polyurethanes for drug delivery: α-hydroxy acids as drug release modifiers. *Int. J. Biol. Macromol.* 80, 683–691, 2015.
33. Sivak, W.N., Pollack, I.F., Petoud, S., Zamboni, W.C., Zhang, J., and Beckman, E.J., Catalyst-dependent drug loading of LDI-glycerol polyurethane foams leads to differing controlled release profiles. *Acta Biomater.* 4, 1263–1274, 2008.
34. Sivak, W.N., Pollack, I.F., Petoud, S., Zamboni, W.C., Zhang, J., and Beckman, E.J., LDI-glycerol polyurethane implants exhibit controlled release of DB-67 and anti-tumor activity *in vitro* against malignant gliomas. *Acta Biomater.* 4, 852–862, 2008.
35. Cherng, J.Y., Hou, T.Y., Shih, M.F., Talsma, H., and Hennink, W.E., Polyurethane-based drug delivery systems. *Int. J. Pharm.* 450, 145–162, 2013.
36. Zhou, L., Yu, L., Ding, M., Li, J., Tan, H., Wang, Z., and Fu, Q., Synthesis and characterization of pH-sensitive biodegradable polyurethane for potential drug delivery applications. *Macromolecules* 44, 857–864, 2011.
37. Chen, Y., Wang, R., Zhou, J., Fan, H., and Shi, B., On-demand drug delivery from temperature-responsive polyurethane membrane. *React. Funct. Polym.* 71, 525–535, 2011.
38. Heo, D.N., Yang, D.H., Moon, H.-J., Lee, J.B., Bae, M.S., Lee, S.C., Lee, W.J., Sun, I.-C., and Kwon, I.K., Gold nanoparticles surface-functionalized with paclitaxel drug and biotin receptor as theranostic agents for cancer therapy. *Biomaterials* 33, 856–866, 2012.

39. Yoonessi, M., Peck, J.A., Bail, J.L., Rogers, R.B., Lerch, B.A., and Meador, M.A., Transparent large-strain thermoplastic polyurethane magnetoactive nanocomposites. *ACS Appl. Mater. Interfaces* 3, 2686–2693, 2011.
40. Ding, M., He, X., Wang, Z., Li, J., Tan, H., Deng, H., Fu, Q., and Gu, Q., Cellular uptake of polyurethane nanocarriers mediated by gemini quaternary ammonium. *Biomaterials* 32, 9515–9524, 2011.
41. Manuscript, A., Kinetics, D.R., Polymeric, D., and Systems, D., Drug release kinetics and transport mechanisms of non-degradable and degradable polymeric delivery systems. *Expert Opin. Drug Deliv.* 7, 429–444, 2010.
42. Adhikari, R., Gunatillake, P.A., Griffiths, I., Tatai, L., Wickramaratna, M., Houshyar, S., Moore, T., Mayadunne, R.T.M., Field, J., McGee, M., and Carbone, T., Biodegradable injectable polyurethanes: Synthesis and evaluation for orthopaedic applications. *Biomaterials* 29, 3762–3770, 2008.
43. Yu, L., Zhou, L., Ding, M., Li, J., Tan, H., Fu, Q., and He, X., Synthesis and characterization of novel biodegradable folate conjugated polyurethanes. *J. Colloid Interface Sci.* 358, 376–83, 2011.
44. Niu, Y., Chen, K.C., He, T., Yu, W., Huang, S., and Xu, K., Scaffolds from block polyurethanes based on poly(ε-caprolactone) (PCL) and poly(ethylene glycol) (PEG) for peripheral nerve regeneration. *Biomaterials* 35, 4266–4277, 2014.
45. Punnakitikashem, P., Truong, D., Menon, J.U., Nguyen, K.T., and Hong, Y., Electrospun biodegradable elastic polyurethane scaffolds with dipyridamole release for small diameter vascular grafts. *Acta Biomater.* 10, 4618–4628, 2014.
46. Ojha, U., Kulkarni, P., and Faust, R., Syntheses and characterization of novel biostable polyisobutylene based thermoplastic polyurethanes. *Polymer (Guildf)* 50, 3448–3457, 2009.
47. Macocinschi, D., Filip, D., Vlad, S., Butnaru, M., and Knieling, L., Evaluation of polyurethane based on cellulose derivative-ketoprofen biosystem for implant biomedical devices. *Int. J. Biol. Macromol.* 52, 32–37, 2013.
48. Lu, Y., and Larock, R.C., Soybean-oil-based waterborne polyurethane dispersions: Effects of polyol functionality and hard segment content on properties. *Biomacromolecules* 9, 3332–3340, 2008.
49. Modesti, M., and Lorenzetti, A., Experimental method for evaluating isocyanate conversion and trimer formation in polyisocyanate-polyurethane foams. *Eur. Polym. J.* 37, 949–954, 2001.
50. Noble, K.-L., Waterborne polyurethanes. *Prog. Org. Coat.* 32, 131–136, 1997.
51. Kuan, H.C., Ma, C.C.M., Chang, W.P., Yuen, S.M., Wu, H.H., and Lee, T.M., Synthesis, thermal, mechanical and rheological properties of multiwall carbon nanotube/waterborne polyurethane nanocomposite. *Compos. Sci. Technol.* 65, 1703–1710, 2005.
52. Saalah, S., Abdullah, L.C., Aung, M.M., Salleh, M.Z., Awang Biak, D.R., Basri, M., and Jusoh, E.R., Waterborne polyurethane dispersions synthesized from jatropha oil. *Ind. Crops Prod.* 64, 194–200, 2015.

53. Zhou, L., Liang, D., He, X., Li, J., Tan, H., Li, J., Fu, Q., and Gu, Q., The degradation and biocompatibility of pH-sensitive biodegradable polyurethanes for intracellular multifunctional antitumor drug delivery. Biomaterials, 33, 2734–2745, 2012.
54. Setty, C.M., Deshmukh, A.S., and Badiger, A.M., Hydrolyzed polyacrylamide grafted maize starch based microbeads: Application in pH responsive drug delivery. *Int. J. Biol. Macromol.* 70, 1–9, 2014.
55. Pradhan, L., Srivastava, R., and Bahadur, D., pH- and thermosensitive thin lipid layer coated mesoporous magnetic nanoassemblies as a dual drug delivery system towards thermochemotherapy of cancer. *Acta Biomater.* 10, 2976–2987, 2014.
56. Wang, J., Gong, C., Wang, Y., and Wu, G., Magnetic nanoparticles with a pH-sheddable layer for antitumor drug delivery. *Colloids Surf. B* 118, 218–225, 2014.
57. Hałupka-Bryl, M., Asai, K., Thangavel, S., Bednarowicz, M., Krzyminiewski, R., and Nagasaki, Y., Synthesis and *in vitro* and *in vivo* evaluations of poly(ethylene glycol)-block-poly(4-vinylbenzylphosphonate) magnetic nanoparticles containing doxorubicin as a potential targeted drug delivery system. *Colloids Surf. B* 118, 140–147, 2014.
58. Schmaljohann, D., Thermo- and pH-responsive polymers in drug delivery. *Adv. Drug Deliv. Rev.* 58, 1655–1670, 2006.
59. Bajpai, A.K., Shukla, S.K., Bhanu, S., and Kankane, S., Responsive polymers in controlled drug delivery. *Prog. Polym. Sci.* 33, 1088–1118, 2008.

13

Synthesis and Characterization of Polymeric Hydrogels for Drug Release Formulation and Its Comparative Study

Nisarg K. Prajapati[1], Nirmal K. Patel[2] and Vijay Kumar Sinha[1]*

[1]Industrial Chemistry Department, V. P. & R. P. T. P. Science College, Sardar Patel University, Vallabh Vidyanagar, India
[2]Institute of Science & Technology for Advanced Studies & Research, Vallabh Vidyanagar, Gujarat, India

Abstract
Presented in this chapter is a study on novel starch-based hydrogel polymer preparation. Hydrogels are advanced polymer systems that hold special advantages for the delivery of drugs and enhancement of biocompatibility of implantable devices. These technologies have the potential to alter drug release and delivery kinetics to better control the level of particular drugs in blood and tissue. The formulations should improve the duration of action and/or reduce undesirable side effects of the selected drug to the extent that they are related to the pharmacokinetics of the specific drugs. In the study presented herein, starch was extracted from potato and then was purified and etherified using monochloropropionic acid. The improved controlled drug release property of the prepared etherified starch resulted in a new high water-holding, biocompatible and biodegradable monomer, 2-(acryloyloxy) propanoic acid, grafted with the etherified starch prepared polymer matrix (hydrogel). The resulting hydrogel was used as drug carrier for loading a specific drug. Scanning electron microscopy (SEM) was used to analyze conformational pore structure of hydrogel; mass spectra for monomer reaction formation; and 13C NMR of monomer for structure conformation. The results were used to study its drug release property and compare the drug release activity of polymeric and standard drugs.

***Keywords*:** 2-(acryloyloxy)propanoic acid, hydrogel, starch, drug loading

**Corresponding author*: drvijaysinha@rediffmail.com

Sanjay K. Nayak *et al.* (eds.) Trends and Applications in Advanced Polymeric Materials, (245–254) 2018 © Scrivener Publishing LLC

13.1 Introduction

Hydrogels are polymeric networks with three-dimensional configuration capable of imbibing high amounts of water or biological fluids. Their affinity to absorb water is attributed to the presence of hydrophilic groups such as –OH, –CONH–, –$CONH_2$–, and –SO3H in polymers forming hydrogel structures [1]. Due to the contribution of these groups and domains in the network, the polymer is thus hydrated to different degrees (sometimes more than 90 wt%), depending on the nature of the aqueous environment and polymer composition [2]. They are insoluble due to the presence of chemical and/or physical cross-links such as entanglements [3].

The term hydrogels implies that the material is distended in water, which is the unique property of hydrogels. Hydrogels are also biodegradable, biocompatible and absorb a large amount of water. These different properties of hydrogel have resulted in their being widely used as a drug carrier. In the present study we prepared a new monomer and grafted it with etherified starch for expanded water absorption properties of etherified starch [4]. Although the etherified starch has good water retention properties, we focused on extending drug release for a long period of time so that the etherified starch grafted with 2,3-dihydroxypropyl 2-hydroxypropanoate monomer would have extended controlled release.

13.2 Materials and Method

Paracetamol was received from Ahmedabad, Gujarat, India; lactose, Mg-stearate talc and acrylic acids were purchased from S.D. Fine Chem, Mumbai; chemicals were analytical grade, purity 99%. The dibasic acids (succinic and sulphuric) and glycerol from Sigma-Aldrich with purity > 99% were used without further purification. All solvents used throughout this study were of the chemically pure grade and were fractionally distilled just before use. The water used in the surfactant tests was purified by water purification system.

Apertures: Mass spectrometry, nuclear magnetic resonance (NMR) spectroscopy and scanning electron microscopy (SEM) analysis.

13.2.1 Preparation of Sodium Salt of Partly Carboxylic Propyl Starch (Na-PCPS)

Carboxylic propyl starch was prepared according to Abdel-Akher *et al.* [5]. In a three-necked flask equipped with condenser, stirrer and gas inlet tube, an affixed amount of starch was suspended in a known volume of isopropyl alcohol. The mixture was stirred for thorough mixing and 50% sodium hydroxide solution was added in a span of half an hour at 40 °C; 3-chloropropionic acid was then added at once. The reaction mixture was stirred for 4 h at a controlled 55 °C temperature through constant temperature bath and the reaction was carried out under nitrogen atmosphere. And then the isopropyl alcohol over the reaction mixture produced 90% aqueous methanol to remove excess of Salt and dried at 35 °C under vacuum.

13.2.2 Preparation of 2-Hydroxy-3-((2-hydroxypropanoyl)oxy) propyl acrylate

Esterification was performed in a 250-mL three-necked flask equipped with a reflux condenser, thermometer and a sampling port. The reaction condition was heated the flask at 80 °C with constant stirring by magnetic stirrer [6]. The reagents used were in a molar ratio of 1 mole of glycerol to 1 mole of lactic acid; the reaction was catalyzed by 0.1% w/w H_2SO_4 (of total). Completion of the reaction: the reaction mixture was washed with water and the glycerol ester extracted with chloroform. Analogous synthetic procedures were performed for the synthesis of 2-hydroxy-3-((2-hydroxypropanoyl)oxy)propyl acrylate. It was cooled at room temperature; then 1 mol of acrylic acid was added with catalyst 0.1% w/w H_2SO_4 (of total) and stirred for 3 to 4 h at room temperature 35 to 40 °C (Figure 13.1).

13.2.3 Graft Copolymerization with PCPS-g-2-hydroxy-3-((2-hydroxypropanoyl)oxy) propyl acrylate (HPA)

A one-step method is used for grafting of PCPS-g-3-(acryloyloxy) 2-hydroxypropyle (2,3-dihydroxypropyle) succinate (AHPS) hydrogel [7]. PCPS (2 g) was added to 35 ml of doubly distilled water in a three-neck reactor equipped with a mechanical stirrer at 45 to 50 °C. Ceric ammonium nitrate was added to initiate the reaction, then a specific amount of monomer (AHPA) was added and the mixture was stirred for 120 min. The pasty mixture was allowed to cool to room temperature and the resulting product was neutralized up to pH 8.0. Then the gelled product was scissored into small pieces and poured into ethanol (200 mL) to dewater for 5 h. The hardened particles were filtered and dried in a vacuum oven (50 °C, 10 h). After grinding, the powdered superabsorbent hydrogel was stored away from moisture, heat and light. Cross-linking graft copolymerization of 2-hydroxy-3-((2-hydroxypropanoyl)oxy) propyl acrylate (HPA) is shown in Figure 13.2.

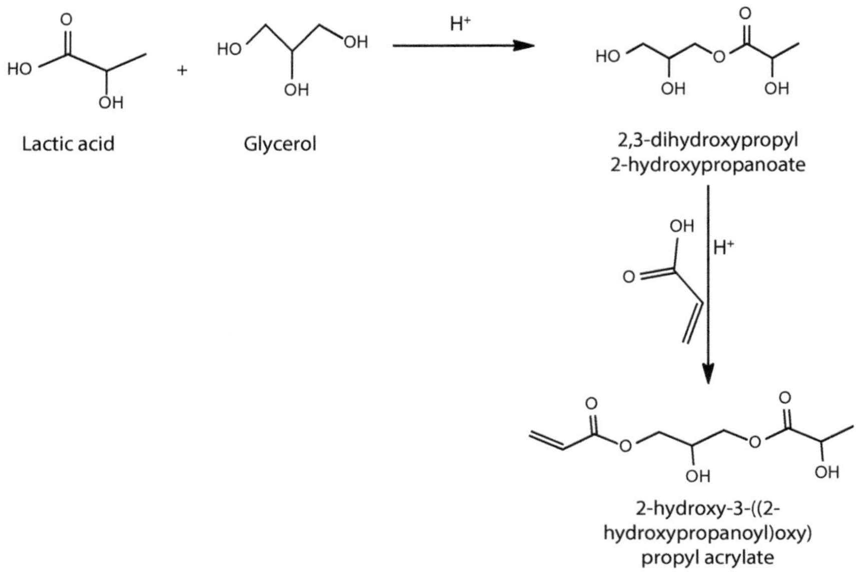

Figure 13.1 The reactive procedures of synthesis on 2-hydroxy-3-((2 hydroxypropanoyl) oxy) propyl acrylate monomer.

13.2.4 Drug Loading in Polymeric Binder

Preparation of powder in the form of small-sized particles was carried out by mortar and pestle. To obtain drug-loaded particles, 250 mg of HPA were accurately weighed and dissolved in 100 ml of water [8]. To this, 500 mg of paracetamol salt was dissolved. After a specific time, water and drug were absorbed by polymeric membrane (HPA) and then the hydrogel particles were filtered and dried. The drug trapped in the HPA particles formulations was analyzed using UV spectrophotometer. A known amount of drug-loaded hydrogel dried powder was dissolved in water to dissolve the particles and volume was made with water. The resultant solution was analyzed to determine the concentration of paracetamol in the samples using a UV-VIS spectrophotometer. The analysis was carried out in triplicate for each batch of drug-loaded hydrogel. The drug loading and entrapment efficiency was calculated using the following formula:

$$\text{Drug Entrapment\%}\frac{\text{Mass of drug in hydrogel}\times 100}{\text{mass of drug used in forulation}}$$

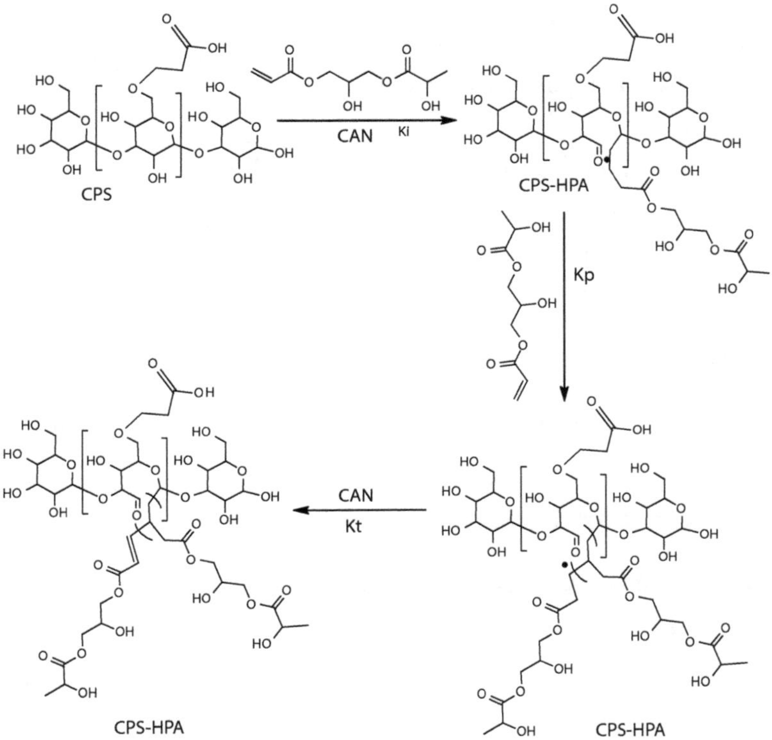

Figure 13.2 The reactive procedures of synthesis on PCPS-g-2-hydroxy-3-((2-hydroxy propanoyl)oxy) propyl acrylate (HPA) copolymer.

13.2.5 Preparation of Matrix Tablets

Solid dosage formulation of paracetamol drug was prepared using the wet granulation method [9]. A batch of 20 tablets was prepared. The ingredients were ground using a mortar and pestle. Various trial batches were prepared using graft copolymer in various concentrations with a fixed quantity of drug. All the ingredients were mixed and passed through a 60-mesh sieve. To the above prepared mixture, isopropyl alcohol in a quantity sufficient to produce wet mass was added with constant mixing; and the wet mass produced was passed through a 60-mesh sieve to produce granules. Then the prepared granules were dried at 55 °C for one hour in an oven. The dried granules were then passed through a 44-mesh sieve. To these granules, 1% Mg stearate and 2% Talc were added and mixed thoroughly. Tablets were prepared by a tablet punching machine. The tablets prepared were of two different types: 1) hydrogel was directly bound with drug and 2) the drug was loaded with hydrogel and then the tablet was prepared. The tablet prepared without drug loading is called PPT (polymeric paracetamol tablet) and with drug loading is called PPT_{Load}.

13.2.6 *In-Vitro* Dissolution Studies of Tablet

The dissolution test was carried out for tablets using 900 ml phosphate buffer pH 7.8 as the medium and rotating the paddle at 50 rpm for 30 min. A suitable volume of the sample was withdrawn and filtered; the first few ml of filtrate were rejected and the remaining were diluted with some solvent. Absorbance of the resulting solution at a maximum of about 249 nm was measured using a UV-1800 Shimadzu spectrophotometer. Similarly, absorbance of a solution of a known concentration of the paracetamol reference standard was measured.

13.3 Result and Discussion

13.3.1 ^{13}C-NMR Spectra Analysis of 2-Hydroxy-3-((2-hydroxypropanoyl)oxy) propyl acrylate

Characteristics of the structure of 3-(acryloyloxy) 2-hydroxypropyle (2,3-dihydroxypropyle) succinate monomer and 2-hydroxy-3-((2-hydroxypropanoyl)oxy) propyl acrylate were analyzed (Figure 13.2). The figure shows carboxylic carbon at 173.57–165.74 ppm and carbon of CH_2 and CH in the region 65.07–68.21 ppm. The peaks observed at 129.92 ppm of CH ethylene, 131.33 ppm of CH_2 ethylene carbon, and 18.81 ppm of CH_3 carbon are all related to the characteristic conformational structure of 2-hydroxy-3-((2-hydroxypropanoyl)oxy) propyl acrylate (Figure 13.3).

13.3.2 XRD Analysis of Starch, CPS, PCPS-g-2-hydroxy-3-((2-hydroxypropanoyl)oxy) propyl acrylate (HPA)

Powder X-ray diffractograms were obtained from starch, CPS and their grafted copolymers. In the X-ray diffractgrams of starch, the peak showing maximum intensity at $2\theta = 18°$ is very well matched with the reported literature value, which indicates that starch is a highly crystalline material. After carboxylproplylation reaction was carried out, crystallites were reduced and the pore size of starch exhibited a peak, which is clearly observed in X-ray diffractograms of CPS (carboxylproply starch). Compared to starch and CPS, the grafting decreases the intensity of the specific peak, i.e., almost no peak is observed in Figures 13.4 and 13.5. The graft copolymerized sample becomes almost amorphous.

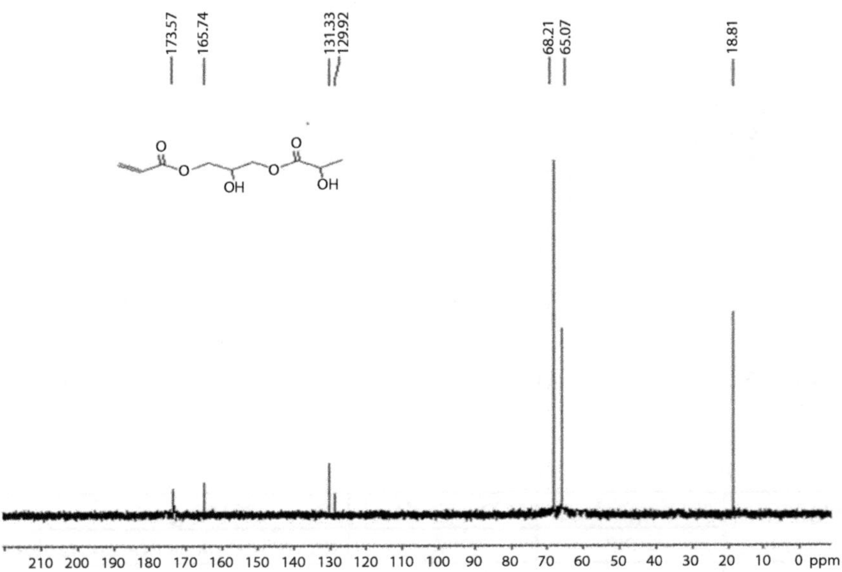

Figure 13.3 ^{13}C-NMR spectra of 2-hydroxy-3-((2-hydroxypropanoyl)oxy) propyl acrylate.

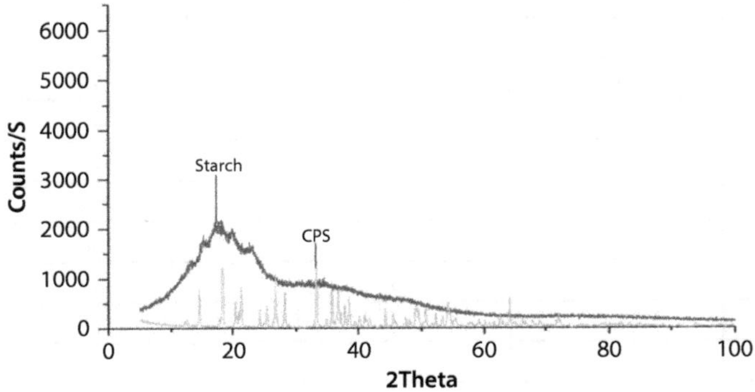

Figure 13.4 X-ray diffraction of starch and carboxylpropyl starch.

13.3.3 *In-Vitro* Study

In the present study, a comparison is given between standard drug and polymeric drug release in a phosphate buffer solution of pH 8. Drug release was observed for 5 h; after 5 h the polymeric paracetamol tablet released 53% of the drug and the polymeric paracetamol-loaded tablet released 33% (Table 13.1; Figure 13.6).

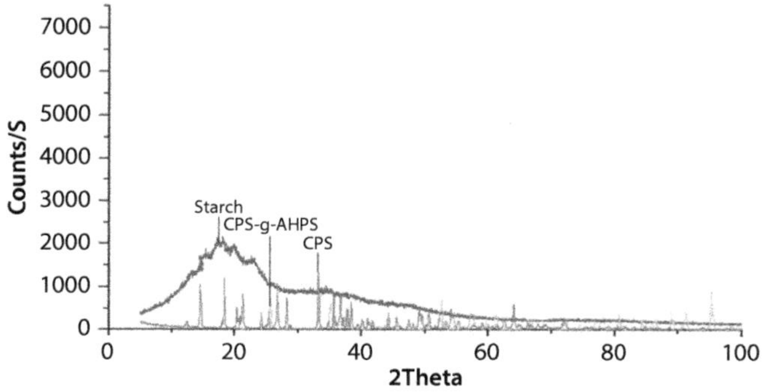

Figure 13.5 X-ray diffraction of CPS-g-3-2-hydroxy-3-((2-hydroxypropanoyl)oxy)propyl acrylate.

Table 13.1 *In-vitro* dissolution studies of tablet.

Time in minutes	% Drug release PPT	% Drug release PPT_{Load}	% Drug release Standard paracetamol tablet
15	13	14	15
30	23	22	20
45	24	25	24
60	28	27	28
75	31	30	33
90	33	32	37
105	34	33	40
120	40	33	45
135	46	33	48
150	50	33	53
165	53	33	58
180	53	33	62
195	53	33	65
205	53	33	68
220	53	33	73
235	53	33	75
250	53	33	80
265	53	33	83
280	53	33	87
295	53	33	92
305	53	33	94
320	53	33	100

PPT: polymeric paracetamol tablet, PPT_{Load}: polymeric paracetamol loaded tablet

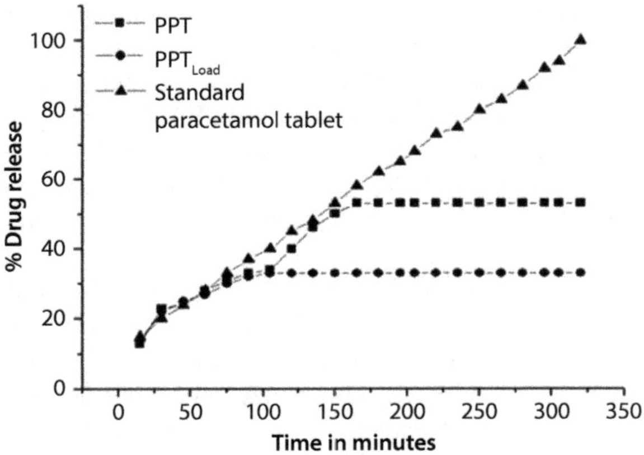

Figure 13.6 Control drug release study of standard tablet, polymeric binding drug and drug-loaded polymeric tablet.

13.4 Conclusion

Sustained-release PPT tablets were prepared. The release rates of standard paracetamol were 100% = 3 h, PTM release 100% = 23 h, and PPT_{Load} release 100% = 46 h in pH 7.8 phosphate buffer, respectively. However, the PTM, PPT_{Load} and standard drug were not significantly different. The PPT_{Load} tablet release was slower than the PPT tablet. The PPT tablet released drug more slowly than the standard tablet, allowing for a high dose to be administered with reduced gastrointestinal side effects and increased compliance of patients; but it did not change the pharmacokinetic behavior.

Acknowledgment

The authors are indebted to the college principal and ISTAR for providing the infrastructural support and laboratory facility. We also acknowledge the support of SICART for timely and accurate analysis of the synthesized product.

References

1. Hamidi, M., Azadi, A., and Rafiei, P., Hydrogel nanoparticles in drug delivery. *Adv. Drug Deliv. Rev.* 14, 60, 1638–1649, 2008.

2. Yu, L., Dean, K., and Li, L., Polymer blends and composites from renewable resources. *Prog. Polym. Sci.* 30, 576–602, 2006.
3. Mansur, H.S., Oréfice, R.L., and Mansur, A.A., Characterization of poly (vinyl alcohol)/poly (ethylene glycol) hydrogels and PVA-derived hybrids by small-angle X-ray scattering and FTIR spectroscopy. *Polymer* 29, 7193–7202, 2004.
4. Farag, S., and Al-Afaleq, E.I., Preparation and characterization of saponified delignified cellulose polyacrylonitrile-graft copolymer. *Carbohydr. Polym.* 48, 1–5, 2002.
5. Roushdi, M., Abdel-Akher, M., Ismail, F. A., & Attia, E. Preparation of mixed derivatives of carboxymethyl starch and determination of substitution degree by a new colorimetric method. Starch-Starke. 1982; 34, 410–413.
6. Koenig, H.S., and Roberts, C.W., Vinylbenzyl ethers of cellulose. Preparation and polymerization. *J. Appl. Polym. Sci.* 18, 651–666, 1974.
7. Wieland, P.C., Stoeckel, N., and Nuyken, O., New syntheses of graft copolymers using the DPE-technique: Cationic graft copolymerization. *J. Macromol. Sci, Part A* 40, 11–19, 2003.
8. Liu, Z., Sun, X., Nakayama-Ratchford, N., and Dai, H., Supramolecular chemistry on water-soluble carbon nanotubes for drug loading and delivery. *ACS Nano* 1, 50–56, 2007.
9. Khullar, P., Khar, R.K., and Agarwal, S.P., Evaluation of guar gum in the preparation of sustained release matrix tablets. *Drug Dev. Ind. Pharm.* 24, 1095–1099, 1998.

14

Enhancement in Gas Diffusion Barrier Property of Polyethylene by Plasma Deposited SiO_x Films for Food Packaging Applications

Purvi Dave*, Nisha Chandwani, S. K. Nema and S. Mukherji

Institute for Plasma Research, Facilitation Center for Industrial Plasma Technologies, Gandhinagar, Gujarat, India

Abstract

The work presented in this chapter provides a better understading of currently used multilayered packaging polymer structures along with how a single layer of a packaging polymer can be modified to have similar properties as multilayer polymer structure. In this study, a single layer of low-density polyethylene (LDPE) packaging film was modfied with the help of SiOx coating deposited by plasma-enhanced chemical vapor deposition (PECVD) coating. Thin SiOx films (100–200 nm thick) deposited by radio frequency (13.56 MHz) capacitively coupled plasma at low pressure using oxygen-HMDSO gas mixture. From Fourier transform infrared spectroscopy (FTIR) results, it was observed that film structure changes from linear type (more organic nature) to network type (inorganic nature) as oxygen concentration was increased in the oxygen-HMDSO gas mixture. Network-type film structure exhibits better oxygen diffusion barrier property. Deposition rate is found decreased with increase of oxygen concentration. In our experiments, deposition of a double layer of SiOx film with different film chemistry has resulted in enhancement of oxygen diffusion barrier property as compared to a single layer of SiOx film. Oxygen diffusion rate (OTR) of polyethylene film was reduced up to 350 $cc/m^{2*}day$ from 3500 $cc/m^{2*}day$.

Keywords: Packaging, oxygen, diffusion, barrier, SiOx, plasma

**Corresponding author*: purvi@ipr.res.in

Sanjay K. Nayak *et al.* (eds.) Trends and Applications in Advanced Polymeric Materials, (255–274) 2018 © Scrivener Publishing LLC

14.1 Introduction

The traditional functions of any type of packaging are to protect food products from degradation processes (primarily produced by environmental factors such as oxygen, light and moisture), to contain the food, and to provide consumers with ingredient and nutritional information. Materials that have conventionally been used in food packaging include glass, metals (aluminum foils and laminates, tinplate, and tin-free steel), paper and paperboards, and plastics. The right selection of packaging material plays an important role in maintaining product quality and freshness during distribution and storage [1–2].

Glass and metal packaging provide a nearly absolute barrier to chemical and other environmental agents, but they are rigid in nature, occupy more space when stored or transported and are heavy in weight; whereas polymer packaging is flexible in nature, occupies less space and is lighter in weight. Polymer packaging possesses many other properties, including strength and stiffness and resistance to food component attack, but in contrast to glass or metal packaging materials, packages made with polymers are permeable to small gas molecules such as oxygen, water vapor, nitrogen, carbon dioxide and to other low molecular weight compounds like aromas [3–4]. The shelf life of products can be increased by using packaging materials that could control or minimize the permeation of different gases towards the internal atmosphere. Barrier properties are mainly correlated with the intrinsic structure of the polymer such as the degree of crystallinity, nature of the polymer, crystalline/amorphous phase ratio, etc. [2]. Controlling the permeability to oxygen and moisture are major challenges to preserve the quality of food products because the presence of oxygen facilitates microbial growth, increases oxidative reactions, and induces the development of off-flavor and color changes. For example, the variation in color produced during the storage of fruit juices can be related to the deterioration of the nutritional and organoleptic properties of the food product. In order to get fine control over the oxygen and moisture diffusion, multilayer flexible packaging is widely used in the food industry. Multilayered films containing layers of different polymers provide desirable properties such as high barrier, mechanical strength, heat sealability, etc. [3–4]

Polyethylene terephthalate (PET), high-density polyethylene (HDPE), polyvinyl chloride (PVC), low-density polyethylene (LDPE), polypropylene (PP), and polystyrene (PS) are the main components of common food packaging plastics. Each food packaging plastic/polymer is used in certain products due to their unique properties. PET has a very good tensile strength and remains transparent even after processing, but melts at low temperature. So this polymer is used for cold beverage packaging. HDPE is used for hard containers or bottles for foods, such as milk, where a strong container is needed but clarity is not required. PVC is most commonly used as clear plastic wrapping due to its low cost and stretching capabilities, as well as its ability to be easily extruded into sheet forms [5].

At present, multilayer (i.e., 3–11 layers) high barrier polymer films are used in the market for packaging various food products. Multiple polymer materials (i.e., PE, PP, Nylon, EVOH, etc.) are used in multilayer film structures for producing flexible packages/pouches, which increase their cost multi-fold. The disadvantage of multilayer packaging materials is that they are not easily recycled and hence are mostly landfilled [1].

Due to its low cost, easy processability, good heat sealing property, compatibility with most categories of foods, fair moisture barrier and very high elogation strength, LDPE is a part of almost all multilayer films ranging from 3 to 11 layers. As LDPE has very poor oxygen barrier property, LDPE monolayer cannot be used as food packaging film.

In order to protect the enviornment, development of surface modification technologies capable of improving the barrier properties of monolayer polymers would be of great help to food processing industries in producing low-cost, environmentally friendly packages. Methods for achiving high barrier properties of monolayer polymers include deposition of thin metal coating such as aluminum, deposition of oxide films such as silicon oxide (SiOx) and aluminum oxide (AlOx), application of nanocomposite materials, incorporating flexes of less permeable material, etc.

Gas diffusion barrier coating on polymeric substrate is an essential part of modern packaging for protection of food and pharmaceutical products against outside influences. Cold plasma technology is an emerging, green process offering many potential applications for food and pharmaceutical packaging as well as for flexible electronics [6–7]. Most high performance semiconducting organic compounds show degraded performance when they are exposed to environmental oxygen and moisture. Thus, gas diffusion barrier coating is again an unavoidable part of such devices [8–9]. The molecular structure of thin polymer film (polymer packaging web) is found in the form of chains oriented like networks. This structure possesses certain porosity as well as gaps through which gas molecules can easily pass through and reach the packed product. Plasma surface modification, such as cross-linking, coating, etc., also reduces such pores and gaps by forming an even, smooth and almost impermeable layer, thus preventing gas molecules from reaching the packed product. Plasma deposited inorganic coating on polymer substrates has been exploited in recent years as an alternative to metalized polymer for packing applications due to its transparency, recyclability, microwavability and excellent barrier properties [10, 11].

The present chapter describes the nature of packaging polymers, mechanism of gas molecules transportation through the packaging polymer films, measurement method of gas transmission rate, deposition of SiOx films on PE films using plasma enhanced chemical vapor deposition method and the results obtained.

14.2 Transport of Gas Molecules Through Packaging Polymers

14.2.1 Packaging Polymer Struture

Bulk polymers are semicrystalline in nature and have both amorphous and crystalline regions, as shown in Figure 14.1. In the case of semicrystalline polymers, the presence of crystallites in the amorphous matrix leads to the different transport properties of gas/vapor molecules compared to amorphous polymers [8–12]. Crystallites are considered to be impenetrable. Penetrant molecules generally enter through the free volume available in the amorphous interlayer separating two crystallites due to segmental mobilities in the amorphous region. Polyethylene has nearly 50% crystalline and 50% amorphous structure.

14.2.2 Transport of Gas Molecules Through Semicrystalline Polymer Films

There are two processes by which gases and vapors may pass through polymeric materials:

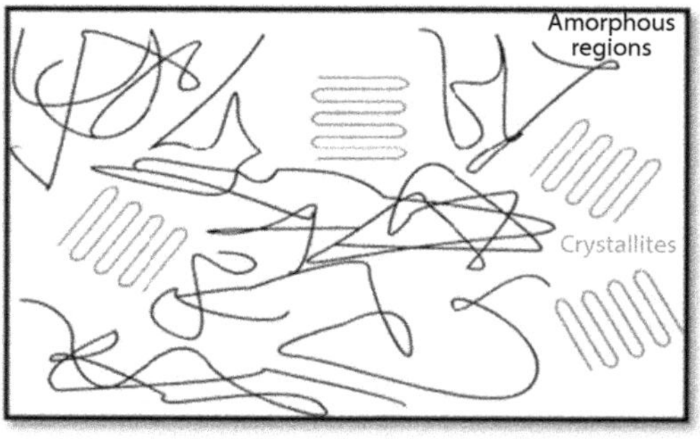

Figure 14.1 Schematic of the semicrystalline nature of polymers.

a. *A pore effect,* in which the gases and vapors flow through microscopic pores, pinholes and cracks in the materials.
b. *A solubility-diffusion effect,* in which gases and vapors dissolve in the polymer at one surface, diffuse through the polymer by virtue of a concentration gradient and evaporate at the other surface of the polymer. This solution-diffusion process is also known as "activated diffusion," also called permeability [13].

In simple terms, movement of gas molecules from one side of a film to another side is called "permeation." This mechanism is explained graphically in Figure 14.2. Diffusion of gas molecules across a film is influenced by the film structure, film permeability to specific gases or vapor, thickness, area, temperature, difference in pressure, or concentration gradient across the film. Diffusion also depends on the size, shape and polarity of the penetrating molecule of the permeant and on the crystallinity, degree of cross-linking and polymer chain segmental motion of the polymer matrix. As reported by Kofinas *et al.* [12], gas molecules are unable to permeate through the polymer crystallites due to the availability of less volume in

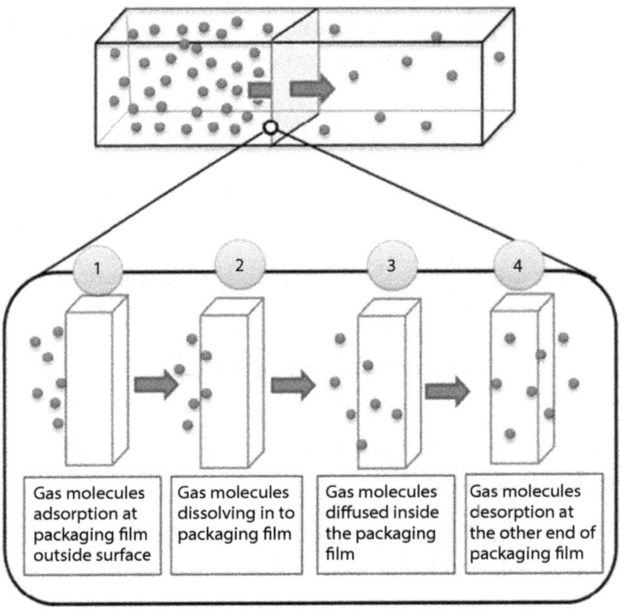

Figure 14.2 Schematic description of the transport of gas molecules through a polymer film.

that region and a large tortuous path between the crystallites. Thus, the gas permeation into semicrystalline polymers is then confined to the amorphous regions. The reduction in permeability value is proportional to the volume fraction of the crystalline phase [4, 10].

In order to explain the activated diffusion (due to concentration gradient) mechanism of gas molecules, let us take an example of two rooms that are separated by a film barrier, as seen in the top part of Figure 14.2. Now, suppose that the pressure in the room on the left is higher than that in the room on the right. In this case, gas flows from the room with the high pressure to that with the low pressure.

The lower part of Figure 14.2 shows an enlargement of the film. Gas transmission goes through the following process. In a first step, gas molecules in the room with the higher pressure condense on the surface of the film. In the second step, the molecules dissolve into the film. The concentration of gas dissolved into the film is proportional to the gas pressure. Driven by a concentration gradient, the dissolved gas molecules diffuse into the room with the lower concentration of gas molecules in the third step and finally the molecules that reach the other side of the film are desorbed from the surface of the film. So gas transmission by the activated diffusion mechanism undergoes the three processes of dissolution, diffusion and desorption.

This process is explained mathematically by Fick's first law of steady state diffusion and governed by the equation:

$$J = -D\frac{\delta c}{\delta x} \tag{14.1}$$

where J is the flux (or rate of transport) per unit area of permeant through the polymer; c is the concentration of the permeant; D is defined as the diffusion coefficient; and $\frac{\delta c}{\delta x}$ is the concentration gradient of the permeant across a thickness δx. D reflects the speed at which the permeant diffuses through the polymer and a unit of area per time [13].

14.2.3 Measurement of Gas Transmission Rate Through a Packaging Film

There are many methods for measuring the gas transmission rate (GTR) of polymer films, out of which the four major methods are:

- Pressure increased method
- Volume increased method

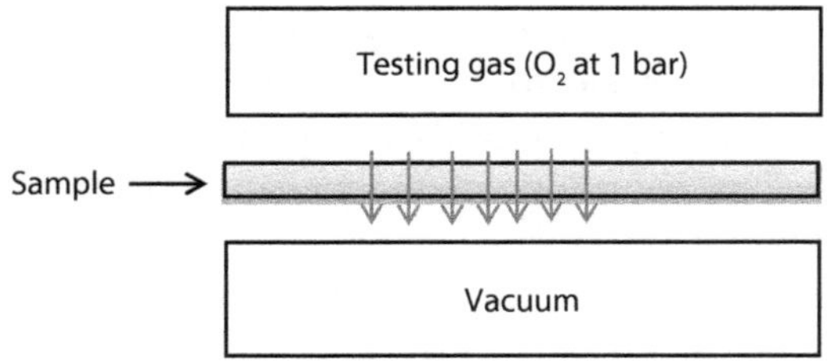

Figure 14.3 Schematic diagram of GTR testing setup using differential pressure method; GTR is calculated using Equation 14.2.

- Concentration increased method
- Detector film method

In the present study, gas transmission rate is measured using the pressure increased method, which satisfies conditions mentioned in ASTM D1434 standard.

In this method, test gas (normally at 1 atm) is introduced on one side of the flat film or sheet which is supported with a filter paper and a sealed 'O' ring. The pressure in the receiving chamber is measured with a mercury manometer. Provided that the pressure on the higher pressure side remains much larger than that on the lower pressure side, the pressure difference remains essentially constant. The test setup is shown in Figure 14.3 [13].

$$GTR = \frac{V}{R \times T \times P \times A} \times \frac{dp}{dt} \frac{cc}{m^2 \times day \times 1bar} \tag{14.2}$$

where V is the volume of the chamber; R is gas constant; T is temperature; P is applied pressure of test gas on higher pressure side; A is the test area; and is differential pressure.

14.3 Experimental

In this study, thin SiOx films were deposited using the PECVD method. Experiments were conducted in a cylindrical-shaped stainless steel process chamber having 60 cm diameter and 30 cm height. Hexamethyldisiloxane (HMDSO) along with oxygen were used as plasma forming gas and were introduced into the process zone via multipoint gas feeding shower head. Plasma was generated at 30 Watt power between two parallel plate electrodes

(35 cm diameter) using radio frequency (13.56 MHz) O_2/HMDSO power source. The gap between electrodes was kept 3.5–5 cm. Partial pressure of oxygen gas was varied in O_2/HMDSO gas mixture while total operating pressure was kept constant at 0.06 mbar in order to study the change in film wettability, film chemistry, deposition rate, oxygen transmission rate, surface morphology and surface roughness as a function of oxygen concentration.

We also deposited double-layer SiOx films with different film chemistry to enhance the oxygen diffusion barrier property of PE filmfurther.

14.3.1 Contact Angle Measurements to Determine Film Wetting Properties

Static water contact angle was measured on each sample after plasma treatment at room temperature using a video contact angle goniometer (OCA 15EC, DataPhysics Instruments). At least three different measurements on each sample surface were conducted and average value for contact angles was calculated.

14.3.2 FTIR-ATR Study to Determine Film Chemistry

Fourier transform infrared–attenuated total reflectance (FTIR-ATR) spectroscopy technique was used to provide information about chemical bonds (such as $Si\text{-}CH_3$, Si-O-Si, Si-OH, etc.) present in the deposited films with various oxygen gas concentrations in O_2/HMDSO gas mixture. This study was performed using a Nicolet 6700 FTIR instrument with multi-bounce HATR accessory. Measurements were performed using a ZnSe ATR crystal with resolution of 4 cm^{-1}.

14.3.3 Film Thickness Measurement

Thickness of the deposited films was measured using a NanoMap-500ES contact mode stylus profilometer.

14.3.4 High Resolution Scanning Electron Microscopy to Determine Film Morphology

Morphology of the bare PE film and SiOx coated PE film was measured using a Carl Zeiss high resolution scanning electron mircoscope.

14.3.5 OTR Measurement to Determine Oxygen Diffusion Barrier Property

Oxygen transmission rate of virgin PE and PE with SiOx barrier layer deposited with different oxygen concentration was measured by differential pressure method using a gas transmission rate tester (GBPI, Model no. N500, China).

14.4 Results

14.4.1 Observations

In these experiments at constant total pressure P_t = 0.06 mbar, oxygen partial pressure was varied in oxygen/HMDSO gas mixture for deposition of SiOx-like films on PE and single-side polished silicon wafer substrates.

14.4.1.1 Wetting Behavior of SiOx Films

Static water contact angles on plasma deposited single-layer SiO_x films for different oxygen concentrations in O_2/HMDSO gas mixture were measured. Contact angle of plasma deposited SiO_x film decreases from 108.44° to 57.02° with increasing oxygen concentration from 0% to 90% in the gas mixture, as shown in Figure 14.4. The experimental set was repeated three times to check the repeatability of the obtained results.

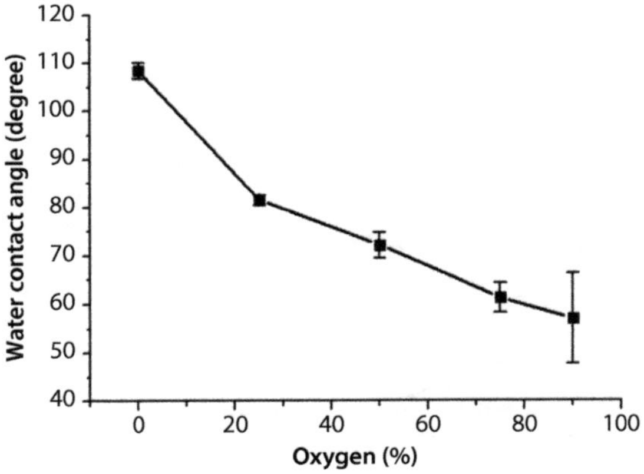

Figure 14.4 Oxygen concentration vs. water contact angle of SiO_x films.

14.4.1.2 Chemistry of SiOx Film

Chemistry of SiOx film is measured by ATR-FTIR spectroscopy. The ATR-FTIR spectra of SiO_x films deposited at different oxygen concentrations in oxygen/HMDSO gas mixture were recorded. In the case of 0% oxygen concentration, the spectra represent the chemistry of plasma polymerized (PP) HMDSO film, which show different modes such as rocking and bending of Si-CH_3 vibrations at 840.52 and 1257.21 cm^{-1} respectively. The spectra also show Si-O-Si stretching and bending vibration at 1031.01 and 800 cm^{-1} respectively. It is observed that as we continued increasing oxygen concentration in the gas mixture, concentration of Si-CH_3 bonds is gradually reduced and the nature of the film becomes more inorganic in type. This is clearly evident in Figure 14.5.

The FTIR spectroscopy results also express that IR peaks associated with Si-O-Si stretching and bending vibrations shift towards the higher wave-number side as we increased oxygen concentration in the oxygen/HMDSO gas mixture. This is shown in Figures 14.6 and 14.7 respectively.

14.4.1.3 Deposition Rate

Deposition rate was calculated at each oxygen concentration from film thickness measured by a contact mode stylus profilometer. The deposition rate was found to decrease with increasing oxygen partial pressure in oxygen/HMDSO gas mixture (Figure 14.8). Maximum deposition rate

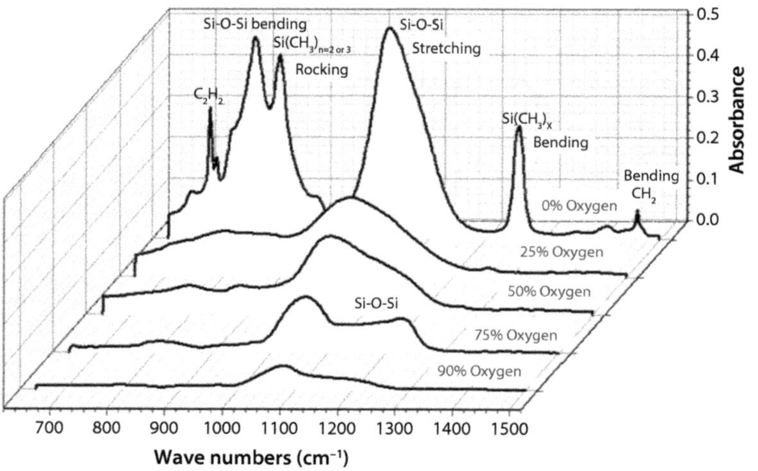

Figure 14.5 FTIR spectra of SiO_x films at various oxygen concentrations.

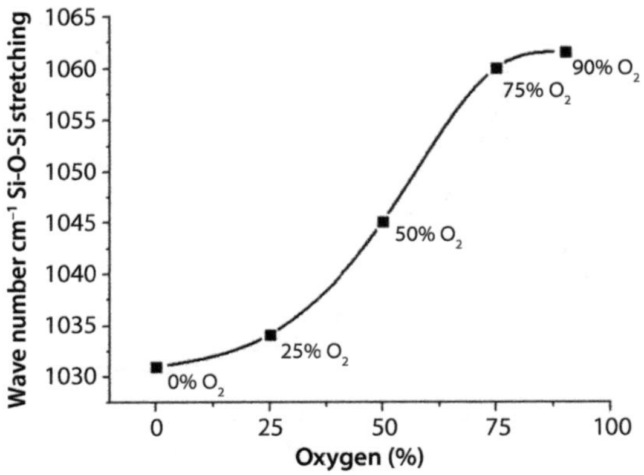

Figure 14.6 Shifting of Si-O-Si stretching peak with increase in oxygen concentration.

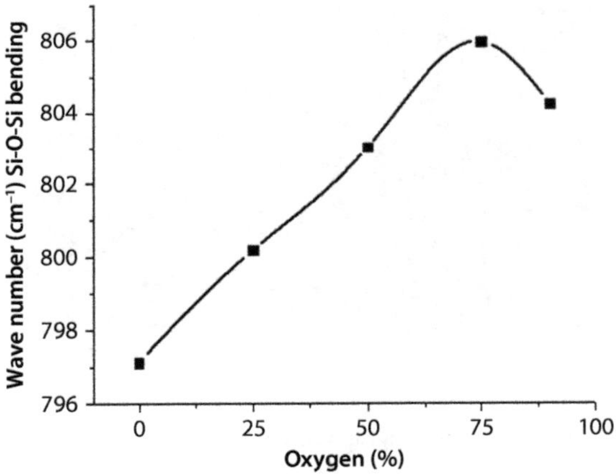

Figure 14.7 Shifting of Si-O-Si bending peak with increase in oxygen concentration.

obtained for 0% oxygen in oxygen/HMDSO gas mixture is 9.66 nm/min; whereas minimum deposition rate obtained for 90% oxygen in oxygen/HMDSO gas mixture is 1.69 nm/min in the present setup.

14.4.1.4 High Resolution Scanning Electron Microscopy

Morphology of virgin and SiOx coated PE surface was analyzed using high resolution scanning electron microscope with SE2 detector (Figures 14.9–14.12).

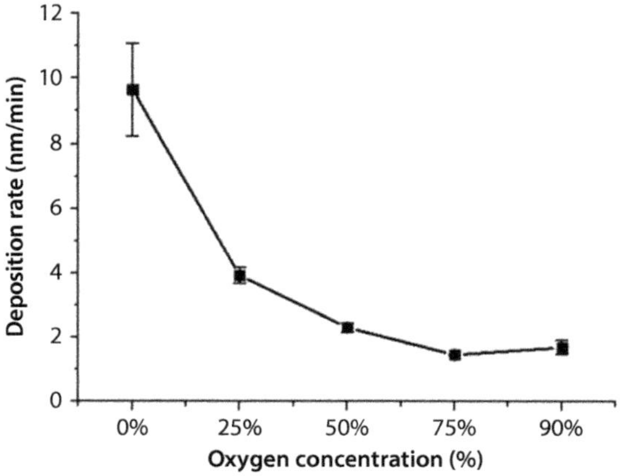

Figure 14.8 Oxygen concentrations vs. deposition rate.

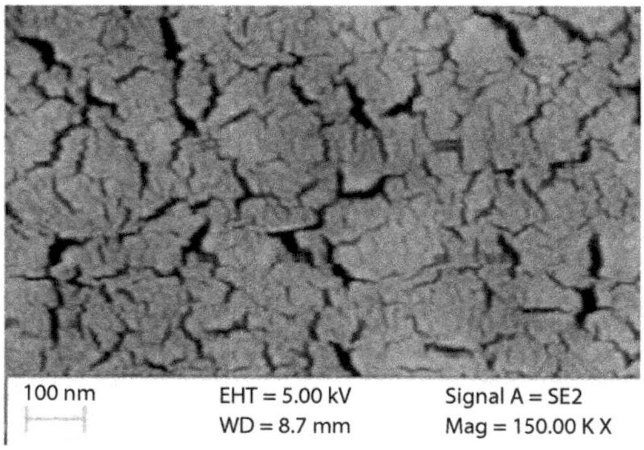

Figure 14.9 Scanning electron micrograph of virgin PE surface.

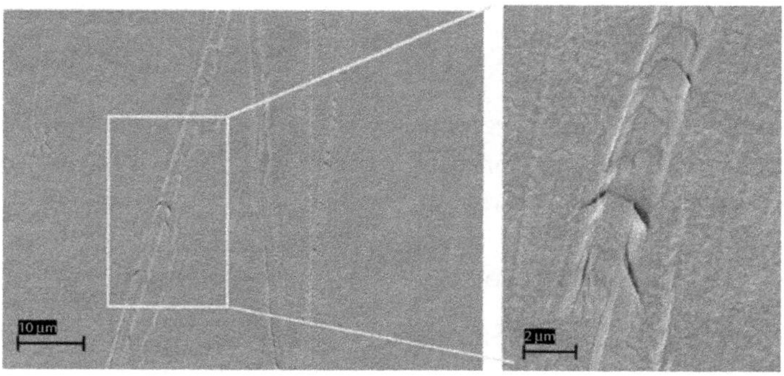

Figure 14.10 Defects in virgin PE surface.

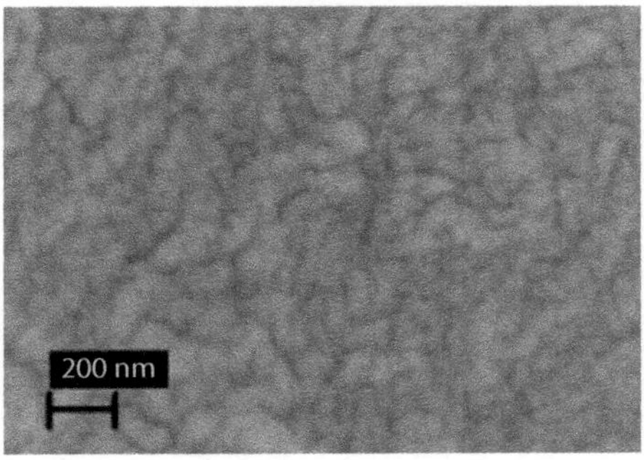

Figure 14.11 SiOx coated PE surface.

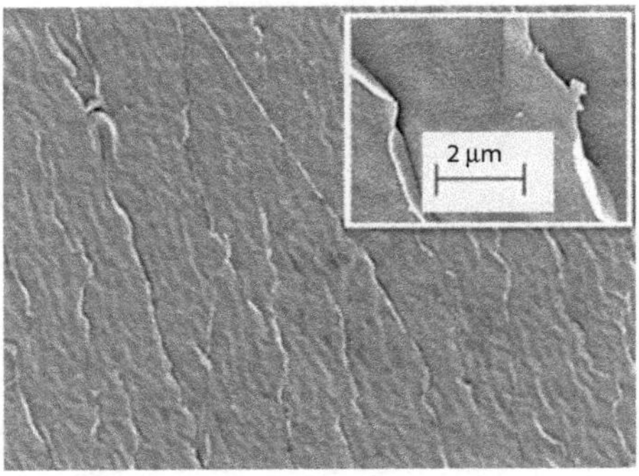

Figure 14.12 Cracks in SiOx coating due to thermal stress.

14.4.1.5 Oxygen Transmission Rate

Oxygen transmission rate (OTR) was measured for virgin PE and SiOx deposited PE film. The OTR value obtained for virgin PE film is 3300 cc/m^{2*}day. The OTR values were found to decrease when oxygen concentration was increased. Minimum OTR value obtained for 75% oxygen concentration was 750 cc/m^2day. Further reduction in OTR values up to 350 cc/m^{2*}day were achieved in the case of double-layer SiOx coated PE film. These results are shown in Figure 14.13.

14.4.2 Discussion

In the reported literature, authors have achieved high barrier properties by depositing thin coating by various methods on PET and PP films for food

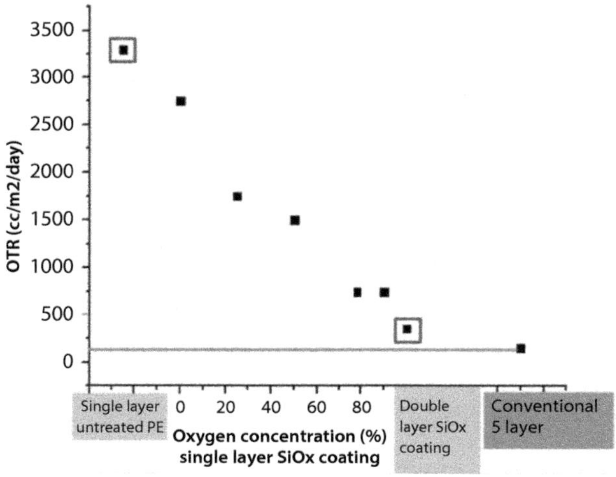

Figure 14.13 Oxygen transmission rate of untreated PE, single-layer SiOx deposited PE, double-layer SiOx deposited PE and conventional 5-layer (different polymers) structure.

packaging applications [14–17]. Studies on single-layer PE films have been reported by few authors [18].

In the SEM image in Figure 14.9, it can be observed that the surface of bare PE film is not homogeneous. It is rough and has many small regions from which a large quantity of oxygen molecules (molecular diameter ~0.2 nm) can enter the PE film. Some manufacturing defects having random scratches of a few micrometers have also been observed in bare PE films, as shown in Figure 14.10.

In our experiments, we aimed at preventing an ingress of oxygen molecules from PE film surface by deposition of thin (100–200 nm) SiOx films by PECVD method.

Results for water contact angle measurements shown in Figure 14.4 convey that plasma deposited film surface exhibits hydrophobic nature with water contact angle value 108.4° when it is deposited with 0% oxygen, i.e., only with HMDSO; we observed that the surface gradually becomes hydrophilic in nature as we increased oxygen concentration in the gas mixture with lowest contact angle value 57.02° when deposited with 90% oxygen. This is because in the absence of oxygen gas, fragmentation of HMDSO in plasma occurs due to electron impact. Carbonated silicon species are generated during such dissociation and they are subsequently deposited on substrate surface without oxidation. This leads to the formation of a hydrophobic surface since the presence of many nonpolar hydrocarbon groups on the film surface contribute to higher water contact angle value. As we increased the oxygen concentration in the gas mixture, fragmented oxygen species started interacting with HMDSO molecules, resulting in its fragmentation in the gaseous phase. Fragmented species were further oxidized by active oxygen species interaction with adsorbed carbon species on the surface. Oxidation of carbonated species resulted in removal of methyl groups and formation of volatile products such as CO, CO_2, and H_2O, which are pumped out during the deposition process. It is reported elsewhere [9] that when the quantity of oxygen gas is < 40%, oxygen molecules are consumed in oxidation of fragmented HMDSO species in the gaseous form and oxidation of surface adsorbed carbon species does not occur completely. Whereas when the quantity of oxygen gas is > 40%, excess oxygen species remain in the vapor phase, which helps in removal of adsorbed carbon species on the surface. So it is understood from the above explanation that, with increasing oxygen concentration in the gas mixture, film chemistry becomes more inorganic in nature and thus surface wettability is improved by lowering the water contact angle value.

The FTIR analysis of plasma deposited SiOx film with different oxygen concentrations clearly shows that the chemical structure of the film is strongly dependent on the oxygen concentration in the oxygen/HMDSO gas mixture in terms of peak locations. The IR spectra of SiO_x film deposited at 0% oxygen, i.e., only with HMDSO, confirm the presence of Si-O-Si bending, Si-CH_3 rocking, Si-O-Si stretching, Si-CH_3 bending and CH_2 bending vibrations at 797.13, 840.51, 1031.00, 1257.23 and 1462.42 cm^{-1} respectively (Figure 14.5). As we increased the oxygen concentration in the gas mixture, the presence of Si-CH_3 and CH_2 vibrations was gradually reduced and were found to completely vanish above 50% concentration. At low oxygen concentration, the presence of Si-CH_3 and CH_2 bond indicate an incomplete oxidation of hydrocarbon radicals existing in the coating [12]. Beyond 50% oxygen concentration, we noticed the appearance of shoulder peak at 1225 cm^{-1} associated with the main Si-O-Si stretching peak. The presence of shouldering in the main peak provides information about the film stoichiometry, i.e., film composition is likely to be SiO_x ($x \sim 2$) [19]. We also observed that the IR peak for Si-O-Si stretching and bending vibrations shift towards the higher wavenumber side as we increased the oxygen concentration in the gas mixture. This is shown in Figures 14.6 and 14.7 respectively.

Peak shift toward higher wavenumber indicates two major characteristics of the type of film growth. First, there is a reduction in molecular weight with increase of oxygen concentration, hence an increase in vibrational frequency. Initially Si-O-Si molecule has six methyl groups attached, so molecular weight is high. As we increased the oxygen concentration, these methyl groups were gradually removed by fragmentation followed by oxidation, thus molecular weight was reduced and so molecule is able to vibrate at higher frequency. Secondly, there are three types of growth mechanism of SiOx films, namely linear, network and cage types, shown schematically in Figure 14.14. As we continued increasing oxygen concentration in the oxygen/HMDSO gas mixture, film growth mechanism changed from linear to cage type [20–21]. Network-type growth structure of SiOx is more dense compared to the other two types and hence exhibits better barrier performance.

Single-layer inorganic/network-type SiOx film on PE surface generates cracks due to dissimilar coefficients of thermal expansion of substrate PE and SiOx film, as shown in Figure 14.12. For this reason, further reduction of OTR values from 750 cc/m^2.day was not obtained in the case of single-layer inorganic SiOx films on PE substrate. In order to overcome these difficulties, a double-layer SiOx film was deposited on PE substrate. The first layer with more organic nature/polymer/linear type SiOx was deposited, which has similar coefficients of thermal expansion as that of PE. Whereas the second layer with inorganic nature/glasslike/network type SiOx was

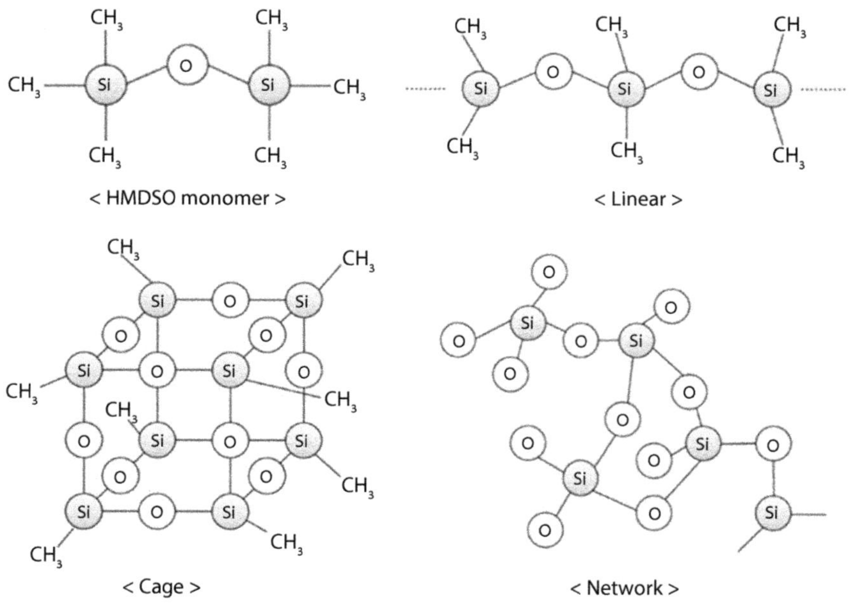

Figure 14.14 Types of SiOx film growth mechanism from HMDSO monomer [20].

deposited. In this configuration, the top network-type SiOx layer gives enhanced barrier properties and the bottom linear-type SiOx layer acts as a buffer layer and does not allow cracks to propagate from the top layer to the substrate. With this configuration we could reduce the OTR values up to 350 cc/m^{2}*day.

Deposition rate is found to gradually decrease with increase in oxygen concentration at constant operating pressure. This decrease in deposition rate may be due to a reduction of the number of HMDSO molecules in the gas phase and/or to the high oxygen reactivity with the organic elements. Assuming that the film growth rate is only proportional to the number of HMDSO molecules introduced into the reactor at constant operating pressure, as well as all other operating parameters, the deposition rate can be considered proportional to the partial pressure of HMDSO [19].

Low oxygen concentration produces porous SiO_x coating due to high carbon content. Hence, the oxygen transmission rate (OTR) is found to be high. As we increased oxygen concentration, carbon content in the film was reduced and OTR values dropped down gradually.

14.5 Conclusion

Silicon oxide (SiOx) films from oxygen/HMDSO gas mixture were deposited at different oxygen concentrations. With increase in oxygen concentration, wettability of the coating surface improved as higher oxygen concentration of nonpolar hydrocarbons in the coating were converted into volatile oxidized products which were pumped out. FTIR results confirmed the presence of hydrocarbon content in the coating at lower oxygen concentration. Si-O-Si peak shifting towards higher wavenumbers indicates that the film chemistry became inorganic in nature when oxygen concentration was increased. Deposition rate was reduced with an increase in oxygen concentration. The OTR values were reduced for film deposited with higher oxygen concentration. Inorganic natured SiOx film on PE substrate generated cracks due to the different coefficients of thermal expansion. Double-layer SiOx deposition enhances the oxygen diffusion barrier property. Due to the intial rough surface of PE and random manufacturing defects of micrometer size, SiOx films with 100–200 nm thickness were unable to fully cover the surface. This may be the limitation for further reduction of OTR values from 350 cc/m^2.day.

References

1. Gherardi, R., Becerra, R., Nerin, C., and Bosetti, O., Development of a multilayer antimicrobial packaging material for tomato puree using an innovative technology. *LWT - Food Sci. Technol.* 72, 361–367, 2016.
2. Siracusa, V., Food packaging permeability behaviour: A report. *Int. J. Polym. Sci.* 2012, 1–11, 2012.
3. Chin, A.W., Polymers for innovative food packaging, Interactive Qualifying Project Report submitted to the Faculty of Worcester Polytechnic Institute, April 29, 2010.
4. Marsh, K., and Bugusu, B., Food packaging—Roles, materials, and environmental issues. *J. Food Sci.* 72, R39–R55, 2007.
5. Ramos, M., Valdés, A., Mellinas, A.C., and Garrigós, M.C., New trends in beverage packaging systems: A review. *Beverages* 1, 248–272, 2015.
6. Pankaj, S.K., Bueno-Ferrer, C., Mishra, N.N., Milosavljevic, V., and O'Donnell, C.P., Application of cold plasma technology in food packaging. *Trends Food Sci. Technol.* 35, 5–17, 2014.
7. Lewis, J.S., and Weaver, M.S., Thin-film permeation-barrier technology for flexible organic light-emitting devices. *IEEE J. Sel. Top. Quantum Electron.*, 10, 45–57, 2004.
8. Galeski, A., Queyroy, S., and Monasse, B., Effect of the molecular structure of semicrystalline polyethylene on mechanical properties studied by molecular dynamics. *J. Appl. Polym. Sci. B.* 125, 4358–4367, 2012.
9. Baudet, C., Grandidier, J.C., and Cangemi, L., A two-phase model for the diffuso-mechanical behaviour of semicrystalline polymers in gaseous environment. *Int. J. Solids Struct.* 46, 1389–1401, 2009.
10. Puleo, A.C., Paul, D.R., and Wong, P.K., Gas sorption and transport in semicrystalline poly (4-methyl-1- pentene). *Polymer* 30, 1357–1366, 1989.
11. Kanehashi, S., Kusakabe, A., Sato, S., and Nagai, K., Analysis of permeability; solubility and diffusivity of carbon dioxide; oxygen; and nitrogen in crystalline and liquid crystalline polymers. *J. Membr. Sci.* 365, 40–51, 2010.
12. Kofinas, P., Cohen, R.E., and Halasa, A.F., Gas permeability of polyethylene/poly (ethylene-propylene) semicrystalline diblock copolymers. *Polymer* 35, 1229–1235, 1999.
13. Robertson, G.L., *Food Packaging: Principles and Practices*, CRC Press, 2012.
14. Körner, L., Sonnenfeld, A., and von Rohr, Ph. Rudolf, Silicon oxide diffusion barrier coatings on polypropylene. *Thin Solid Films* 518, 4840–4846, 2010.
15. Wu, C.-Y., Liao, R.-M., Lai, L.-W., Jeng, M.-S., and Liu, D.-Q., Organosilicon/silicon oxide gas barrier structure encapsulated flexible plastic substrate by using plasma-enhanced chemical vapor deposition. *Surf. Coat. Tech.* 206, 4685–4691, 2012.
16. Iwamori, S., Gotoh, Y., and Moorthi, K., Characterization of silicon oxynitride gas barrier films. *Vacuum* 68, 113–117, 2003.

17. Bahroun, K., Behm, H., Mitschker, F., Awakowicz, P., Dahlmann, R., and Hopmann, Ch., Influence of layer type and order on barrier properties of multilayer PECVD barrier coatings. *J. Phys. D: Appl. Phys.* 47, 015201, 2014.
18. Hedenqvist, M.S., and Johansson, K.S., Barrier properties of SiOx-coated polymers: Multilayer modelling and effects of mechanical folding. *Surf. Coat. Technol.* 172, 7–12, 2003.
19. Sahli, S., Segui, Y., and Ramdani, S., R.f. plasma deposition from hexamethyl disiloxane-oxygen mixtures. *Thin Solid Films* 250, 206–212, 1994.
20. Lin, T.-Y., and Lee, C.-T., Organosilicon function of gas barrier films purely deposited by inductively coupled plasma chemical vapour deposition system. *J. Alloys Compd.* 542, 11–16, 2012.
21. Huan, W., Lizhen, Y., and Qiang, C., Investigation of microwave surface-wave plasma deposited SiOx coatings on polymeric substrates. *Plasma Sci. Tech.* 16, 37, 2014.

15

Synthesis and Characterization of Nanostructured Olivine $LiFePO_4$ Electrode Material for Lithium-Polymer Rechargeable Battery

K. Rani[1], M. Abdul Kader[1] and S. Palaniappan[2*]

[1]Central Institute of Plastics Engineering & Technology, Guindy, Chennai, India
[2]Polymers & Functional Materials Division, Indian Institute of Chemical Technology, Tarnaka, Hyderabad, India

Abstract

Renewable power generation systems are currently preferred for clean power generation. However, due to their intermittent and unpredictable nature, energy storage is inevitable to ensure an uninterrupted supply of power at all times. Although there are many possible options for storage of electrical energy, the most popular and technologically mature option is the use of batteries. Among the rechargeable (secondary) batteries, lithium-based rechargeable batteries have many advantages. Lithium-ion batteries are popular worldwide for sourcing power to wireless telephones, laptop computers, and other electronic devices. Current lithium-ion batteries mostly consist of $LiCoO_2$ and layered graphite as electrode materials. Possible alternatives to $LiCoO_2$ and graphite are being intensively investigated throughout the world. Lithium iron phosphate ($LiFePO_4$) is a promising cathode material for lithium-ion batteries due to its versatility. In this chapter, we discuss the storage device; common rechargeable batteries, in particular lithium battery; the comparison between lithium-ion and lithium-polymer batteries; commercial manufacturers; advantages and limitations of the lithium-ion/polymer batteries; the importance of $LiFePO_4$ material, its synthesis, characterization, structure and use as cathode along with polymer electrolyte in lithium-polymer rechargeable batteries.

***Keywords*:** Lithium-polymer battery, $LiFePO_4$ cathode, polymer electrolyte, electronic application

**Corresponding author*: palaniappan@iict.res.in

Sanjay K. Nayak *et al.* (eds.) Trends and Applications in Advanced Polymeric Materials, (275–304) 2018 © Scrivener Publishing LLC

15.1 Introduction

15.1.1 Energy Storage: Rechargeable Batteries

In recent years, limited mineral oil resources and global warming threats have continued to exert pressure on an already stretched world energy infrastructure. Hence, building a low carbon society has been a topic for new research attempts worldwide. If a new energy economy is to emerge, it must be based on cheap and sustainable energy harvesting and uninterrupted supply. One of the most energy-consuming activities is travel, and here battery devices can potentially provide a solution, especially as they can be used to store electrical energy from sustainable sources such as solar and wind power.

Energy storage refers to a process of converting electrical energy from a power network into a form that can be stored for converting back to electrical energy when needed. Such a process enables electricity to be produced at times of either low demand, low generation cost or from intermittent energy sources, and to be used at times of high demand, high generation cost or when no other means of generation is available. Energy storage has numerous applications, including portable devices, transport vehicles and stationary energy resources.

Energy storage technologies includes compressed air energy storage system, battery, flow battery, fuel cell, solar fuel, superconducting magnetic energy storage system, flywheel, capacitor, super capacitor, and thermal energy storage system. Some of them are currently available and some are still under development.

A rechargeable battery consists of one or more electrochemical cells. It is also known as "secondary cell" because its electrochemical reactions are electrically reversible. Rechargeable batteries come in many different shapes and sizes, ranging from button cells to megawatt grid systems. Rechargeable batteries have lower total cost and lesser environmental impact than non-rechargeable (disposable) batteries. Some rechargeable battery types are available in the same form as disposables. Rechargeable batteries have higher initial cost but can be recharged very cheaply and used many times.

A battery is composed of several electrochemical cells that are connected in series and/or in parallel to provide the required voltage and capacity, respectively. Each cell consists of a positive and a negative electrode (both sources of chemical reactions) separated by an electrolyte solution containing dissolved salts, which enable ion transfer between the two electrodes. Once these electrodes are connected externally, the chemical reactions proceed in tandem at both the electrodes, thereby liberating electrons and enabling the current to be tapped by the user. The amount of electrical energy, expressed either per unit of weight (W h kg^{-1}) or per unit of volume (W h l^{-1}), that a battery is able to deliver is a function of the cell potential (V) and capacity (A h kg^{-1}), both of which are linked directly to the chemistry of the system. Among the various existing technologies, because of their high energy density and design flexibility, Li-based batteries currently outperform other systems, accounting for 63% of worldwide sales in portable batteries. This explains why they receive the most attention at both fundamental and applied levels.

Although rechargeable battery provides efficient storage of electrical energy as chemical energy, the storage capacity of the individual battery cell is limited, and the construction of large stationary batteries containing many cells for use at a centralized facility is currently prohibitively expensive. A more efficient and cost-effective way to use batteries for electrical energy storage has been in portable electronic devices, e.g., cell phones, laptop computers, power tools, etc. Other than these the plug-in hybrid electric vehicle (PHEV) is particularly attractive for large-scale electrical energy storage since the electric energy can be stored in commuter vehicles during off-peak demand without contributing to distributed air pollution. However, the cost of the battery for even the PHEV presently limits the commercial potential of this alternative. The greater convenience and higher energy density of the chemical energy stored in gasoline compared to that in battery electrodes make it difficult for a large, all-electric vehicle to compete with the modern family car powered by the internal-combustion engine, and the oil network on which our modern transportation system mainly depends is well-established. Nevertheless, the rechargeable battery offers the most promising possibility of displacing a large fraction of our oil consumption by providing distributed storage of electricity generated by centralized and/or dispersed alternative energy sources. Lithium-ion rechargeable battery is considered to be the most promising alternate candidate for internal combustion energy because of its superior energy density compared to conventional rechargeable batteries.

Common rechargeable batteries:

- *Lead-acid battery:* Lead acid batteries hold the largest market share of electric storage products. A single cell produces about 2V when charged.
- *Nickel–cadmium battery* (Ni-Cd): Cadmium is a toxic element, and was banned for most uses by the European Union

in 2004. Nickel-cadmium batteries have been almost completely replaced by nickel-metal hydride (NiMH) batteries.

- Nickel–metal hydride battery (NiMH): First available in 1989, the NiMH battery was the first commercial battery. It is now a common consumer and industrial type.
- Lithium-ion battery: The choice in many consumer electronics with one of the best energy-to-mass ratios and a very slow self-discharge when not in use.
- Lithium polymer battery: These batteries are light in weight and can be made in any shape as desired.

15.1.1.1 Lithium Battery

A lithium-ion battery belongs to the family of rechargeable-type battery in which lithium (Li) ions move from the negative electrode to the positive electrode during discharge and back when charging (Figure 15.1). Li-ion batteries use intercalculated lithium compound as one electrode material, compared to the metallic lithium used in a non-rechargeable lithium battery. The electrolyte, which allows for ionic movement, and the two electrodes are the constituent components of a lithium-ion battery cell [1].

Three primary components of a lithium-ion battery are the positive and negative electrodes and electrolyte. Generally, the negative electrode of a conventional lithium-ion cell is made from carbon. The positive electrode is a metal oxide, and the electrolyte is a lithium salt in an organic solvent

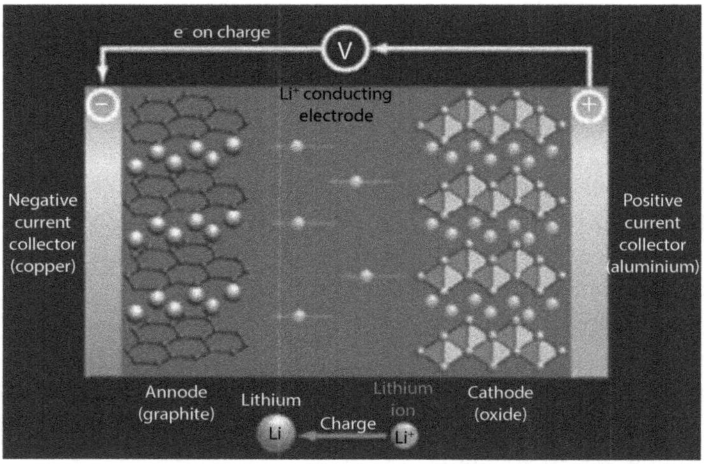

Figure 15.1 Li-ion cell operation: During charging, lithium ions intercalate into the anode; the reverse occurs during discharge [1].

[2]. The electrochemical roles of the electrodes will reverse between anode and cathode, depending on the direction of current flow through the cell.

The most commercially popular negative electrode is graphite. The positive electrode is generally one of three materials: a layered oxide (such as lithium cobalt oxide), a polyanion (such as lithium iron phosphate) or a spinel (such as lithium manganese oxide) [3].

The electrolyte is typically a mixture of organic carbonates such as ethylene carbonate or diethyl carbonate containing complexes of lithium ions [4]. These nonaqueous electrolytes generally use non-coordinating anion salts such as lithium hexafluorophosphate ($LiPF_6$), lithium hexafluoroarsenate ($LiAsF_6$), lithium perchlorate ($LiClO_4$), lithium tetrafluoroborate ($LiBF_4$) and lithium triflate (CF_3LiSO_3).

Lithium-ion batteries are common in household electronics. They are one of the most popular types of rechargeable batteries for portable electronics, with a high energy density, tiny memory effect [5] and low self-discharge. Beyond consumer electronics, lithium-ion batteries are also growing in popularity for military, battery electric vehicle and aerospace applications [6]. For example, lithium-ion batteries are becoming a common replacement for the lead acid batteries that have been historically used for golf carts and utility vehicles. Instead of heavy lead plates and acid electrolyte, the trend is to use lightweight lithium-ion battery packs that can provide the same voltage as lead-acid batteries; therefore no modification to the vehicle's drive system is required.

15.1.1.2 Comparison between Li-Polymer Battery and Liquid Battery

The advantages of a lithium-ion polymer battery compared to a liquid battery are:

- *Excellent safety performance:* As the aluminum blister outer package is different from the metal-shelled liquid lithium power cell, no safety problems arise, as it cannot detonate, only expand.
- *Super thin design:* The design is suitable for all kinds of super-thin electrical appliances, but liquid lithium ion will have a technical bottleneck as thickness less than 3.6 mm is not admissible.
- *Light weight:* The lithium polymer cell is 40% lighter than the same specification of the steel shell liquid lithium cell, and 20% lighter than aluminium shell liquid lithium cell.

- *High capability:* The polymer capacity is 10–15% greater than the same specification of steel shell and aluminium shell liquid lithium cell.
- *Low internal resistance:* Presently the product impedance can be made below 35 Ohm. Because of low impedance, the usage time can differ by 20–50% with the same capacity.
- *Customized shape:* The thickness and shape can be made according to customer requirements. Some special shapes, such as arc-shape battery, can also be made.
- *Excellent discharge performance:* Li-polymer applies gelled electrolyte with more stable discharge characteristics and higher discharge plateau.

15.1.1.3 Commercial Production

A brief history of the commercial production of lithium-ion battery (LIB) is given below [7]:

- 1991 – Sony and Asahi Kasei released the first commercial lithium-ion battery.
- 1996 – John Goodenough and coworkers proposed lithium iron phosphate ($LiFePO_4$) and other phospho-olivines (lithium metal phosphates with the same structure as mineral olivine) as positive electrode materials.
- 2002 – Yet-Ming Chiang and his group at MIT showed a substantial improvement in the performance of lithium batteries by boosting the material's conductivity by doping it with aluminium, niobium and zirconium. The exact mechanism causing the increase in conductivity became the subject of widespread debate.
- 2004 – Chiang again increased performance by utilizing iron(III) phosphate particles of less than 100 nanometers in diameter. This decreased particle density almost one hundred-fold, increased the positive electrode's surface area and improved capacity and performance. Commercialization led to a rapid growth in the market for higher capacity LIBs, as well as a patent infringement battle between Chiang and John Goodenough.
- 2011 – Lithium-ion batteries accounted for 66% of all portable secondary (i.e., rechargeable) battery sales in Japan.

- 2012 – John Goodenough, Rachid Yazami and Akira Yoshino received the 2012 IEEE Medal for Environmental and Safety Technologies for developing the lithium-ion battery.
- 2014 – Commercial batteries from Amprius Inc. reached 650 Wh/L (a 20% increase) using a silicon anode and were delivered to customers. The National Academy of Engineering recognized John Goodenough, Yoshio Nishi, Rachid Yazami and Akira Yoshino for their pioneering efforts in the field.

Rechargeable Li-ion batteries are key components of the portable, entertainment, computing and telecommunication equipment required by today's information-rich, mobile society. Despite the impressive growth in sales of batteries worldwide, the science underlying battery technology is often criticized for its slow advancement. This is true whatever the technology considered (e.g., nickel-cadmium, nickel-metal hydride or ion). Certainly, although energy storage cannot keep pace with the rate of progress in the computer industry, the past decade has produced great advances in chemistry and engineering within the emerging technologies of Ni-MeH and Li-ion batteries. These cells are now supplanting the well-known Ni-Cd batteries.

Today's lithium polymer cells have improved design. Their defining feature is a flexible, foil-type polymer laminate exterior; they still contain organic solvent. There is a wide array of different technologies used to produce these batteries, some of which have liquid electrolytes, some are gelled by inclusion of polymers in the electrolyte, some with layers laminated together, and others may have their electrodes stacked rather than wound. The chemistry is the same as Li-ion. Basically, it is a Li-ion battery in a soft pack.

15.1.1.4 Advantages of Lithium Polymer Batteries

Lithium-polymer batteries have several advanges, which are:

- They can be made very thin (down to around half a millimeter), however, much of the space is wasted by the packaging. At the bottom of this range, cells typically range from 2 to 6.5 mm thick.
- Their length and width can be made quite large. Cell capacities can range anywhere from 50 mAh for a small cell, such as for a Bluetooth headset, up to 10Ah or more for an electric vehicle battery.

- Absence of metal container allows more flexibility to change sizes per customer requirement. This flexibility exists for several reasons. First, from the standpoint of component supply, the laminate material is just slit to different widths rather than requiring a can supplier to produce new tooling to manufacture some can size; also, the heat sealing process is easily modified compared to crimping for cylindrical cells. Laser welding of prismatic cells can also be easily adapted, but large, flat aspect ratios are problematic for assembly.

15.1.1.5 Limitations of Lithium-Polymer Batteries

Lithium-polymer batteries have some limitations, which are:

- Lithium polymer cells are more expensive per watt hour compared to other types of Li-ion cells for several reasons. First, the high quality laminate material and special tabs that allow sealing against the bag are expensive. Secondly, the lower speed of manufacturing increases both labor and overhead costs. Finally, while lower production runs allow for size flexibility, they result in lower yields and higher prototyping costs.
- The soft packaging on polymer cells is easily punctured and has more swelling than metal cans.
- Compared to a cylindrical cell, the lithium-polymer cell has less volumetric energy density. This is because cylindrical cells do not bulge due to their extremely strong shape, so very high electrode densities can be used. Also, the selection of materials is easier because the small amount of gas produced by a cylindrical cell has no effect on its performance or shape.

15.1.2 Cell Manufacturers Using Lithium Iron Phosphate

Below is the list of manufacturers [8] of lithium battery using lithium iron phopate as electrode material:

- A123 (cylindrical)
- Advanced Lithium Electrochemistry Co., Ltd./Aleees Eco Ark Co., Ltd. (cylindrical, large prismatic)
- B&K Rechargeable Battery Inc. (Shenzhen)/Enersource (26650, large prismatic)

- BYD Co., Ltd. (Shenzhen)
- CBAK Energy Technology, Inc. (A123, cylindrical)
- Discover Energy Corp. (large prismatic)
- DLG Battery Co., Ltd. (cylindrical)
- Dongguan Victory Battery Technology Co., Ltd.
- Ecity Power Co., Ltd. (cylindrical, large cylindrical, large prismatic)
- European Batteries Ltd. (large prismatic)
- Electric Vehicle Power System Technology Co., Ltd. (EVPST) (cylindrical, pouch, large prismatic)
- Fullriver Battery New Technology Co., Ltd. (Fent Battery) (cylindrical, large prismatic)
- GAIA/LTC (Lithium Technology Corp.) (large cylindrical)
- GBP Battery Co., Ltd. (prismatic)
- Gold Peak Group/GP Batteries/EVB Technology (HK) Ltd.
- General Electronics Battery Co., Ltd. (large prismatic)
- Heter Electronics Group/Shandong Heter Battery Technology Co. (cylindrical, large prismatic)
- Hipower New Energy Group Co., Ltd. (cylindrical, large prismatic)
- Huanyu Power Source Co., Ltd. (large prismatic)
- Inanovation Inc. (pouch)
- International Battery (large prismatic)
- Ionex Energy Storage Systems, Inc. (large prismatic)
- Jiangxi Jingang Energy Sources Technology Co., Ltd./ Kingcell Shenzhen (cylindrical)
- JND Battery Co., Ltd. (GBP) (prismatic)
- Joolee Battery Co., Ltd. (Shenzhen) (cylindrical, pouch, large prismatic)
- Kayo Battery (Shenzhen) Co., Ltd. (cylindrical, large cylindrical)
- K2 Energy (cylindrical)
- Leclanché (large prismatic)
- LiFeBatt Inc. (large cylindrical, prismatic)
- Meidy Battery (cylindrical, large cylindrical)
- Microvast, Inc. (cylindrical)
- Mottcell Battery Technology Co., Ltd. (Shenzhen) (cylindrical, large prismatic)
- Optimum Battery (cylindrical)
- PHET (Pihsiang Energy Technology Co., Ltd.) (cylindrical, large cylindrical, prismatic)

- PKCell Battery Co., Ltd. (Shenzhen) (cylindrical, large prismatic)
- Saft Groupe S.A. (large cylindrical)
- Sawtry Technology Ltd.
- Sky Energy (Luoyang) Co., Ltd. (CALB) (large prismatic)
- Samsung/Bosch SB LiMotive
- Shandong Realforce Enterprises Co., Ltd.
- Shenzhen Universal Battery Technology Co., Ltd. (PTI)
- Sunhigh Battery Co., Ltd.
- Tenergy Corp.
- Thundersky (large prismatic)
- Tongyu Technology Ltd. (cylindrical)
- Top Energy
- Valence
- Vigor Power Battery Co., Ltd. (Shenzhen)
- Voltronix USA, Inc. (large prismatic)
- YikLik Energy Co., Ltd. (Shenzhen) (cylindrical, small prismatic, large prismatic)
- Yintong Energy Co., Ltd. (Zhuhai)
- Yok Energy Ltd.
- Zhejiang GBS Energy Co., Ltd.
- Zhejiang Xinghai Energy Technique Co., Ltd. (Headway)
- Zhuhai Sanchuan Electronic Technology Co., Ltd. (Seiden)

15.1.3 Lithium Iron Phosphate ($LiFePO_4$)

Lithium iron phosphate ($LiFePO_4$) battery [5] is a type of Li-ion rechargeable battery for high power applications in electric vehicles, power tools, remote control (RC) hobbies, etc. A lithium iron phosphate ($LiFePO_4$ or LFP) cell features high discharging current, is nonexplosive and has a long cycle life (> 2000@0.2C rate, IEC Standard); however, its energy density is lower than a normal Li-ion cell. LFP cell has 3.2V nominal working voltage and power will be cut off at 3.6–3.8V per cell during charging. Please see Table 15.1 for a comparison between LFP and other rechargeable batteries. An LFP battery has the lowest cost if its long cycle life is taken into consideration.

The $LiFePO_4$ batteries have shown a huge improvement over the lead acid type in weight, capacity and shelf life. They are the safest type of lithium batteries as they will not overheat, and even if punctured they will not catch on fire. The cathode material, $LiFePO_4$, is not hazardous, and so poses no negative health hazards or environmental hazards. Due to the oxygen being bonded tightly to the molecule, there is no danger of the

Table 15.1 Performance comparison of $LiFePO_4$ battery with other batteries.

Chemistry	Voltage	Energy density	Working temperature (°C)	Cycle life	Safety	Environmental	Cost based on cycle life x wh of SLA[1]
$LiFePO_4$	3.2V	> 120 Wh/kg	–20–60	> 2000 (0.2C rate, IEC Standard)	Safe	Good	0.15–0.25 lower than SLA
Lead acid	2.0V	> 35Wh/kg	–20–40	> 200	Safe	Not good	1
NiCd	1.2V	> 40Wh/kg	–20–50	> 1000	Safe	Bad	0.7
NiMH	1.2V	> 80 Wh/kg	–20–50	> 500	Safe	Good	1.2–1.4
$LiMn_xNi_yCo_zO_2$	3.7V	> 160 Wh/kg	–20–40	> 500	better than LiCo	OK	1.5–2.0
$LiCoO_2$	3.7V	> 200 Wh/kg	–20–60	> 500	Unsafe w/o PCM[2]	OK	1.5–2.0

[1]Sealed lead acid; [2]Protection circuit module

battery erupting into flames like there is with lithium-ion. The chemistry is so stable that $LiFePO_4$ batteries will accept a charge from a lead-acid configured charger. Though less energy dense than lithium-ion the lithium-polymer battery has advantages due to the presence of iron and phosphate which are abundant and cheaper to extract, so costs are much more reasonable. The life expectancy of $LiFePO_4$ is approximately 5–7 years.

Lithium iron phosphate is a natural mineral of the olivine family (triphylite). Its use as a battery electrode was first described as a cathode material for rechargeable lithium batteries in the published literature by John Goodenough's research group at the University of Texas in 1996 [9, 10]. Due to its low cost, nontoxicity, excellent thermal stability, safety characteristics, electrochemical performance, specific capacity (170 mA·h/g, or 610 C/g) and the nautral abundance of iron, $LiFePO_4$ has gained some market acceptance [11, 12].

15.1.3.1 Synthesis of $LiFePO_4$

The precursor mixture for the preparation of $LiFePO_4$ usually consists of a stoichiometric amount of an iron salt, a lithium compound and a phosphate. For the solid-state process, the starting materials are commonly Fe(II)-acetate or Fe(II)-oxalate, Li_2CO_3 and mostly ammonium phosphate. These materials can decompose at a relatively low temperature without generation of undesirable solid phases. In general, the starting mixture first decomposes at about 200 °C to release NH_3, then the oxalates are decomposed at 300–400 °C to release CO_2; after regrinding, they are calcined at a temperature ranging from 400–800 °C for 8–24 h. The solid routes always need a lengthy sintering process at high temperature to achieve a pure phase. If a mechanical alloying process is used, the sintering time can be significantly reduced. In order to achieve an ideal electrochemical performance, after short-time milling a long-time thermal treatment is usually needed; whereas a long-time milling just needs a subsequent short-time thermal treatment. In addition, ball milling helps both homogeneity and reduction of the particle size. In most cases, Fe(III) compounds, such as Fe_2O_3 and $FePO_4$, are much cheaper and more stable than Fe(II) salts, but Fe(III) precursors need a reductive gas like Ar + 15% H_2 to form the ferrous $LiFePO_4$. With an Fe(III)compound as the iron source, the carbon-thermal reduction method is used in which a calculated amount of carbon added to the precursor effectively reduces Fe(III) to Fe(II) at high temperature. In this case, carbon serves as the reducing agent; no additional reductive gas is needed. However, the carbon-thermal reduction always needs a very high temperature to get a high-quality $LiFePO_4$/C composite.

Some undesired impurities, such as phosphide, Fe_2O_3 and $Li_3Fe_2(PO_4)_3$ may be generated by a temperature that is too high and always leads to uncontrollable particle growth and agglomeration. If a polymer is used as a carbon source, complete Fe^{3+} reduction can occur at a temperature below 700 °C, even as low as 400 °C. In this case, the reduction of Fe^{3+} is due to hydrogen gas coming from decomposition of the polymer, not to carbon-thermal reduction. A carbon-containing compound or another compound containing a doping element can be added into the precursor either before the second grinding step or at the beginning to obtain $LiFePO_4$/C or doped $LiFePO_4$ samples. It should be noted that the electronic conductivity of *in-situ* carbon coating is strongly dependent not only on the sintering temperature, but also on the type of organic precursor. As a conventional method, solid-state synthesis is limited for controlling particle growth and agglomeration. The final product always has large particles with irregular morphology. Compared to the solid-state approach, solution chemistries (sol-gel, coprecipitation, spray pyrolysis and hydrothermal reaction) are well known for their advantages in tailoring the size and morphology of the nanostructure, which makes them powerful routes to prepare $LiFePO_4$.

The solution route starts from a precursor in liquid solution. The precursor solution can provide intimate mixing of the starting ingredients at the atomic level, thus allowing fine particles of high purity to be produced by rapid homogeneous nucleation. In addition, a wide range of reactants for the solution method are available. Some compounds, such as iron chloride and iron sulfate, which are not suitable for a solid-state reaction, can be used successfully in the solution method because Cl^- and SO_4^{2-} can be dissolved in the solution and removed with the solution. With this method, some impurities can also be removed with the solution, which reduces the requirement of purity for the reactants and thus lowers fabrication costs. To reduce Fe(III) in solution, some reducing agents, such as oxalic acid or ascorbic acid, can be used. Similar to the solid route, solution synthesis using either sol-gel, coprecipitation or hydrothermal reaction also needs a sintering process at high temperature for a certain period of time. Hydrothermal synthesis has been successfully used to transform transition metal oxides into high-quality nanostructures. It is the most common way to prepare nano-sized $LiFePO_4$. Precursors used in the synthesis via various synthetic routes are given in Tables 15.2 to 15.6.

15.1.3.2 Structure of $LiFePO_4$

According to Yuan *et al.* [37], $LiFePO_4$ has the ordered-olivine structure, space group Pnma. The oxygen atoms are located in a slightly distorted,

Table 15.2 Solid-state synthesis.

Li precursor	Fe precursor	P precursor	Product	Ref.
Li_2CO_3	$FePO_4(H_2O)_2$		$LiFePO_4/C$	[13]
Li_2CO_3	$FeC_2O_4.2H_2O$	$NH_4H_2PO_4$	$LiFePO_4$	[14]
$LiOH.2H_2O$	$FeC_2O_4.2H_2O$	$NH_4H_2PO_4$	$LiFePO_4/C$	[15]
$LiOH.2H_2O$	$FeC_2O_4.2H_2O$	$NH_4H_2PO_4$	$Li_{0.99}FeNb_{0.01}PO_4$	[16]
$LiOH.\ H_2O$	$FeC_2O_4.2H_2O$	$(NH_4)_2HPO_4$	$LiFePO_4$	[17]

Table 15.3 Carbothermal reduction method.

Li precursor	Fe precursor	P precursor	Product	Ref.
Li_2CO_3	$FeSO_4.7H_2O$	$NH_4H_2PO_4$	$LiFePO_4/C$	[18]
Li_2CO_3	Fe_2O_3	$NH_4H_2PO_4$	$LiFePO_4/C$	[19]
Li_2CO_3	$FePO_4$	---------	$LiFePO_4/C$	[20]
LiH_2PO_4	Fe_2O_3	-----------	$LiFePO_4/(C+Fe_2P)$	[21]
Li_2CO_3	$Fe(NO_3)_3.9H_2O$	$NH_4H_2PO_4$	$LiFePO_4/C$	[22]

Table 15.4 Microwave heating method.

Li precursor	Fe precursor	P precursor	Product	Ref.
CH_3COOLi	$FeC_2O_4.2\ H_2O$	$NH_4H_2PO_4$	$LiFePO_4$	[23]
Li_3PO_4	$FeSO_4.7H_2O$	H_3PO_4	$LiFePO_4/C$	[24]
$LiOH.\ H_2O$	$FePO_4.4H_2O$	-----------	$LiFePO_4/C$	[25]
Li_3PO_4	$Fe(PO_4)_2.8H_2O$	------------	$LiFePO_4/C$	[26]
Li_3PO_4	$Fe(PO_4)_2.8H_2O$	------------	$LiFePO_4/C$	[27]

Table 15.5 Sol-gel method.

Li precursor	Fe precursor	P precursor	Product	Ref.
Li_3PO_4	$C_6H_5FeO_7$	H_3PO_4	POROUS $LiFePO_4/C$	[28]
$LiCO_3$	$FeC_2O_4.2\ H_2O$	$NH_4H_2PO_4$	POROUS $LiFePO_4/C$	[29]
$LiCO_3$	$FeC_2O_4.2\ H_2O$	$NH_4H_2PO_4$	POROUS $LiFePO_4/C$	[29]
$LiNO_3$	$FeC_2O_4.2\ H_2O$	$NH_4H_2PO_4$	$LiFePO_4/C$	[30]
$Li(COO\ CH_3)$	$Fe\ (COO\ CH_3)_2$	H_3PO_4	$LiFePO_4/C$	[31]

Table 15.6 Hydrothermal method.

Li precursor	Fe precursor	P precursor	Product	Ref.
CH_3COOLi	$FePO_4.2H_2O$	-------------	$LiFePO_4/C$	[32]
LiOH. H_2O	$FeSO_4.7H_2O$	H_3PO_4	$LiMg_{0.02}Fe_{0.98}PO_4$	[33]
LiOH. H_2O	$FeSO_4.7H_2O$	H_3PO_4	$LiFePO_4$/MWCNTs	[34]
LiOH. H_2O	$FeSO_4.7H_2O$	H_3PO_4	$LiFePO_4/C$	[35]
Li_3PO_4	$FeSO_4.7H_2O$	-------------	$LiFePO_4/C$	[36]

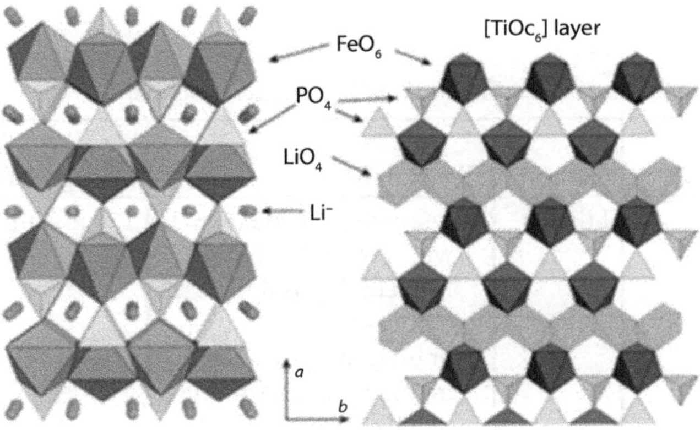

Figure 15.2 Structure of olivine $LiFePO_4$ [1].

hexagonal close-packing arrangement. The phosphorus atoms occupy tetrahedral sites; the iron and lithium atoms occupy octahedral 4a and 4c sites, respectively. Each FeO_6 octahedron is linked with four FeO_6 octahedra through common corners in the b-c plane, forming zigzag planes. The LiO_6 octahedra form edge-sharing chains along the *b*-axis. One FeO_6 octahedron has common edges with two LiO_6 octahedra. PO_4 groups share one edge with a FeO_6 octahedron and two edges with LiO_6 octahedra. The delithiated phase $FePO_4$ has essentially the same structure as $LiFePO_4$ but the $FePO_4$ host framework deforms slightly to orthorhombic symmetry (Figure 15.2). The structural similarity between $LiFePO_4$ and $FePO_4$ not only avoids capacity degradation resulting from severe volumetric changes during the charge-discharge process, but also effectively compensates the volume changes of the carbon anode during lithiation and delithiation.

This explains to a large extent the truly excellent electrochemical cyclability of the system.

15.1.3.3 Work on $LiFePO_4$ Cell Systems

Most of the research concentrating on cathode materials is based on $LiFePO_4$, for example, $LiFePO_4$-C [38–47], Li-polyaniline [48, 49], Li-polypyrrole [50–53], Li-C-polypyrrole [54, 55], and other materials [56–58]. Most groups are working on polymer electrolytes such as polyethylene oxide-polyethylene glycol [59–63], poly(vinylidene fluoride-co-hexafluoropropylene) [64–68], polyethylene oxide [69, 70], other polymer electrolytes [71–78] and ionic liquids [79–83]. The literature reports on the performance of the $LiFePO_4$-polymer systems are reported in Table 15.7.

Table 15.7 Literature reports on $LiFePO_4$-polymer systems.

Sl. no	Anode	Cathode	Electrolyte	Charge rate	Ref.
1	Li	$LiFePO_4$	poly(ethylene oxide) (PEO) with Li+ and N-butyl-N-methylpyrrolidinium (PYR14+) salts of bis (trifluoro methanesulfonyl)imide (TFSI)	154 mAh/g	[84]
2	Li	$LiFePO_4$	mesoporous lithium aluminate hexa decyltrimethyl ammonium bromide (CTAB) as the template	140 mAh/g at 60 °C	[85]
3	Li	$LiFePO_4$	(PVdF-HFP-based) gel and composite microporous membranes 0.5-M LiPF3(CF2CF3) (3) (lithium fluoroalkylphosphate, LiFAP) in EC:DEC (1:1 v/v)	157 and 148 mAh/g	[86]
4	Li	$LiFePO_4$	PYR14TFSI + LiTFSI		[87]
5	Graphite	$LiFePO_4$	EC-DEC and EMI(FSI)-LiFSI	160 mAh/g at C/24	[88]

Table 15.7 Cont.

Sl. no	Anode	Cathode	Electrolyte	Charge rate	Ref.
6	Li	$LiFePO_4$	1.0 M lithium bis(trifluoromethane) sulfonamide in various volume proportions of glutaronitrile, a dinitrile, and butyronitrile, a mononitrile (10/0; 8/2; 6/4; 4/6; 2/8; 10/0).	100 cycles at a 0.5C rate	[89]
7	Li	$LiFePO_4$	viscous PEGMEM-co-SMA copolymer, lithium salt and cellulose matrix	156.9 mAh/g	[90]
8	Sn-C	$LiFePO_4$	polyethylene oxide-based electrolyte	120 mAh/g at a voltage of 2.8 V	[91]
9	Li	$LiFePO_4$	guanidinium-based ionic liquid with 0.8 mol kg(-1) lithium bis(trifluoromethane sulfonylimide)	142 and 150 mAh/g at the 100th cycling at 25 and 50 °C	[92]
10	Li	$LiFePO_4$	N-methyoxymethyl-N-methyl piperidinium bis(trifluoromethane sulfonyl) imide (PP1.1O1TFSI) and polyethylene oxide		[93]
11	$Li4Ti_5O_{12}$	$LiFePO_4$	liquid N-butyl-N-methylpyrrolidinium bis(fluorosulfonyl) imide	140 mAh/g constant for more than 150 cycles	[94]
12	graphite	$LiFePO_4$	CH_3-(OC_2H_4)(3)-CH_3 (TG)/LiN(SO_2F)(2) (LiFSI)		[95]
13	Li	$LiFePO_4$	gel polymer electrolyte containing ionic liquid		[96]
14	$LiFePO_4$	$LiFePO_4$	poly(ethylene oxide) PEO-based solid polymer electrolyte		[97]

15.2 Experimental

It is clear from the above discussion that lithium-polymer battery using $LiFePO_4$ as cathode material is important work for future secondary batteries for the market. The following work have been carried out.

15.2.1 Synthesis

$LiFePO_4$ materials were synthesized via three different methods:

1. Sol-Gel
2. Hydrothermal
3. Glycothermal

Sol-Gel method: As a starting precursor one part of iron(III) citrate and two parts of citric acid were dissolved at 60 °C in water. Separately, an equimolar water solution of LiH_2PO_4 was prepared from lithium phosphate and phosphoric acid. Clear solutions were mixed together and dried at 60 °C for 24 h. After thorough grinding with a mortar and pestle, the obtained xerogel was fired in an inert atmosphere at 700 °C. Finally, the product was taken out and ground into powder.

Hydrothermal method: As a starting precursor, one part of iron(III) citrate and two parts of citric acid were dissolved at 60 °C in water. Separately, an equimolar water solution of LiH_2PO_4 was prepared from lithium phosphate and phosphoric acid. Clear solutions were mixed together and transferred to an autoclave, heated at 200 °C for 6 h. The solid portion was filtered and dried in an oven. After thorough grinding with a mortar and pestle, the obtained xerogel was fired in an inert pure atmosphere at 700 °C. Finally, the product was taken out and ground into powder.

Glycothermal method: As a starting precursor, iron(III) citrate and two parts of citric acid were dissolved at 60 °C in glycerol. Separately, an equimolar water solution of LiH_2PO_4 was prepared from lithium phosphate and phosphoric acid. Clear solutions were mixed together and dried at 60 °C for 24 h. The solutions were heated in autoclave at 200 °C for 6 h and the solid portion was filtered and dried in an oven. After thorough grinding with a mortar and pestle, the obtained xerogel was fired in an inert pure atmosphere at 700 °C. Finally, the product was taken out and ground into powder.

15.3 Characterization

X-ray diffraction (XRD) experiments were carried out using an X-ray mini diffractometer (Rigaku, MiniFlex IIC, Japan) equipped with a Cu Kα source (wavelength 1.5406 Å) at a scan speed of 0.045° min^{-1}. Morphology of cathode samples was investigated by scanning electron microscopy (SEM: Hitachi, model no. S-3400N)) and the images were taken at 15 to 20 kV. The polymer powder sample was gold-sputtered on a carbon disc with the help of double-sided adhesive tape

The elemental composition of the cathode material was analyzed using an energy dispersive X-ray spectrometer (EDX) (Horiba, model no. 7021-H) attached to the SEM using an accelerating voltage of 15 kV. To ensure reliable average values, EDX analyses were conducted for each material component on at least 6 different scan areas of each sample with a minimum of 3 spot analyses per scan area. As this characterization technique has limitations in detecting light elements, we could not detect any peak corresponding to Li element.

Charge-discharge cycling measurements were performed using an electrochemical workstation (ZIVE MP5, WonATech, S. Korea). Coin-type cells (Type 2016) were assembled by using graphite sheet anode and 1 M lithium trifluoromethanesulfonate (lithium triflate) in ethylene carbonate/propylene carbonate (1:1 in volume) electrolyte with Celgard separator. The electrospun microporous membrane gel polymer electrolyte from PVDF-co-HFP was soaked in above said electrolyte solution. The coin cell was assembled in a glove box filled with argon gas. $LiFePO_4$ was used as cathode material coated on Cu current collector. Cells were cycled at room temperature (25 °C) in the voltage range of 2–3V by applying constant current of 0.5 mA. Charge and discharge were carried out at 1C rating.

15.4 Results and Discussion

15.4.1 Morphology

The morphology and microstructure of $LiFePO_4$ prepared by three different methods as described in the experimental section are shown in Figure 15.3. It can be seen from the figure that there is a variation in the size and shape of electrode material that depended on the preparation method, as the environmental conditions for synthesis of electrode material were different. This is due to the fact that in the sol-gel method the material was synthesized without stringent conditions. As a result, the morphology of $LiFePO_4$ became porous with pore density preferably more than 50%, which is clearly seen at higher magnification (not shown in figure). Moreover, $LiFePO_4$ prepared by sol-gel method showed honeycomb-like morphology with the pore dimension of 60 to 120 nm. This porous structure is helpful for insertion/intercalation of lithium ion

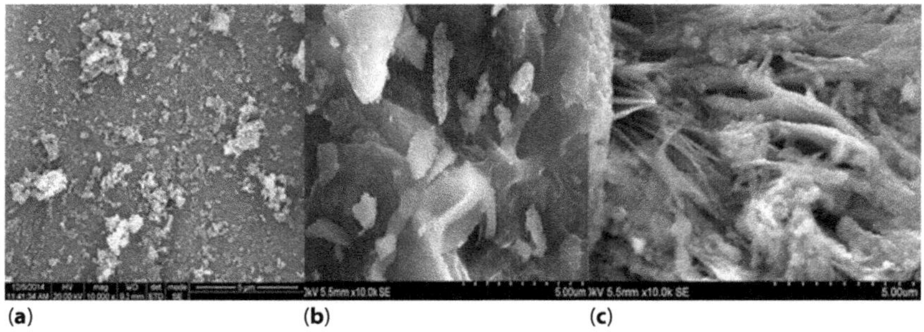

Figure 15.3 SEM images of $LiFePO_4$ prepared via (a) sol-gel, (b) hydrothermal and (c) glycothermal methods.

during the electrochemical process. However, the hydrothermal method, with high pressure and temperature conditions, yielded particles with layers/flake-like structure with channels in the nanoscale range, which is beneficial for conducting lithium ions. Interestingly, the glycothermal method produced $LiFePO_4$ with a mix morphology, having fiber as well as flakes with the diameter of fiber in the nanoscale region, as the condition of synthesis was still severe due to the high viscosity of glycerol compared to water in the hydrothermal process. Again, this kind of morphology is advantageous for both ion transport and intercalation of lithium ion.

15.4.2 E-DAX

Energy dispersive X-ray analysis (E-DAX) of $LiFePO_4$ was performed for all the samples and a representative image is depicted in Figure 15.4. It shows the presence of constituent elements corresponding to the iron, phosphorus and oxygen. Lithium, being a light metal, could not be detected by E-DAX. Based on the peak heights, the compositions of $LiFePO_4$ prepared by three different methods are listed in Table 5.8.

15.4.3 Charge-Discharge Characteristics

The dynamic performance of a battery in charge and discharge is the speed at which current can be put into and taken from storage. The terminal voltage rises and falls during charging and discharging. The charge and discharge dynamics of batteries can be characterized by measurements of voltage under constant charge/discharge current inputs. The charge can be controlled by current or voltage. Charging usually consists of periods of constant current (CC) and/or constant voltage (CV) charging. A CC

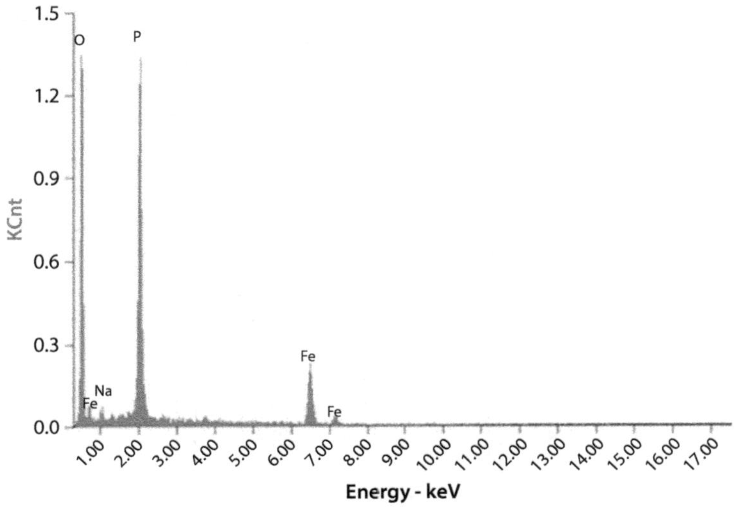

Figure 15.4 EDAX of $LiFePO_4$ prepared via sol-gel method indicating the composition as $Li_{0.05}Fe_{0.35}(PO_4)_{0.6}$.

Table 15.8 E-DAX compositions of $LiFePO_4$ prepared by three different methods.

Method of preparation	Elemental composition (%)	Composition
Sol-Gel	Fe = 20.47; P = 29.0; O = 48.80	$Li_{0.02}Fe_{0.41}(PO_4)_{0.58}$
Hydrothermal	Fe = 24.0; P = 16.02; O = 60.01	$Li_{0.02}Fe_{0.47}(PO_4)_{0.52}$
Glycothermal	Fe = 6.58; P = 27.5; O = 65.96	$Li_{0.02}Fe_{0.32}(PO_4)_{0.67}$

charge is initially applied to bring the voltage up to the CV level. If the CV charge is at the beginning, the current will be too high and an excessive temperature rise will happen. Galvan static charge-discharge experiments were carried out for button cells at a constant current of 0.5 mA and the results are shown in Figure 15.5. The capacity of the cell was obtained as 120, 132 and 140 $mAhg^{-1}$ for the sample prepared via hydrothermal, glycothermal and sol-gel methods respectively.

15.4.4 XRD Studies on $LiFePO_4$

The formation of $LiFePO_4$ was found out from X-ray diffraction spectrophotometer. The X-ray diffraction pattern of the sample prepared via sol-gel method is given in Figure 15.6, which shows d space values at 1.47, 1.61, 2.0, and 2.32. These peaks correspond to the material $LiFePO_4$.

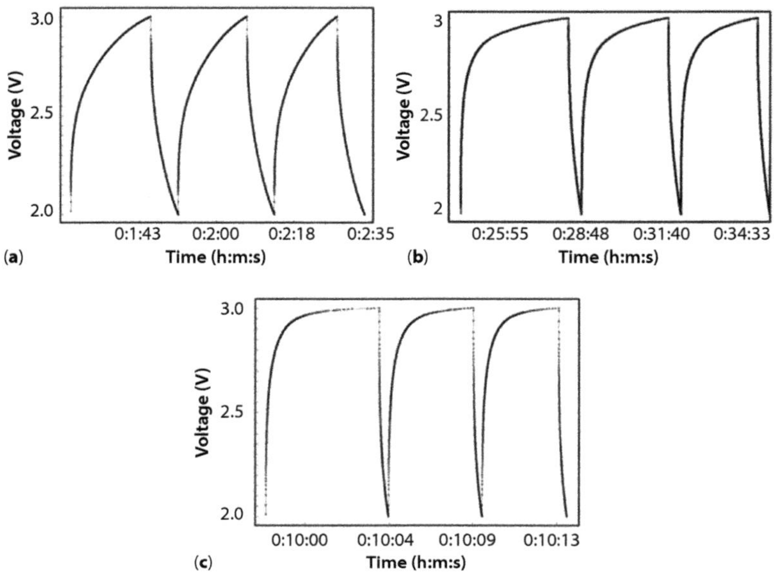

Figure 15.5 Charge-discharge experiments on the cell prepared from $LiFePO_4$ anode and graphite cathode with PVDF-HFP gel polymer electrolyte. $LiFePO_4$ was prepared by (a) sol-gel, (b) hydrothermal and (c) glycothermal methods.

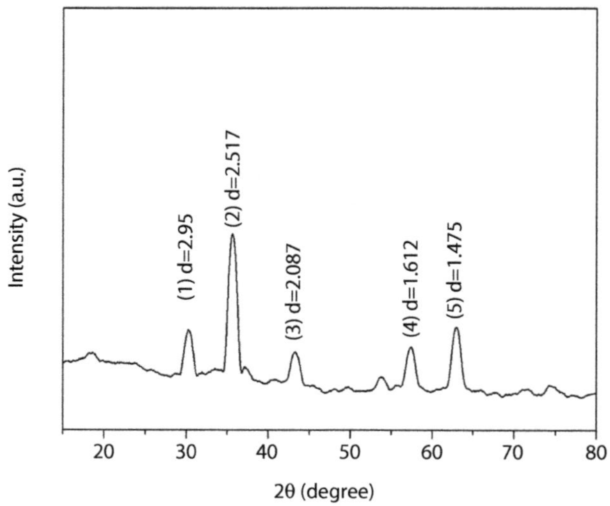

Figure 15.6 XRD pattern of $LiFePO_4$ prepared via sol-gel method.

15.5 Conclusion

Lithium-polymer batteries are the new generation type of lithium batteries with numerous advantages over their predecessors. Huge market potential exists for lithium-polymer rechargeable batteries. They have a high voltage of battery life, a steady discharge rate and are non-polluting. Lithium-polymer battery also eliminates safety hazards which are prevalent in lithium-ion battery. Additionally, its design has become more flexible, convenient, lightweight and handy. The performance of this type of battery has reached or exceeded the technology parameter of liquid lithium-ion battery with a better reliability. Lithium-polymer cells can be made very thin and their length and width can also be made with quite large cells with capacities ranging anywhere from 50 mAh for a small cell, such as for a Bluetooth headset, up to 10 Ah or more for an electric vehicle battery. The lack of metal allows more flexibility to change sizes per customer requirements. From a component supply stand point, the laminate material is just slit to different widths rather than requiring a can supplier to produce new tooling to manufacture some can size; also, the heat sealing process is easily modified compared to crimping for cylindrical cells. Prismatic cells, laser welding can also be easily adapted, but large, flat aspect ratios are problematic for assembly. Development of $LiFePO_4$ cathode and polymer electrolyte with lithium/graphite anode is very much required for energy storage system. Although great progress has been achieved in improving the electrochemical performances of $LiFePO_4$ cathode material by carbon coating, minimization of practical particle size and supervalent cation doping, there are still some challenging problems that need to be solved before attempting to commercialize $LiFePO_4$ for the next generation of lithium-ion battery for electric vehicles. As far as the development of $LiFePO_4$ material is concerned, one has to consider low temperature synthesis with no impurities, increasing the electrical conductivity without impurities by using carbon or conducting polymers and a low-cost device. In this study, $LiFePO_4$ was prepared via three different methods, i.e., sol-gel, hydrothermal and glycothermal methods. Formation of $LiFePO_4$ was confirmed from XRD spectrum. These samples were used as electrodes in button cell configuration with graphite electrode, which showed capacity of 120, 132 and 140 mAh g^{-1} for the sample prepared via hydrothermal, glycothermal and sol-gel methods respectively.

Acknowledgments

The authors (KR and MAK) thank the Department of Science and Technology, New Delhi, India for the financial support to carry out this project (Project No. SR/NM/NS-1001/2010(G).

References

1. Long, R.T., Kahn, M., and Mikolajczak, C., Lithium-ion battery hazards, *Fire Protection Engineering*, w.sfpe.org/page/2012_Q4_2, 2012.
2. Silberberg, M., *Chemistry: The Molecular Nature of Matter and Change*, pp. 935–971, McGraw-Hill Education, 2006.
3. Thackeray, M.M., Thomas, J.O., Whittingham, M.S., Science and applications of mixed conductors for lithium batteries. *MRS Bull.* 25, 39, 2011
4. National Power Corp., Lithium ion batteries, http://www.pssurvival.com/PS/Batteries/Lithium/Lithium-Ion_Battery-2016.pdf, 2010.
5. $LiFePO_4$/$LiFeMnPO_4$ Batteries, http://www.batteryspace.com/$LiFePO_4$/$LiFeMnPO_4$-Batteries.aspx, 2016.
6. Electrovaya, Tata Motors to make electric Indica [Press release], Electrovaya Inc., 2008.
7. Lithium ion batteries, https://en.wikipedia.org/wiki/Lithium-ion_battery, 2017.
8. http://www.lifepo4-info.com
9. Padhi, A.K., Nanjundaswamy, N.S., and Goodenough, J.B., A novel cathode material for rechargeable batteries. *Electrochimical Society Meeting Abstracts* 96, 73, 1996.
10. Padhi, A.K., Nanjundaswamy, N.S., and Goodenough, J.B., Phospho-olivines as positive-electrode materials for rechargeable lithium batteries. *J. Electrochem. Soc.* 144, 1188, 1997.
11. Gorman, J., Bigger, Cheaper, safer batteries: New material charges up lithium-ion battery work, https://www.sciencenews.org/article/bigger-cheaper-safer-batteries-new-material-charges-lithium-ion-battery-work, 2002.
12. House of Batteries, Building safer Li-Ion batteries, http://www.houseofbatteries.com/articles.php?id=27, 2017.
13. Zaghib, K., Ravet N., Gauthier, M., Gendron, F., Mauger, A., Goodenough, J.B., and Julien, C.M., Optimized electrochemical performance of $LiFePO_4$ at 600C with purity controlled by SQUID magnetometry. *J. Power Sources* 163, 560, 2006.
14. Li, M., Xie, K., Li, D., and Pan, Y., Synthesis of $LiFePO_4$ by one-step annealing under the vacuum condition. *J. Mater. Sci.* 40, 2639, 2005.
15. Lai, C., Xu, Q., Ge, H., Zhou, G., and Xie, J., Improved electrochemical performance of $LiFePO_4$/C for lithium–ion batteries with two kinds of carbon sources. *Solid State Ionics* 179, 1736, 2008.
16. Mi, C.H., Zhang, X.G., and Li, H.L., Electrochemical behaviors of solid $LiFePO_4$ and Li0.99Nb0.01 $FePO_4$ in Li_2SO_4 aqueous electrolyte. *J. Electroanal. Chem.* 602, 245, 2007.
17. Takahashi, M., Tobishima, S.H., Takei, K., and Sakurai, Y., Characterization of $LiFePO_4$ as the cathode material for rechargeable lithium batteries. *J. Power Sources* 97–98, 508, 2001.
18. Zhu, B.Q., Li, X.H., Wang, Z.X., and Guo, H.J., Novel synthesis of $LiFePO_4$ by aqueous precipitation and carbothermal reduction. *Mater. Chem. Phys.* 98, 373, 2006.

19. Liu, H.P., Wang, Z.X., Li, X.H., Guo, H.J., Peng, W.J., Zhang, Y.H., and Hu, Q.Y., Synthesis and electrochemical properties of olivine $LiFePO_4$ prepared by carbothermal reduction method. *J. Power Sources* 184, 469, 2008.
20. Wang, L., Liang, G.C., Ou, X.Q., Zhi, X.K., Zhang, J.P., and Cui, J.Y., Effect of synthesis temperature on the properties of $LiFePO_4$/C composites prepared by carbothermal eduction. *J. Power Sources* 189, 423, 2009.
21. Liu, H., Xie, J., and Wang, K., Synthesis and characterizations of $LiFePO_4$/ (C+Fe_2p) composite cathodes. *Solid State Ionics* 179, 1768, 2008.
22. Yu, F., Zhang, J., Yang, Y., and Song, G., Reaction mechanism and electrochemical performance of $LiFePO_4$/C cathode materials synthesized by carbothermal method. *Electrochim. Acta* 54, 7389, 2009.
23. Wang, L., Huang, Y., Jiang, R., and Jia, D., Preparation and characterization nano-sized $LiFePO_4$ by low heating solid-state coordination method and microwave heating. *Electrochim. Acta* 52, 6778, 2007.
24. Beninati, S., Damen, L., and Mastragostino, M., MW-assisted synthesis of $LiFePO_4$ for high power applications. *J. Power Sources* 180, 875, 2008.
25. Guo, X.F., Zhan, H., and Zhou, Y.H., Rapid synthesis of $LiFePO_4$/C composite by microwave method. *Solid State Ionics* 180, 386, 2009.
26. Song, M.S., Kang, Y.M., Kim, J.H., Kim, H.S., Kim, D.Y., Kwon, H.S., and Lee, J.Y., Simple and fast synthesis of $LiFePO_4$/C composite for lithium rechargeable batteries by ball- milling and microwave heating. *J. Power Sources* 166, 260, 2007.
27. Song, M.S., Kang, Y.M., Kim, Y.I., Park, K.S., and Kwon, H.S., Nature of insulating-phase transition and degradation of structure and electrochemical reactivity in an olivine-structured material, $LiFePO_4$. *Inorg. Chem.* 48, 8271, 2009.
28. Dominko, R., Goupil, J.M., Bele, M., Gaberscek, M., Remskar, M., Hanzel, D., and Jamnik, J., Impact of $LiFePO_4$/C composites porosity on their electrochemical performances. *J. Electrochem. Soc.* 150, A8508, 2005.
29. Kim, J.K., Choi, J.W., Chauhan, G.S., Ahn, J.H., Hwang, G., Choi, J.B., and Ahn, H.J., Enhancement of electrochemical performance of lithium iron phosphate by controlled sol-gel synthesis. *Electrochim. Acta* 53, 8258, 2008.
30. Lin, Y., Gao, M.X., Zhu, D., Liu, Y.F., and Pan, H.G., Effects of carbon coating and iron phosphides on the electrochemical properties of $LiFePO_4$/C. *J. Power Sources* 184, 444, 2008.
31. Bhuvaneswari, M.S., Bramnik, N.N., Ensling, D., Ehrenberg, H., and Jaegermann, W., Synthesis and characterization of carbonnano fiber/$LiFePO_4$ composites for Li-ion batteries. *J. Power Sources* 180, 553, 2008.
32. Yang, S., Song, Y., Zavalji, P.Y., and Whittingham, M.S., Reactivity, stability and electrochemical behaviour of lithium iron phosphates. *Electrochem. Commun.* 4, 239, 2002.
33. Ou, X., Liang, G., Wang, L., Xu, S.H., and Zhou, X., Effects of magnesium doping on electronic conductivity and electrochemical properties of $LiFePO_4$ prepared via hydrothermal route. *J. Power Sources* 184, 543, 2008.

34. Xu, J., Chen, G., and Li, X., Electrochemical performances of $LiFePO_4$ cathode material coated with multi-wall carbon nanotubes. *Mater. Chem. Phy.* 118, 9, 2009.
35. Zhang, C., Huang, X., Yin, Y., Dai, J., and Zhu, Z., Hydrothermal monodispersed $LiFePO_4$ cathode material in alcohol-water mixed solutions. *Ceram. Int.* 35, 2979, 2009.
36. Lee, J., and Teja, A.S., Synthesis of $LiFePO_4$ micro and nanoparticles in supercritical water. *Mater. Lett.* 60, 2105, 2006.
37. Yuan, L.X., Wang, Z.H., Ahang, W.X., Hu, X.L., Chen, J.T., Huang, Y.H., and Goodenough J.B., Development and challenges of $LiFePO_4$ cathode materials for lithium-ion batteries. *Energy Environ. Sci.* 4, 269, 2010.
38. Nien, Y.H., Carey, J.R., and Chen, J.S., Physical and electrochemical properties of $LiFePO_4$/C composite cathode prepared from various polymer-containing precursors. *J. Power Sources* 193, 822, 2009.
39. Jin, B., Jin, E.M., Park, K.-H., and Gu, H.-B., Electrochemical properties of $LiFePO_4$-multiwalled carbon nanotubes composite cathode materials for lithium polymer battery. *Electrochem. Commun.* 10, 1537, 2008.
40. Jin, E.M., Jin, B., Park, K.H., Gu, H.B., Park, G.C., and Kim, K.W., Electrochemical characteristics of lithium iron phosphate with multi-walled carbon nanotube for lithium polymer batteries. *J. Nanosci. Nanotechnol.* 8, 5057, 2008.
41. Murugan, A.V., Muraliganth, T., and Manthiram, A., Comparison of microwave assisted solvothermal and hydrothermal syntheses of $LiFePO_4$/C nanocomposite cathodes for lithium ion batteries. *J. Phys. Chem. Part C* 112, 14665, 2008.
42. Hanai, M., Imanishi, H., and Takeda, Y., Enhancement of electrochemical performance of lithium dry polymer battery with $LiFePO_4$/carbon composite cathode. *J. Power Sources* 178, 789, 2008.
43. Jin, E.M., Jin, B., Jun, D.K., Park, K.H., Gu, H.B., and Kim, K.W., A study on the electrochemical characteristics of $LiFePO_4$ cathode for lithium polymer batteries by hydrothermal method. *J. Power Sources* 178, 801, 2008.
44. Kim, J.K., Choi, J.W., Cheruvally, G., Shin, Y.J., Ahn, J.H., Cho, K.K., Ahn, H.J., and Kim, K.W., Electrochemical properties of lithium iron phosphate cathode material using polymer electrolyte. *Phys. Scripta* T129, 66, 2007.
45. Zhang, C.H., Yao, L., and Qiu, Y.P., Synthesis and characterization of $LiFePO_4$-carbon nanofiber-carbon nanotube composites prepared by electrospinning and thermal treatment as a cathode material for lithium-ion batteries. *J. Appl. Polym. Sci.* 133, 43001, 2016.
46. Vu, A., and Stein, A., Multi-constituent synthesis of $LiFePO_4$/C composites with hierarchical porosity as cathode materials for lithium ion batteries. *Chem. Mater.* 23, 237, 2011.
47. Yang, K.R., Lin, Z.J., Hu, X.B., Deng, Z.H., and Suo, J.S., Preparation and electrochemical properties of a $LiFePO_4$/C composite cathode material by a polymer-pyrolysis reduction method. *Electrochim. Acta* 56, 2941, 2011.

48. Chen, W.M., Qie, L., Yuan, L.X., Xia, S.A., Hu, X.L., Zhang, W.X., and Huang, Y.H., Insight into the improvement of rate capability and cyclability in $LiFePO_4$/polyaniline composite. *Electrochim. Acta* 56, 2689, 2011.
49. Chen, W.M., Huang, Y.H., and Yuan, L.X., Self-assembly $LiFePO_4$/polyaniline composite cathode materials with inorganic acids as dopants for lithium-ion batteries. *J. Electroanal. Chem.* 660, 108, 2011.
50. Park, K.S., Schougaard, S.B., and Goodenough, J.B., Conducting-polymer/iron-redox-couple composite cathodes for lithium secondary batteries. *Adv. Mater.* 19, 848, 2007.
51. Dinh, H.C., Yeo, I.H., Cho, W.I., and Mho, S.I., Advanced organic and inorganic material electrochemistry power source. *ECS Trans.* 28, 167, 2010.
52. Fedorkova, A., Wiemhofer, H.D., Orinakova, R., Orinak, A., Stan, M.C., Winter, M., Kaniansky, D., and Alejos, A.N., Improved lithium exchange at $LiFePO_4$ cathode particles by coating with composite polypyrrole-polyethylene glycol layers. *J. Solid State Electrochem.* 13, 1867, 2009.
53. Wang, J.Z, Chou, S.L, Chen, J., Chew, S.Y., Wang, G.X., Konstantinov, K., Wu, J., Dou, S.X., and Liu, H.K., Paper-like free-standing polypyrrole and polypyrrole-$LiFePO_4$ composite films for flexible and bendable rechargeable battery. *Electrochem. Commun.* 10, 1781, 2008.
54. Huang, Y.H., Park, K.S., and Goodenough, J.B., Improving lithium batteries by tethering carbon-coated $LiFePO_4$ to polypyrrole. *J. Electrochem. Soc.* 153, A2282, 2006.
55. Boyano, I., Blazquez, J.A., de Meatza, I., Bengoechea, M., Miguel, O., Grande, H., Huang, Y.H., and Goodenough, J.B., Preparation of C-$LiFePO_4$/polypyrrole lithium rechargeable cathode by consecutive potential steps electrodeposition. *J. Power Sources* 195, 5351, 2010.
56. Xie, H.M., Wang, R.S., Ying, J.R., Zhang, L.Y., Jalbout, A.F., Yu, H.Y., Yang, G.L, Pan, X.M., and Su, Z.M., Optimized $LiFePO_4$-polyacene cathode material for lithium-ion batteries. *Adv Mater.* 18, 2609, 2006.
57. Murugan, A.V., Muraliganth, T., and Manthiram, A., Rapid microwave-solvothermal synthesis of phospho-olivine nanorods and their coating with a mixed conducting polymer for lithium ion batteries. *Electrochem. Commun.* 10, 903, 2008.
58. Shi, J.Y., Yi, C.W., and Kim, K., An investigation of $LiFePO_4$/poly(3,4-ethylenedioxythiophene) composite cathode materials for lithium-ion battery. *Bull. Korean Chem. Soc.* 31, 2698, 2010.
59. Masuda, Y., Nakayama, M., and Wakihara, M., Fabrication of all solid-state lithium polymer secondary batteries using PEG-borate/aluminate ester as plasticizer for polymer electrolyte. *Solid State Ionics* 178, 991, 2007.
60. Bakenov, Z., Nakayama, M., and Wakihara, M., A nonflammable lithium polymer battery with high performance for elevated temperature applications. *Electrochem. Solid State Lett.* 10, A208, 2007.
61. Nara, H., Momma, T., and Osaka, T., Feasibility of an interpenetrated polymer network system made of di-block copolymer composed of polyethylene

oxide and polystyrene as the gel electrolyte for lithium secondary batteries. *Electrochem.* 76, 276, 2008.
62. Mohan, V.M., Qiu, W.L., Shen, J., and Chen, W., Electrical properties of poly(vinyl alcohol) (PVA) based on LiFePO4 complex polymer electrolyte films. *J. Polym. Res.* 143, 143, 2010.
63. Nair, J.R., Juarez, D.C., Perez, V.C., Tirado, J.L., Ahmad, S., and Gerbaldi, C., Truly quasi-solid-state lithium cells utilizing carbonate free polymer electrolytes on engineered $LiFePO_4$. *Electrochim. Acta* 199, 172, 2016.
64. Li, X., Cheruvally, G., Kim, J.K., Choi, J., Ahm, J.H., Kim, K., and Ahn, H.J., Polymer electrolytes based on an electrospun poly(vinylidene fluoride-co-hexafluoropropylene) membrane for lithium batteries. *J. Power Sources* 167, 491, 2007.
65. Xie, H., Tang, Z.Y., Li, Z.Y., He, Y.B., Liu, Y., and Wang, H., PVDF-HFP composite polymer electrolyte with excellent electrochemical properties for Li-ion batteries. *J. Solid State Electrochem.* 12, 1497, 2008.
66. Nah C, Jeong K-U, Lee Y.S, LeeS.H, Abdul Kader M.M, Leed H-K and Ahn J.H., Polymer electrolyte membranes composed of an electrospun poly(vinylidene fluoride) fibrous mat in a poly(4-vinylpyridine) matrix, *Polym Int.*, 62, 375, 2013.
67. Kim, J.K., Cheruvally, G., Li, X., Ahn, J.H., Kim, K.W., and Ahn, H.J., Preparation and electrochemical characterization of electro spun, microporous membrane-based composite polymer electrolytes for lithium batteries. *J. Power Sources* 178, 815, 2008.
68. Aravindan, V., Senthilkumar, V., Nithiananthi, P., and Vickraman, P., Characterization of poly(vinylidenefluoride-co-hexafluoroprolylene) membranes containing nanoscopic AlO(OH)(n) filler with $Li/LiFePO_4$ cell. *J. Renew. Sustain. Energy* 2, 033105, 2010.
69. Wang, X.L., Mei, A., Li, X.L., Lin, Y.H., and Nan, C.W., Novel polymer electrolytes based on triblock poly(ethylene oxide)-poly(propylene oxide)-poly(ethylene oxide) with ionically active SiO_2. *J. Power Sources* 171, 913, 2007.
70. Choi, J.W., Cheruvally, G., Kim, Y.H., Kim, J.K., Manuel, J., Raghavan, P., Ahn, J.H., Kim, K.W., Ahn, H.J., Choi, D.S., and Song, C.E., Poly(ethylene oxide)-based polymer electrolyte incorporating room-temperature ionic liquid for lithium batteries. *Solid State Ionics* 178, 1235, 2007.
71. Reddy, C.V.S., Chen, M., Jin. W., Zhu. Q.Y., Chen, W., and Mho, S.I., Characterization of (PVDF+LiFePO4) solid polymer electrolyte. *J. Appl. Electrochem.* 37, 637, 2007.
72. Hu, L.F., Tang, Z.L., and Zhang, Z.T., New composite polymer electrolyte comprising mesoporous lithium aluminate nano sheets and PEO/LiClO4. *J. Power Sources* 166, 226, 2007.
73. Mohan, V.M., Qiu, W.L., Shen, J., and Chen, W., Electrical properties of poly(vinyl alcohol) (PVA) based on $LiFePO_4$ complex polymer electrolyte films. *J. Polym. Res.* 17, 143, 2010.

74. Li, Y., and Fedkiw, P.S., Nanocomposite gel electrolytes based on fumed silica for lithium-ion batteries. *J. Electrochem. Soc.* 154, A1140, 2007.
75. Miyashiro, H., Kobayashi, Y., Nakamura, T., Seki, S., Mita, Y., and Usami, A., Basic properties of all-solid-state lithium polymer secondary batteries using $LiFePO_4$. *Electrochem.* 74, 321, 2006.
76. Raghavan, P., Zhao, X., Manuel, J., Chauhan, G.S., Ahn, J.-H., Ryu, H.-S., Ahn, H.-J., Kim, K.-W., Nah. C., Electrochemical performance of electrospun poly(vinylidene fluoride-co-hexafluoropropylene)-based nanocomposite polymer electrolytes incorporating ceramic fillers and room temperature ionic liquid. *Electrochim. Acta* 55, 1347, 2006.
77. Angulakshmi, N., Kumar, T.P., Thomas, S., and Stephan, A.M., Ionic conductivity and interfacial properties of nanochitin-incorporated polyethylene oxide-$LiN(C_2F_5SO_2)_2$ polymer electrolytes. *Electrochim. Acta* 55, 1401, 2010.
78. Ferrari, S., Quartarone, E., Mustarelli. P., Magistris, A., Fagnoni, M., Protti, S., Gerbaldi, C., and Spinella, A., Lithium ion conducting PVDF-HFP composite gel electrolytes based on N-methoxyethyl-N-methylpyrrolidinium bis(trifluoromethanesulfonyl)-imide ionic liquid. *J. Power Sources* 74, 321, 2010.
79. Cheruvally, G., Kim, J.K., Choi, J.W., Ahn, J.H., Shin, Y.J., Manuel, J., Raghavan, P., Kim, K.W., Ahn, H.J., Choi, D.S., Song C.E., Electrospun polymer membrane activated with room temperature ionic liquid: Novel polymer electrolytes for lithium batteries. *J. Power Sour.*, 172, 863, 2007.
80. Choi, J.W., Cheruvally, G., Kim, Y.H., Kim, J.K., Manuel, J., Raghavan, P., Ahn, J.H., Kim, K.W., Ahn, H.J., Choi, D.S., and Song. C.E., Poly(ethylene oxide)-based polymer electrolyte incorporating room-temperature ionic liquid for lithium batteries. *Solid State Ionic* 178, 1235, 2007.
81. Zhang, H., Cheng, X.R., Ma, Q., Feng, W.F., Zheng, L.P., Nie, J., Huang, X.J., Armand, M., and Zhou, Z.B., New ionic liquids based on a super-delocalized perfluorinated sulfonimide anion: physical and electrochemical properties. *Electrochim. Acta* 207, 66, 2016.
82. Shin, J.H., Henderson, W.A, Tizzani, C., Passerini, S., Jeong, S.S., and Kim, K.W., Characterization of solvent-free polymer electrolytes consisting of ternary PEO-LiTFSI-PYR_{14} TFSI. *J. Electrochem. Soc.* 153, A1649, 2006.
83. Wu, F., Chen, N., Chen, R.J., Zhu, Q.Z., Ban, J., and Li, L., "Liquid-in-solid" and "solid-in-liquid" electrolytes with high rate capacity and long cycling life for lithium-ion batteries. *Chem. Mater.* 28, 848, 2016.
84. Lane, G.H., Best, A.S., Farlane, D.R., Hollenkamp, A.F., and Forsyth, M., An azo-spiro mixed ionic liquid electrolyte for lithium metal-$LiFePO_4$ batteries. *J. Elecrochem. Soc.* 157, A876, 2010.
85. Aravindan, V., Vickraman, P., Sivashanmugam, A., Thirunakaran, R., and Gopukumar, S., LiFAP-based PVdF-HFP microporous membranes by phase-inversion technique with Li/$LiFePO_4$ cell. *Appl. Phys. A* 97, 811, 2009.
86. Shin, J.H., Basak, P., Kerr, J.B., and Cairns, E.J., Rechargeable Li/$LiFePO_4$ cells using N-methyl-N-butyl pyrrolidinium bis(tri fluoro methane sulfonyl)

imide-LiTFSI electrolyte incorporating polymer additives. *Electrochim. Acta* 54, 410, 2008.

87. Guerfi, A., Duchesne, S., Kobayashi, Y., Vijh, A., and Zaghib, K., $LiFePO_4$ and graphite electrodes with ionic liquids based on bis (fluoro sulfonyl)imide (FSI)(-) for Li-ion batteries. *J. Power Sources* 175, 866, 2008.
88. Rohan, R., Kuo, T.C., Lin, J.H., Hsu, Y.C., Li, C.C., and Lee, J.T., Di nitrile-mono nitrile-based electrolyte system for lithium-ion battery application with the mechanism of reductive decomposition of mono nitriles. *J. Phy. Chem.* 120, 6450, 2016.
89. Zhang, J.F., Ma, C., Xia, Q.B., Liu, J.T., Ding, Z.P., Xu, M.Q., Chen, L.B., and Wei, W.F., Composite electrolyte membranes incorporating viscous copolymers with cellulose for high performance lithium-ion batteries. *J. Membr. Sci.* 497, 259, 2016.
90. Hassoun, J., Lee, D.J., Sun, Y.K., and Scrosati, B., A lithium ion battery using nanostructured Sn-C anode, $LiFePO_4$ cathode and polyethylene oxide-based electrolyte. *Solid State Ionics* 202, 36, 2011.
91. Li, M.T., Yang, L., Fang, S.H., Dong, S.M., Jin, Y.D., Hirano, S., and Tachibana, K., Li/$LiFePO_4$ batteries with gel polymer electrolytes incorporating a guanidinium-based ionic liquid cycled at room temperature and 50 degrees C. *J. Power Sources* 196, 6502, 2011.
92. An, Y.X., Cheng, X.Q., Zuo, P.J., Liao, L.X., and Yin, G.P., The effects of functional ionic liquid on properties of solid polymer electrolyte. *Mater. Chem. Phys.* 128, 250, 2011.
93. Kim, G.T., Jeong, S.S., Joost, M., Rocca, E., Winter, M., Passerini, S., and Balducci, A., Use of natural binders and ionic liquid electrolytes for greener and safer lithium-ion batteries. *J. Power Sources* 196, 2187, 2011.
94. Seki, S., Takei, K., Miyashiro, H., and Watanabe, M., Physicochemical and electrochemical properties of glyme-$LiN(SO_2F)_2$ complex for safe lithium-ion secondary battery electrolyte. *J. Electrochem. Soc.* 158, A769, 2011.
95. Yun, Y.S., Song, S.W., Lee, S.-Y., Kim, S.H., and Kim, D.W., Lithium metal polymer cells assembled with gel polymer electrolytes containing ionic liquid. *Curr. Appl. Phys.* 10, E97, 2011.
96. Damen, L., Hassoun, J., Mastragostino, M., and Scrosati, B., Solid-state rechargeable Li/$LiFePO_4$ polymer battery for electric vehicle application. *J. Power Sources* 195, 6902, 2010.
97. Cui, W.Y., An, M.Z., and Yang, P.X., Preparation of an ionic liquid gel polymer electrolyte and its compatibility with a $LiFePO_4$ cathode. *Acta Phys. Chim. Sin.* 26, 1233, 2010.

Index

Also of Interest

Polymer Nanocomposites Based on Inorganic and Organic Nanomaterials

Edited by Smita Mohanty, Sanjay K. Nayak, B.S. Kaith and Susheel Kalia
Published 2015 ISBN: 978-1-118-38509-8

Nanomaterials and related polymer nanocomposites have gained enormous interest in many fields in recent years because bio and synthetic nanomaterials display dramatically improved properties when blended with polymer matrices. In view of the advancing technology employing nanocomposites, researchers have focused on developing high performance polymer nanocomposites and their myriad applications.

In this volume, *Polymer Nanocomposites Based on Inorganic and Organic Nanomaterials:*

- Presents all aspects of the different classes of nanomaterials, from synthesis to application.
- Investigates the use and feasibility of developing nanocomposites with these nanomaterials as reinforcements.
- Encompasses synthesis and properties of cellulose nanofibers, bacterial nanocellulose, carbon nanotubes/nanofibers, graphene, nanodiamonds, nanoclays, inorganic nanomaterials and their nanocomposites for high-end applications such as electronic devices, energy storage, structural and packaging.
- Details green and carbon-based nanocomposites.
- Provides insight into the various modification techniques for improving the functionality of nanomaterials apart from their compatibility with the base matrix.

www.scrivenerpublishing.com

Lightning Source UK Ltd.
Milton Keynes UK
UKHW022012061119
353032UK00003B/53/P